"Smell is perhaps the most potent and the least understood of the senses. It infuses all that we do, sending subtle and not-so-subtle messages that we do not just react to but think with. It is a vital part of how the spaces we inhabit are enlivened and made into what we count as real. It follows that smell can be actively designed and this book acts as both a primer and a guide to research in this vibrant area which holds such great potential for understanding how we understand space and, by extension, ourselves."

Professor Sir Nigel Thrift DL FBA, Schwarzman Scholars Executive Director

"From the imagined smell of Cardiff's sewers to the real (and fading) smells of Seoul Olympic stadium and the Arctic, *Designing with Smell* surprises, challenges, and inspires its readers to reconsider what they know about smell and its role in our built and natural environments."

Holly Dugan, The George Washington University,
author of The Ephemeral History of Perfume

DESIGNING WITH SMELL

Designing with Smell aims to inspire readers to actively consider smell in their work through the inclusion of case studies from around the world, highlighting the current use of smell in different cutting-edge design and artistic practices. This book provides practical guidance regarding different equipment, techniques, stages and challenges which might be encountered as part of this process.

Throughout the text there is an emphasis on spatial design in numerous forms and interpretations – in the street, the studio, the theatre or exhibition space, as well as the representation of spatial relationships with smell. Contributions originate across different geographical areas, academic disciplines and professions. This is crucial reading for students, academics and practitioners working in olfactory design.

The late **Victoria Henshaw**'s last post was as a Lecturer in Urban Design and Planning at the University of Sheffield, UK. Her research focused upon experiences and design of cities and buildings according to the senses, and smell in particular. Victoria led smellwalks in cities around the world, wrote the blog *Smell and the City*, and was author of *Urban Smellscapes: Understanding and Designing City Smell Environments* (2013). Victoria's work featured in broadcast, print and electronic media, including BBC Radio 4, *The New York Times*, *New Scientist*, *The Guardian*, *The Times*, *The Independent* and *Wired* magazine amongst others.

Kate McLean is Programme Director for Graphic Design at Canterbury Christ Church University, UK, and a part-time PhD candidate in Information Experience Design at the RCA, London. Her research through design investigates possibilities for rendering 'eye-visible' and communicating the space between a smell's source and the human nose through practices of smellwalking and smellscape mapping in cities around the world. She has published in *Leonardo* and *Communication Design* and her smellmap research has received international coverage, most recently in the *New Yorker*, BBC Radio 4 and the Australian Broadcasting Corporation.

Dominic Medway is a Professor of Marketing in the Institute of Place Management at Manchester Metropolitan University, UK. Dominic's work is primarily concerned with the complex interactions between places, spaces and those who manage and consume them, reflecting his academic training as a geographer. He has published extensively in a variety of leading academic journals, including *Environment & Planning A*, *Tourism Management*, *Journal of Environmental Psychology*, *Cities*, *European Journal of Marketing* and *Marketing Theory*.

Chris Perkins is Reader in Geography at the University of Manchester, UK, where he has taught since 1998, having previously run the university's map libraries. His research interests lie at the interface between mapping technologies and social and cultural practices, with ongoing research into performative aspects of contemporary mapping behavior, an interest in sensory mapping, and an emerging interest in play. Chris was Chair of the International Cartographic Association Maps and Society Research Commission, and is the author of numerous books and academic articles.

Gary Warnaby is Professor of Retailing and Marketing in the Institute of Place Management at Manchester Metropolitan University, UK. His research interests focus on the marketing of places and retailing. Results of this research have been published in various academic journals in both the management and geography disciplines, including *Environment and Planning A*, *Journal of Business Research*, *Journal of Marketing Management* and *European Journal of Marketing*. He was co-author of *Relationship Marketing: A Consumer Experience Approach*, co-editor of *Rethinking Place Branding*, and has contributed to numerous edited books.

DESIGNING WITH SMELL

Practices, Techniques and Challenges

Edited by Victoria Henshaw, Kate McLean, Dominic Medway, Chris Perkins and Gary Warnaby

Routledge
Taylor & Francis Group
NEW YORK AND LONDON

First published 2018
by Routledge
711 Third Avenue, New York, NY 10017

and by Routledge
2 Park Square, Milton Park, Abingdon, Oxon, OX14 4RN

Routledge is an imprint of the Taylor & Francis Group, an informa business

Library of Congress Cataloging-in-Publication Data
Names: Henshaw, Victoria, editor.
Title: Designing with smell : practices, techniques and challenges / edited by Victoria Henshaw, Kate McLean, Dominic Medway, Chris Perkins and Gary Warnaby.
Description: New York : Routledge, 2017. | Includes bibliographical references.
Identifiers: LCCN 2017017985| ISBN 9781138955530 (hardback) | ISBN 9781138955547 (pbk.)
Subjects: LCSH: Smell in design. | Smell—Social aspects.
Classification: LCC NK1520 .
D473 2017 | DDC 745.4—dc23
LC record available at https://lccn.loc.gov/2017017985

ISBN: 978-1-138-95553-0 (hbk)
ISBN: 978-1-138-95554-7 (pbk)
ISBN: 978-1-315-66627-3 (ebk)

Typeset in Bembo
by Keystroke, Neville Lodge, Tettenhall, Wolverhampton

Dedicated to Victoria Henshaw, the original editor of this collection

CONTENTS

ILLUSTRATIONS

Figures

Tables

CONTRIBUTORS

Jo Barratt works in branding and graphic design. His popular podcast, *Life in Scents*, explores guests' lives through their scent memories. Jo has also worked with brands and performers to bring an awareness of the sense of smell to what they do, particularly in the music industry, resulting in scented events, performances and exhibition design.

Michael Bull is the Director leading Arup's global Environmental Consulting Practice. After research at Imperial College, where he gained his PhD, he has spent most of his career working in environmental consulting specializing in air quality and odor. He has published and spoken widely on these matters, looking to raise awareness of the capabilities of air quality and odor assessment. His work on odors includes the compilation of a detailed odor emission database, assessing the use of modeling for odor assessment and how well models can predict complaints, as well as undertaking odor assessments from a wide range of different activities.

Olivia Alice Clemence embarked upon her research into scent and design whilst studying at Goldsmiths College, London. She received a first class honors degree in Design, and has subsequently created her own distilling kit, developed her own scents, and exhibited them at the London Southbank Centre, Milan Design Week and Kensington Palace. More recently her practice has explored relationships between scent and food, and she has been working alongside a chef to distil food scents prior to cooking, to explore the potential of adding the essence back to the dishes. Olivia also works at the Institute of Making at University College London.

Anna D'Errico is a neuroscientist at the Max-Planck-Institut für Biophysik, Frankfurt and at Goethe Universität Frankfurt, and an independent performer. She obtained her PhD in Physiology and Neuroscience at the University of Pavia, Italy. As a performer and dancer she studied several techniques, and has practiced dancetheater since 2004, following the theoretical and practical approach of theater anthropology. She works on the physiology of olfaction and on the aesthetical dimension and practical applications of odor in theatrical and performance space. She is a contributor to various science websites and magazines in Italy and Switzerland, and is the author of *Il senso perfetto*, a bilingual (Italian/English) blog about smell and olfaction.

Jim Drobnick is a critic, curator and Associate Professor of Contemporary Art and Theory at OCAD University, Toronto. He has published on the visual arts, performance, the senses and post-media practices in anthologies such as *The Multisensory Museum* (2014), *Senses and the City* (2011) and *Art, History and the Senses* (2010), and in journals such as *Angelaki*, *Canadian Theatre Review*, *High Performance*, *Parachute* and *Performance Research*. He edited the anthologies *Aural Cultures* (2004) and *The Smell Culture Reader* (2006), and recently co-edited special thematic issues of *Public* (2012) and *The Senses & Society* (2012). In 2012, he co-founded the *Journal of Curatorial Studies*. He is a co-founder of DisplayCult, a curatorial collaborative (www.displaycult.com). He is working on an upcoming book on smell in contemporary art.

Nada Endrissat is Assistant Professor at Bern University of Applied Sciences, Switzerland. She studied Psychology at Free University of Berlin, and received her PhD in Management from the University of Basel in Switzerland. Her current research interests include organizing and artistic practices in the creative industries.

Stuart Eve is an Honorary Research Associate at the Institute of Archaeology, University College London. He specializes in using mixed reality techniques to aid in landscape-based archaeological investigations. By creating innovative delivery mechanisms that combine sight, sound and smell, he aims to get closer to the experiences of people in the past. He is also a founding partner of L-P: Archaeology, a commercial archaeology company specializing in using digital techniques to aid in the archaeological process.

Julia Feuer-Cotter is a PhD student in geography at the University of Nottingham with an interest in the roles of smellscapes for human–environment interactions, and specifically the realities and imagination of Arctic infrastructure. Her work explores everyday sensory practices through an open, creative and practice-based approach in the form of community-based workshops, storytelling, and interacting with recreated smellscapes. Her research aspires to contribute to our understanding of diverse Arctic micro-geographies, in which the experiences of the region's inhabitants play the foremost role, and to explore the complex connections between sensed reality and geographical imagination.

Brian Goeltzenleuchter is an artist based in San Diego. His recent projects investigate the use-value of cultural objects and institutions. In 2001 Brian received his MFA from the University of California, San Diego. He was Associate Professor of Art at Central Washington University. He is currently a Research Fellow at the Institute for Public and Urban Affairs at San Diego State University, and Artist-In-Residence at the Institute for Art and Olfaction in Los Angeles. His work has been screened, performed and exhibited throughout the United States, Canada, Austria, Italy, China, Croatia and the Netherlands.

Simon Guy is Professor of Design and Society and Dean of the Faculty of Arts and Social Sciences at Lancaster University. His research explores the co-evolution of art, design and development in the pursuit of sustainable urban futures. His work is inter-disciplinary and collaborative, most recently working with sculptor Wolfgang Weileder on an Arts and Humanities Research Council project, *Jetty*, which explored the role of public art in generating public debate about urban sustainability, and led to the publication of the edited volume *Catalyst: Art, Sustainability and Place in the Work of Wolfgang Weileder* (2015).

Alexandra Horowitz is a Professor of Animal Behavior and Cognition at Barnard College, Columbia University, and the author of *Inside of a Dog: What Dogs See, Smell, and Know* (2009) and *On Looking: Eleven Walks with Expert Eyes* (2013). The Horowitz Dog Cognition Lab at

Barnard College conducts research on a wide range of topics, including dog olfaction, inter-species play behavior and attributions of secondary emotions to dogs. In addition to many scholarly articles relating to dog behavior and cognition, she writes regularly for *The New York Times* and other publications.

David Howes is Professor of Anthropology at Concordia University, Montreal, and the Co-Director of the Concordia Sensoria Research Team (CONSERT). He has conducted field research on the cultural life of the senses in Papua New Guinea, northwestern Argentina, and the southwestern United States. He recently completed a project exploring the sensory life of things in the Pitt Rivers Museum, Oxford. Recent publications include: *Ways of Sensing: Understanding the Senses in Society* (2013), co-authored with Constance Classen; the edited collection *A Cultural History of the Senses in the Modern Age, 1920–2000* (2014); and a special issue of the journal *The Senses and Society* on "Sensory Museology" (2014).

Victoria J. E. Jones is an artist and researcher whose main interest is the ways people connect with one another through common, universal experiences. She is particularly focused on how people understand and make sense of place. Victoria creates multi-sensory artworks that intersect place philosophy, installation, performance and participatory art. A graduate of Fine Art at Camberwell College of Art, Victoria's work has been shown in the UK and internationally.

Jian Kang is Professor of Acoustics at the University of Sheffield. He is distinguished by his work in architectural and environmental acoustics, evidenced by numerous engineering consultancy projects, funded research projects and publications. His work on acoustic theories, design guidance and products has brought major improvements to soundscape design in urban areas. He chairs the Technical Committee for Noise of the European Acoustics Association, the WUN (Worldwide Universities Network) Environmental Acoustics Network; and EU COST Action on Soundscape of European Cities and Landscapes. He was awarded the John Connell Award 2011, and UK IOA Tyndall Medal 2008.

EunSook Kwon, a professor and director, has developed the curriculum of the new Industrial Design program at the Gerald D. Hines College of Architecture, University of Houston, since 2003. She has published and lectured throughout the world about creative design thinking and design strategy development. She has been a consultant to numerous corporations, and has developed educational programs for government agencies and corporations. She was appointed Director General of Seoul Design Olympiad 2008, a design and cultural event with 86 programs. Two million visitors participated in the 21 days of design exhibitions as part of a festival. She was selected twice in DesignIntelligence's list of 'Most Admired Educators' in architecture, industrial design and interior design in 2011 and 2015. In 2015, DesignIntelligence recognized Kwon for "her insight to nurture and grow the program with her vision and passion."

Victoria L. Lygum is Assistant Professor at the Department of Geosciences and Natural Resource Management at the University of Copenhagen. Her expertise is within Landscape Architecture, with a focus on evidence-based design. Her current research interests include healing gardens at crisis shelters for women and child survivors of domestic violence. She has also been involved in a research project about a healing forest garden for patients with stress-related illnesses, where she had a focus on designing with smell. She is EDAC-certified (Evidence-based Design Accreditation and Certification) from the Center for Health Design.

Gwenn-Aël Lynn, after having spent the previous decade focusing on sculpture and installation, in the summer of 2000 collaborated with a dancer and a musician on a performance where the scent of beeswax was prevalent. In 2002, Gwenn-Aël created *Olfactory Path*, his first olfactory installation. In 2005 he was awarded an artist residency in the Netherlands, where he created *Interactive Sound and Smell Installation*, his first work where technology played a prominent role. Since then he has continued to build interactive installations and performances that articulate scent, sound and technology to problematize the relationship between identity, culture and the political.

Melanie McBride is a Canadian educator and a researcher with Ryerson University's Responsive Ecologies Lab (RE/Lab) in Toronto. She is currently a PhD candidate in York University's joint program in Communication and Culture. Melanie's doctoral research investigates ecologically situated approaches to inquiry-based learning with aroma materials and sources. She can be found online at melaniemcbride.net

Charles McGinley is the Technical Director of St. Croix Sensory. He has a BS in Chemical Engineering from the University of Minnesota (1970) and an MBA from St. Thomas University in St. Paul, Minnesota. As founder of St. Croix Sensory, Charles provides technical direction to the laboratory. His vast experience in mass transfer and air distribution systems is utilized in designing scenting systems.

Michael McGinley is the Laboratory Director of St. Croix Sensory. He has a BS in Chemical Engineering from the University of Minnesota (1995) and an MS in Environmental Health Engineering from Johns Hopkins University School of Public Health. His work at St. Croix Sensory involves overseeing daily laboratory operations, including environmental air and product odor evaluations and directing research in olfactory art and design.

Jason Nolan is Director of the Responsive Ecologies Lab, and Associate Professor in Early Childhood Studies at Ryerson University, Toronto. His research focuses on multisensory technologies, adaptive design for children with disabilities, children's play, privacy and autonomy. His work has appeared in journals such as *Information, Communication & Society*, *New Media & Society*, *Surveillance and Society* and *Canadian Children*. He was co-editor of *The International Handbook of Virtual Learning Environments* (2006).

Claus Noppeney is a professor at Bern University of Applied Sciences. Claus' research investigates the role of the senses, and in particular smell, for organizing practices in the creative industries. This interest has recently focused on the context of scent development, with three Swiss National Science Foundation-funded projects into materiality, sense-making and processes of culturalizing in the perfume industry.

Joanna Norman is a postgraduate lecturer at Kingston University, London, and a consultant/founding director of Pandora (www.pandoraltd.com), providing independent training and consultancy within the fragrance industry. She graduated from the Royal College of Art with an MPhil, where she researched the link between the luxury industries of fashion and fragrance. Joanna was trained professionally in London and Paris at leading supplier International Flavours and Fragrances, encompassing all aspects of fragrance development. Recent projects include *Marty the Mighty Nose*, introducing primary school teachers to extend learning using the sense of smell, and chairing the Jasmine Awards Judges, pursuing excellence in fragrance journalism.

Ashraf Osman is the founder-director of Artinect, a consultancy that connects art and people, virtually and physically, and Scent Art Net in Switzerland, a think-tank focusing on the cultural aspects of smell. He is also an advisor for the Institute for Art and Olfaction in Los Angeles and a judge for the Experimental category of its Art and Olfaction Awards. He recently curated a year-long Olfactory Exhibition series at Kunstmuseum Thun in Switzerland (in cooperation with Scent Culture Institute). He is currently working on a traveling project, in partnership with Givaudan, titled *Scents of Exile*.

Debra Riley Parr has a PhD from Boston University and is Associate Professor of Art and Design History at Columbia College Chicago. She has presented several conference papers on scent, conducted workshops on olfactory experience, and curated an exhibition entitled *Volatile!* on poetry and scent for the Poetry Foundation. She is currently writing a book on scent in art and design practice, and was recently awarded a research travel grant to work in residency at the Chemical Heritage Foundation in Philadelphia.

Nicola Pozzani is a Lecturer at Kingston University, London, Faculty of Art, Design & Architecture, and at the Y Institute, Bern University of the Arts, Switzerland. He is the Founder of S Sense (www.ssense-perfume.com), specializing in cross-sensory fragrance education and innovation. Nicola studied Science of Perfume under Master Perfumer Jean-Claude Ellena at the Università dell'Immagine of Milan before working for global perfume companies Symrise and LVMH. Together with Perfumers Les Christophs he has created the Duft-Visions perfume workshops in Berlin, and he has been a guest speaker for the Scent Marketing Institute in NYC and the British Society of Perfumers, amongst others.

Jonathan Reinarz is Professor of the History of Medicine and Director of the Social Studies in Medicine Unit at the University of Birmingham. He has published extensively on the history of hospitals and medical education, and in 2017 was elected President of the European Association for the History of Medicine and Health. Jonathan has additional research interests in the senses and smell, specifically as author of *Past Scents: Historical Perspectives on Smell* (2014).

Kathleen Riach is Associate Professor in Management at Monash Business School Australia. Her main research explores the unspoken, affective experiences of work, with a focus on ageing, bodies and the senses. Her work has been published in internationally renowned journals across the social sciences, including *Organization Studies*, *Urban Studies* and *Sociology*, and has been featured in a variety of national and international media and policy arenas, including the United Nations and the BBC. She is currently working on a project exploring the menopause at work.

Claire Sokell Thompson is the founder of the sensory brand agency Vetyver. Her work in strategic and creative direction has spanned most sectors: retail, travel and leisure, luxury, digital and telecommunications. During her career she has developed in-depth and innovative work in less traditional areas of sonic, naming and intellectual property, tone of voice, cultural training, digital, customer journeys and brand experience, and in the development of fragrances.

Miriam Songster is an artist and designer who engages with the themes of minimalism, site-specificity and the multi-faceted nature of sensory perception. Her practice includes sculpture and video as well as installation- and scent-focused immersive works. Miriam's work has been funded by the Robert Rauschenberg Foundation and the New York Foundation for the Arts, and shown in the US, Germany, Iceland and China. She recently contributed to *Protein Journal*, and has received widespread press coverage from such diverse publications as *Hyperallergic*, *Forbes* and *The Japan Times* for the installation *GhostFood*. She is based in Brooklyn, New York.

Charles Spence is Professor in Experimental Psychology and Head of the Crossmodal Research Laboratory at the University of Oxford. He is interested in how our brains process the multisensory information that fills our daily lives, and the implications for the way in which we design everything from household products to mobile phones, and from the food we eat to the places in which we work and live. Over the years Charles has consulted for a number of multinational companies, has published over 500 scientific journal articles, and has received major awards, including the 2008 Ig Nobel Prize for Nutrition.

Prask Sutton studied Physics and Astronomy as an undergraduate at University College London, and the History, Philosophy and Public Understanding of Science as a postgraduate at the London School of Economics and Political Science. Shortly thereafter he embarked on a career in advertising. He is currently Co-Founder and Chief Innovation Officer at Nock, a London-based agency innovating through creative solutions and interaction design within the marketing and advertising sector. Specializing in multisensory and creative technologies, he has been both a pioneer and advocate of olfactory marketing, behind the most cutting-edge scent delivery technologies and methodologies currently used in Out of Home. Outside of Nock, Prask has founded and co-founded a number of tech startups as well as acting as non-executive director to companies both within and outside the marketing and advertising industries. As a futurist, he also speaks on the subject of emergent and yet to be realized technologies and their potential impact on society, humankind and the Universe at large.

Malcolm Tait is Senior Lecturer in Town and Regional Planning at the University of Sheffield. His research focuses on the means by which places are planned, designed and made. In doing so, he seeks to uncover how the relations between those actors involved in the processes of place-making are constructed and mediated through institutional forms. His previous research has investigated the design and planning of urban villages and how communities develop within them, the relations of trust between those involved in design control, and most recently, the processes by which environments are designed for older people.

Jean-Paul Thibaud, sociologist and urban planner, is a researcher at CRESSON (Centre for Research on Sonic Space & Urban Environment). His field of research covers the theory of urban ambiences, ordinary perception in the urban environment, the social anthropology of sound and the sensory ethnography of public places. He has directed the CRESSON research laboratory, and is currently Co-director of the International Ambiances Network (www.ambiances.net). Jean-Paul has published numerous papers on urban ambiences, and has co-edited various books on this field of research.

Kevin Thwaites graduated in Landscape Architecture at Leeds Metropolitan University in 1983. Since then, he has worked in private practice and higher education, completing a PhD in 1999. Prior to moving to the University of Sheffield in 2003, he was Course Leader for Undergraduate Programmes in the Landscape Architecture Department, Faculty of Health and Environment, Leeds Metropolitan University. He now researches and teaches in the Faculty of Social Sciences at the University of Sheffield, where his work is focused on the development of theory and practice in experiential landscape, and socially restorative urbanism. His research interests include the philosophy and theory of landscape design, the relationship between spatial organization and experience, and its impact on design languages and processes.

Caro Verbeek is a researcher and curator. She did an MA in Art History and in Curatorial Studies, and is currently a lecturer at VU University, Amsterdam. She teaches the preliminary course 'The Other Senses' at the Royal Academy of Art, The Hague (KABK) and is an embedded researcher at the Rijksmuseum, Amsterdam, and International Flavours & Fragrances. Verbeek creates olfactory tours for museums. She is also the author of *Something in the Air: Scent in art* (2015) on the history of olfactory art.

Samantha Warren is a Professor of Management, Employment and Organization at Cardiff Business School, UK. She has been interested in sensory methodologies since using photography to research organizational aesthetics, space and fun in contemporary workplaces in 2001. She is a co-founder of the inVisio visual research network, and co-editor of *The Routledge Companion to Visual Organization* (2014). Her interest in the olfactory arose through collaboration with Kathleen Riach on a British Academy-funded project to explore smell in office environments. This work was recently published in *Human Relations*.

Melissa Westbrook is an Australian-born curator and art historian. Founder of an international art movement called The Neo-Outsiders, she has a distinctive voice in the insider/outsider art debate, and actively promotes self-determination. Her unique exhibitions, which incorporate adult and children's pieces, champion the idea that art of all styles – and from all ages – can inspire us. Through her exhibition projects, Melissa devised a series of multisensory art classes. Known as *The Literary Workshops*, they use art, literature, drama, music, touch and smell to encourage participants to engage all of their senses in their creative processes.

Jieling Xiao is a doctoral candidate at the University of Sheffield, and is currently working on a cross-cultural smellscape study in public transport hubs in the UK and China. This work provides insights to improve sensory environmental quality by designing smellscapes from an in-depth understanding of human perceptions. Jieling obtained a master's degree in Urban Design from University College London, and a bachelor's degree in Architecture at Wuhan University of Technology, China. She has a particular interest in place-making through human sensory experiences and everyday life culture in cities.

PREFACE

> That this is what we fear – no sight, no sound,
> No touch or taste or smell, nothing to think with,
> Nothing to love or link with . . .
>
> *from the poem* Aubade *by Philip Larkin*

Victoria Henshaw was a pioneer in the field of sensory urbanism, adding her passionate voice to a critique of rational planning and ocular-centric design that marginalizes the sensual experience of cities. For Victoria, it was not enough to think about and look at cities as expert observers, situated in an architectural office, a planning department or a university. Instead, her work calls for us to engage with cities as explorers, to open ourselves to a richer urban experience and ultimately to become everyday designers who own, occupy and order our own environments. Victoria was just such an engaged ethnographer, who treated every urban encounter as a new perceptual experience that gives fresh meaning to the city.

Victoria developed, distilled and disseminated this sensory ethnography through her academic positions at the Universities of Salford, Manchester and Sheffield, and also through her professional life as an active member of the design community, working for over 13 years as a practitioner in urban development and management, where she was a founding member and Chair of the South Yorkshire Design and Architecture Centre. This crossing of professional and academic boundaries was key to her work, through which she sought to bring together and influence both theoretical debate and design practice. Her doctoral research involved undertaking 'smellwalks' with built environment professionals such as architects, urban designers, planners and engineers, investigating relationships between smell and place perception, and identifying factors by which such actors might begin to incorporate a more proactive approach to odor in city design and management. She subsequently organized smellwalks in cities across the UK, mainland Europe, North America and Montreal, Canada, and led the delivery of a cross-disciplinary project, *Smell and the City*, at the University of Manchester. This research culminated in her book *Urban Smellscapes: Understanding and Designing with Smell in the City* (2014), which established her as a leading international authority in the field and provides the methodological framework and intellectual agenda for this collection.

In the conclusion to her book, Victoria noted the limited exploration of smellscape experience and design internationally, and the need for more work in exploring the transition from theory to practice. This book answers her call, and provides the next chapter in this debate and a rich set of studies to complement and extend Victoria's work. If we are serious about designing what Victoria termed "more humanistic places for the future," then we should celebrate and connect with Victoria's urban smellscapes through this book and through our own ethnographies. I can think of no better tribute to an inspiring and much missed sensory scholar.

Professor Simon Guy
Lancaster University, August 2016

1

INTRODUCTION

Kate McLean, Dominic Medway, Chris Perkins and Gary Warnaby

The Smell of the Mountains

In January 2013, Victoria Henshaw chose the city of Grenoble, in the foothills of the Alps, as her temporary home to transform her PhD thesis into her first book, *Urban Smellscapes*. Victoria loved the smell of mountains in the winter (most especially when she was at the top of a snow-capped peak with a snowboard attached to her feet), where meteorology, geology and geography combine to form an olfactory purity and clarity devoid of any lingering sweetness or residual pungency. The scent of cold mountains is paradoxically both full and empty, reassuring as an olfactory entity of its own, but equally adept as a neutral container co-existing with the more transitory whiffs of foliage, humanity and vernacular activities. The smell of the mountains has a similar presence to the mountains themselves, a background odor anchored in space and permanent in time, a dependable and reliable, self-assured scent. *Urban Smellscapes* emerged in this montane environment, as a masterly text depicting the background to the urban olfactory world from a humanist perspective. Urban smellscapes around the world are subject to odor monitoring and environmental control, and the book showed how people living in or visiting these spaces might reclaim some agency in the determination of the smellscape. As a text it served as an inspiration for future involvement in decision-making, and will certainly change the design of the cities of our future. Victoria proposed employing a pragmatic philosophy to effect this transition from theory to practice:

> The design of urban smellscapes must therefore acknowledge and allow for difference, both in recognising that odours will be perceived differently by different people, and in thinking about the variety of odours that may be detected in a city or neighbourhood.
>
> *Henshaw, 2014, p. 204*

These 'background' smells are arguably analogous to base notes in a scent; they form the structure to which the lighter episodic smells attach themselves prior to volatilization. *Urban Smellscapes* can be read as the 'background' text to the book we are now introducing, which builds on the theoretical background to highlight the practical, episodic and case based olfactory practice, and also extends the range beyond the urban focus of its predecessor. *Designing with Smell* as an edited collection was conceived by Victoria as a practical 'hands-on' guide and a natural successor to the call in *Urban Smellscapes* to design "more humanistic places for the future" (Henshaw, 2014,

p. 224). Pulling together professionals, practitioners and academics from a variety of countries and disciplines highlights Victoria's unique ability to connect with those working in this field, and is testament to the respect in which she is held across many disciplines and also beyond the academy from practitioners and artists and across different language communities.

The call for contributions for *Designing with Smell* went out in early 2014, and several of the authors featured in this collection swiftly submitted their abstracts. Tragically, just six months later, Victoria discovered she was terminally ill and suggested that we, the *Smell and the City* team who formed as a group in 2012, might continue her work in ensuring that students and professionals interested in the practicalities of designing with smell had a body of experience, guidance and reference to inspire and lead them. We took the project on, and where necessary adapted Victoria's plans in the light of changing submissions and activities in the field, but the practical ethos remains. So does the trans-disciplinary and multi-national nature of the proposal, with chapters from disciplines as diverse as human geography, management science, art and design, history, psychology and landscape architecture and with case evidence from the Americas, Europe and Asia. There is clearly scope for more work from the global South and Australasia. You can also explore the very different perspectives that emerge from consultants to the perfume industry, and practitioners focusing on branding and environmental monitoring, as against chapters from artists discussing their practice, or educators exploring how smell practice might be learnt. These practical observations are set against more academic chapters, and *Designing with Smell* offers much more than an academic take on its central themes. Three years after Victoria's early drafts of the proposal, the project has come to fruition.

The Themes

The 25 central chapters of *Designing with Smell* (which exclude this Introduction and the Afterword) are organized into seven parts, each with an introduction reflecting on the ideas presented by the part authors. This thematic organization offers one cut into a diverse field.

Part 1 highlights the significance of olfactory art, with contributions discussing practice in contexts across Europe and North America. These chapters bring together a concern with multimodal artistic encounters focusing on artistic interventions highlighting often contentious aesthetic olfactory installations. This aesthetic challenges the dominance of perfumed and pleasant smells: sewers and sweat are frequently foregrounded, and the olfactory art described here also focuses on the relations of smell to the places it evokes and the spaces in which it is exhibited. This art frequently escapes from the confines of the gallery to work as part of an educational encounter in a school (Melissa Westbrook, Chapter 4), or in a skittle ally of a pub (Victoria J.E. Jones, Chapter 2), or as theatrical performances (Gwenn-Aël Lynn, Chapter 3).

Such artistic endeavor often strongly depends upon a means of translating smell into another medium, in order to represent aspects of the olfactory experience in ways that might be communicated. Part II highlights different aspects of this representational practice, with contributions from diverse fields and backgrounds. The cartographic challenges faced by Kate McLean in her smell mapping (Chapter 7) are set against storytelling amongst First Nation communities of North America explored by Julia Feuer-Cotter (Chapter 6). These in turn are juxtaposed to Miriam Songster's artistic practice relating taste, smell and eating practices in an exhibit in various North American cities (Chapter 5). Stories of the Arctic are brought home to English cities; the experiential smellscape of Singapore is set against a different cartographic imaginary reflecting the canals of Amsterdam, and the global is rendered into taste and smells for New York participants in the *GhostFood* project.

Part III focuses much more explicitly on the worlds of urban design that were central to Victoria's first book. Here, the focus is on the practicalities of design and monitoring, and the chapters present evidence from the UK, Korea, the Netherlands, Greece and China. Landscape architects play an important role in these design processes, and Chapter 9 from Victoria L. Lygum highlights how the natural world might be enrolled to enhance wellbeing, but also to enrich our urban experiences. A more regulatory and scientific approach is central in Michael Bull's exploration of the monitoring of smellscapes (Chapter 11). Planting in cities offers great potential and Jieling Xiao, Malcolm Tait and Jian Kang focus on best practice in the design of natural elements of our urban fabric (Chapter 8). By way of contrast, a cultural experience can also be enhanced through intervention in the smellscape, evidenced in EunSook Kwon's focus (Chapter 10).

Smell is also increasingly important for those interested in the branding and design of retail and service environments in the city. This commercial theme is explored in Part IV with chapters illustrating global variations in practice, but also contrasting academic with practitioner views of the field. Prask Sutton explores best practice in outdoor advertising (Chapter 13), whilst Claire Sokell Thompson focuses upon how smells might best be enrolled into the development of brands (Chapter 14). These two practical chapters are set against much more academic treatment of smell in organizational environments from Samantha Warren and Kathleen Riach (Chapter 15), and a review of the wider context of designing smell into the consumer experience provided by Dominic Medway and Gary Warnaby (Chapter 12).

Part V focuses on pedagogic contexts for smell. Disciplinary practices are contrasted, from Victoria Henshaw's focus on architecture and planning (Chapter 16), to the emphasis on art and design practice documented by Ashraf Osman, Claus Noppeney and Nada Endrissat (Chapter 17), through to design fashion and art presented by Joanna Norman and Nicola Pozzani (Chapter 18) and Melanie McBride and Jason Nolan's focus on situated learning in the design professions (Chapter 19). The European and North American practices described here highlight how more than vision is crucial in the education and training for professionals acquiring skills in the management of the built environment. Together these chapters also demonstrate that research alone is insufficient: it needs to be disseminated and enacted in and through participatory curricular structures.

The four chapters in Part VI focus on changes in the deployment of smell, with an emphasis on historical practice and upon the role of smell in the theatrical professions. On the one hand the smells of the past can be unearthed: either by archaeologists such as Stuart Eve, in Chapter 21 about the smellscape of the battle of Waterloo, or by historians of art, as in Caro Verbeek's consideration of avant-garde and futurist encounters with smell (Chapter 20). On the other hand, the final two chapters in this part highlight the more performative potential of smell by exploring theatrical contexts. Michael and Charles McGinley emphasize the practical issues involved in olfactory staging (Chapter 22), whilst Anna D'Errico charts perceptions of smell in different cases of multi-sensorial performances (Chapter 23).

The three chapters in Part VII each focus on different aspects of making odors, bringing the book back to a key theme underpinning many chapters. The artists, designers and academics writing in this part highlight the role of technologies in this process, but also reflect on the conceptual implications of creating smells. Olivia Alice Clemence starts with a consideration of DIY-based approaches to smell distillation (Chapter 24). Brian Goeltzenleuchter shifts focus to the frequently antiseptic art museum and explores the technological possibilities of making smell-based exhibits in these kinds of contexts (Chapter 25). Debra Riley Parr emphasizes the digital possibilities of making smells in her exploration of recent changes in practice (Chapter 26).

We conclude with Jim Drobnick's thoughtful words. Drobnick highlights three aspects of the different contributions to our anthology: their protean focus on diverse contexts; the need for critical thinking about smell which they together evoke; and the frequent recognition that the intersection of smell and design demands a consideration of the ethical implications of intervening in the smellscape. He ends with a recognition of the spectacular growth in research and literature in this field, highlighting the incredible diversity of olfactory research, but also its dynamism, versatility and performativity.

Moving on from the Mountains . . .

So this collection of articles and papers deals with "smells, of scents, perfumes, odours, whiffs and niffs; of aromas, bouquets and fragrances; and also . . . of effluvia, reeks, foetors, stenches, and – stinks" (McKenzie, 1923, p. 1). It is Victoria Henshaw's second and final book. It serves as a tribute to an inspiring voice, and a celebration of her wonderful capacity to make things happen. We urge you to read these contributions and think about the richly diverse ways in which they enable Victoria's vision for a human-centered re-imaging of our smellscapes!

References

Henshaw, V. (2014). *Urban smellscapes: Understanding and designing city smell environments*. London: Routledge.
McKenzie, D. (1923). *Aromatics and the soul: A study of smells*. London: Heinemann.

PART I

Olfactory Art

INTRODUCTION

David Howes

"To what organic sense do we owe the least and which seems to be the most dispensable?" the 18th-Century German philosopher Immanuel Kant (1978) asked:

> The sense of smell. It does not pay us to cultivate it or to refine it in order to gain enjoyment; this sense can pick up more objects of aversion than of pleasure (especially in crowded places) and, besides, the pleasure coming from the sense of smell cannot be other than fleeting and transitory.
>
> *p. 46*

Kant's disqualification of smell, not only as an aesthetic sense but also as a cognitive sense, cast a long shadow. It is echoed, for example, in the art psychologist Rudolf Arnheim's assertion in *Visual Thinking* (1969, p. 19) that: "one can indulge in smells and tastes, but one can hardly think in them." This is in contrast to the senses of sight and hearing, the aesthetic and cognitive or 'intellectual' vocation of which has always been treated as given in the Western tradition. The curious thing is that smell was also seen as an intellectual sense in premodernity. For example, 'nose-wise' (now obsolete) could mean either 'clever' or 'keen-scented,' and the words 'sagacious' and 'sage' derive from Latin words meaning 'to have a good sense of smell' (Classen, 1993, pp. 36–60). Moreover, on account of its identification with the breath (Latin *spiritus*), smell was widely understood to be the most spiritual of the senses (Classen, 1998, ch. 2). Thus, Kant was out of line.

Recent years have witnessed a reversal of Kant's withering judgment of smell, first in the case of the 2012 exhibition at the New York Museum of Arts and Design entitled *The Art of Scent, 1889–2012*, curated by Chandler Burr. This exhibition's stated objective was to "situate olfactory art within the larger historical context of the visual arts." Burr accordingly assembled 12 perfumes, from *Jicky* by Aimé Guerlain to *Angel* by Olivier Cresp and *Osmanthe Yunnan* by Jean-Claude Ellena in a bid to put perfumery on a par with painting. Then, in 2015, as the first in a series of five exhibitions on the senses in art, the Museum Tinguely in Basel put on *Belle Haleine: The Scent of Art*. This show featured a range of avant-garde works of art, from Marcel Duchamp's *Air de Paris* (1919) and Piero Manzoni's *Merda d'artista* (1961) to such contemporary works as Clara Ursitti's *Eau Claire* (1992–1993), composed of vaginal secretions, Ernesto Neto's *While nothing happens* (2008), redolent of turmeric, saffron and other spices, and Carsten Höller's pheromone-spewing scent machine, *Hypothèse de grue* (2013). Sissel Tolaas' installation *The Smell of Fear – The Fear of*

Smell (2006–2015) figured as the centerpiece of the exhibition. For it, Tolaas collected the sweat of 20 paranoiac men during periods of crisis and incorporated its smell into the walls of a 10 × 10 m room, with each subject's smell identified by a number. To experience the individual smells the visitor had to scratch the walls, but (as I can attest) there was also a composite odor that had a distinct edge. Tolaas' work raises the intriguing question of whether 'bad' odors can make 'good' art, a question also addressed by Victoria J. E. Jones in her olfactory exploration of city sewer fat in Chapter 2.

These two exhibitions suggest that the idea of an olfactory aesthetics or 'art of odors' has come of age (much like in the case of sound art – a field which has also undergone significant expansion in recent years). The apparent novelty of this development should not, however, be allowed to distract attention from the long, if hidden, history of smell as a vehicle for the aesthetic imagination. Already in the late 18th Century, Kant's British contemporary Horace Walpole speculated on the possibility of creating a poetry of odors. "How charming it would be to smell an ode from a nosegay!" (quoted in Classen, 1998, p. 110). And, in the 19th Century, inspired by Baudelaire, a number of artists experimented with creating art forms for the so-called lower senses, and with combining different sensory stimuli in their theatrical and musical productions (Classen, 2014). Constance Classen has excavated much of this hidden history of olfactory and indeed multi-sensory aesthetics in *The Color of Angels: Cosmology, Gender and the Aesthetic Imagination*. Her work is essential reading for all those interested in investigating the aesthetic potential of all those senses that Kant othered – not only smell, but texture, gustatory taste and even music (see Vercelloni, 2016) – and in crossing sensory borders in the arts. The work of Gwenn-Aël Lynn (Chapter 3), which artfully blends scents and sounds, speaks to this, as does Melissa Westbrook's experiment in multi-modal art education in the *Alice 3.0* workshops (Chapter 4).

Some might argue that it is unnecessary to call for an olfactory aesthetics since perfumery already exists to fit this bill. Classen (1998) roundly rejects this suggestion:

> To limit olfactory aesthetics to the productions of the perfume industry . . . would be like limiting the visual arts to the field of fashion design. There is a whole world of vital olfactory imagery and meaning which cannot be, and is not meant to be, encompassed in a perfume bottle.
>
> *p. 151*

Classen (1998) goes on to point out that the most fundamental deterrent to the development of an olfactory art has, paradoxically, precisely to do with meaning – that is, with the cognitive *under-development* of olfaction: "Not only do we lack a well-elaborated code of odors in the West, we are often unable to recognize even the most familiar odors when these are separated from their source" (p. 152). To continue with the example of perfumery, scents have hedonic value, but scant semantic value. A further deficiency is that perfumes scarcely matter, there is hardly any substance to them. That is, in the West, perfume is normally consumed in the form of a textureless, mildly tinted liquid, and perfumery is basically a unimodal art. To put this in perspective, consider the case of Indian perfumery. In India, perfume normally takes the form of a paste, and has a plethora of sensible qualities. Consider the following description of *candana* (sandalwood): "Light, unctuous, not dry, smearing oil like ghee, of pleasant smell, suitable for the skin, mild, not fading, tolerant of warmth, absorbs great heat, and pleasant to touch – these are the qualities of sandalwood" (quoted in McHugh, 2012, p. 187). Preparations involving sandalwood are thus valued not only for their smell, but also for their light white or yellow color (sight), cooling potency (temperature), pleasing feel (touch) and longevity (other aromatics fade), in addition to being exotic and expensive. This explains their vaunted status alongside gemstones in royal treasuries, their use to cut fever, incite pleasure, and for visual body decoration. Perfumery in India is thus a multimodal art, and the semantics of smell have been elaborated to a high degree. There are many lessons for designing with smell to be learned from the Indian example (see McHugh, 2012).

The meaning problem, which must be overcome in the interest of olfactory art, is compounded by the sentimentalization problem. Here we are referring to what is commonly known as the 'Proust Effect,' in honor of the author of *À la Recherche du Temps Perdu*. 'Proust Syndrome' would be a more accurate descriptor for this so-called effect, given all the limiting effects of the idea on our smell life, which is ironic, since Proust is widely thought to be one of smell's greatest champions. Everyone is familiar with the incident involving the madeleine cake dipped in tea, even those who have never read Proust, so there is no need to repeat it here. Suffice it to say that the taste of the madeleine precipitated a moment of 'total recall' in the author, suffused with the sentiments of childhood, and gave rise to the doctrine of smell being the sense of memory and emotions. It is the charter myth of that doctrine (even though the actual medium of Proust's revelation was a taste).

While the madeleine incident might seem like a celebration of smell, it was actually a demotion, which compounded the Kantian devaluation of olfaction on cognitive and aesthetics grounds. No good for thinking, at least smell is good for emoting and remembering, the doctrine suggests.[1] And so smell has come to be known as the 'affective sense,' with the result that the gap between it and the intellectual and aesthetic senses proper (i.e. sight and hearing) grows ever wider.

This is not to suggest that smell is not good for stimulating recall or triggering emotions (it is in the West), only that that is not all that smell is good for. For example, Francis Galton (1894) invented a smell arithmetic, which proves that smell can have cognitive value (it just didn't catch on). To arrive at a proper understanding of the aesthetic and cognitive potential of smell we must look *outside* the Western tradition, to those traditions where it does not carry all the assumptions as to its 'essential' nature and all the disqualifications that Kant and Proust and legions of psychologists and neuroscientists have saddled it with.

Take the case of India again. Whereas in the West, following Proust, the dominant understanding of smell is as a temporal sense (the sense of recollection) and it is supposed to act inwardly (resonating through the corridors of memory), in India, smell is considered a spatial sense in that it acts outwardly, and is a force either for attraction or repulsion (on the analogy of the way a bee is drawn to a flower by its scent, or the smell of carrion is found repellent). As a spatial sense, smell is not primarily concerned with recollection. For example, based on an extensive analysis of Sanskrit literature, James McHugh (2012) found that "smells are no more prominent than other sensory stimuli when it comes to memory (e.g. in contexts of remembering and longing for absent lovers). . . . It is not the case that when smells are present memories are automatically triggered" (p. 14).

While not particularly implicated in processes of recollection, being a spatial sense, smell is good for navigation. By way of illustration, McHugh (2012) points to a scene in a well-known Sanskrit play where a king sniffs his way around the royal pleasure garden on his way to a secret tryst in the dead of night (p. 28). In the scene, smell has replaced sight (disabled by darkness) as the sense that orients a person in space, and leads them to their destination.

Interestingly, all three of the chapters in this part have to do with navigation and smell's potential as a spatial sense (over its being a temporal sense). Forget Proust. Furthermore, all three, but Melissa Westbrook's chapter in particular, are concerned with the development of olfactory cognition or thinking in smell. Forget Kant. Finally, all three treat smell as an entrée to making multimodal art, and by so doing help us exit the confines of conventional Western aesthetics (while bringing us closer to Indian understandings of the art of scenting).

Acknowledgments

This essay draws on a program of research into the cultural life of the senses which I have been pursuing for a number of years with the support of the Social Sciences and Humanities Research Council of Canada and the Fonds de Recherche du Québec – Société et Culture. An earlier version of part of this essay was presented at the interdisciplinary symposium held in conjunction with the exhibition *Belle Haleine: The Scent of Art* at the Museum Tinguely, Basel in April 2015 and published in the catalogue of the same title. I am grateful to the Director of the Museum Tinguely, Roland Wetzel, for permission to reuse this material here.

Note

1 As should be apparent, this critique is addressed not at Proust, but rather the reception of Proust in the psychologistic and neuroscientific literature.

References

Arnheim, R. (1969). *Visual thinking*. Berkeley, CA: University of California Press.
Classen. C. (1993). *Worlds of sense: Exploring the senses in history and across cultures*. London: Routledge.
Classen, C. (1998). *The color of angels: Cosmology, gender and the aesthetic imagination*. London: Routledge.
Classen, C. (2014). Art and the senses: From the Romantics to the Futurists. In C. Classen (Ed.), *A cultural history of the senses in the age of empire, 1800–1920* (pp. 185–210). London: Bloomsbury.
Galton, F. (1894). Arithmetic by smell. *Psychological Review*, *1*, 61–62.
Kant, I. (1978). *Anthropology from a pragmatic point of view*. Carbondale, IL: Southern Illinois University Press.
McHugh, J. (2012). *Sandalwood and carrion: Smell in Indian religion and culture*. Oxford: Oxford University Press.
Vercelloni, L. (2016). *The invention of taste: A cultural account of desire, delight and disgust in fashion, food and art*. London: Bloomsbury.

2

SMELL THE CITY

A Participatory Art Installation

Victoria J. E. Jones

In the summer of 2013, I was given rare access to the fat deposits in the sewers of both London and Cardiff, UK. Following the news coverage of the discovery of a giant fatberg in London in August 2013, the sewer workers reporting that various locations across London have distinctive sewer smells intrigued me. As part of an artist in residency with the South London Underground Department of Geo-potation and Effluent, I was able to gain access to sewer fat samples across London and two locations in Cardiff.

Smell the City was an exhibition of these samples (Figure 2.1). The samples had been sanitized in laboratory conditions, but not deodorized. Visitors could therefore find out for themselves what Mayfair, Brick Lane and the West End, amongst other locations, smelt like. Except that was not quite it.

Set up in a pub as a fake museum outreach project, the participatory installation artwork *Smell the City* presented a series of smells masquerading as sewer fat samples. The work was constructed as a foil to gain insight into how humans make 'sense of place' (Relph, 1976, p. 48; Cresswell, 2004, p. 7) when encountering the unfamiliar. Olfactory stimuli played a central point in the

FIGURE 2.1 *Smell the City*: an olfactory exploration of city sewer fat.

Image courtesy of Diana Oliveira, 2013.

makeup of this work, in an attempt to evoke associated smell memories. The inspiration for the work, its setup, the theory underpinning it and insights from the work's visitors may be of interest to practitioners creating smells and spaces/places.

Installation Art

I make installation art as defined by Bishop (2005): "the type of art into which the viewer physically enters . . . often described as 'theatrical,' 'immersive' or 'experiential'"(p. 6). Bestor (2003) outlines that "In installation art, the space itself, and the artefacts that are enclosed within it, are the work of art" (p. 16). I view the interplay between my work and the audience in line with Bestor's assertion that "It is a world that the artist has created, and of which the audience is intrinsically a part" (p. 16). As a conceptually led artist, I utilize a range of materials and mediums to make works. Several themes reoccur within my works: space and place, simulation, memory and narrative.

Intention

> People understand themselves as selves through the stories they tell and the stories they feel part of . . . whilst narratives are personal, they are also social.
>
> *Smith, 2007, p. 391*

The *Smell the City* installation was created as an experiential story. The intention was to present a narrative that could be touched and interacted with as well as told. In this instance the story could be experienced, added to and shared by people meeting in a social setting – a pub. The idea was that the audience would perceive it as a bit of fun, and that the smells would act as a foil for them to relate stories from their personal narrative. Ideas of disgust or any negative connotations of sewers did not factor in the making of the work. I had assumed that people would think it was a joke and not take it seriously. In the event, disgust as well as delight did feature in the audience responses.

Back-story

The residency story was as faked as the sewer fat I exhibited. In the summer of 2013 I was not at a geo-potation depot in London exploring sewer fat, but on holiday in Greece with an adult child with severe tonsillitis and an antibiotic allergy. Between dusty trips to the doctor and pharmacy, I whiled away my time surfing the Internet. I found myself fascinated by the British news reports of fatbergs gathering in the sewers around Greater London. 'Fatberg' is a word used to define large-scale fatty deposits found in sewers caused by a build-up of fat from household and commercial waste that is entangled with non-biodegradable items such as wiping cloths, tampons and condoms (Carson, Zimmer & Solomon, 2014). The resultant fatty mass coats the inside of the drains, preventing the free flow of waste through the sewer system (Atkinson, 2014). In August 2013 the British media focused on a fatberg found underneath Kingston upon Thames in South London. The news reports centered on the removal of the fatberg by sewer workers. The workers reportedly described the fatberg as being "the size of a double decker bus" and like "mushy stalactites" (Vidal, 2013; Webb, 2013).

Fatbergs

Jane Bennett in her book *Vibrant Matter* (2010, p. 2) describes how she is drawn to certain objects or materials because they are vibrant matter with an innate agency which she calls "thing power."

I was drawn to the idea of fatbergs in the same manner. It was not just the essentially disgusting materialization of human waste that fascinated me, rather the narrative being built around it that drew my attention. The 'fatberg' story had been given to the media by Thames Water as a way of drawing attention to the behavior of its customers around waste disposal (Thames Water, 2014, p. 26). Graphic images of the fatberg appeared across the media. It resembled a giant lump of lard, pale in color and not at all fecal in appearance. It reminded me of the lard (clarified pork fat) which was much used in baking by my late grandmother. The images of workers removing the fat looked like they were digging in a lard mine. And yet, however disgusting it was supposed to be, it somehow became less vile because people were interacting with it and telling stories about it. Indeed, fatbergs were somehow comical.

As my interest had been piqued by fatbergs, I felt that at some future point I would explore them in my artwork. I had not ascribed a smell to fatbergs. The media images showed the matter looking like lard, so I mentally noted that it must smell innocuous like lard. My associative memory of lard and baking was overriding any rational thoughts about the actual smells of sewers. In line with most people in Western society, I am removed from my own bodily waste by effective water and sewage systems and so never have to consider them too fully (Urry, 2000, p. 96).

The opportunity to explore the notion of fatbergs came a few months later when I was commissioned to make a work by a Cardiff-based arts festival, *Made in Roath*. The commissioners asked that I make a work under the title *The Museum of Mystery* and that it be sited in a public house (pub).

For me artworks afford an opportunity for concepts that I have been thinking about, researching or already exploring to collide and manifest. In this instance it became apparent that this would be an opportunity to explore ideas I had around the human propensity to exchange personal stories and the memory recall of place through olfactory stimuli. The intention was to make a work that was humorous in nature, and would evoke conversations around place and personal narrative. I did not intend that people would think it was real, although it was presented as a museum outreach project. That it was unlikely for health and safety reasons that a pub would hold an exhibition of sewer fat led me to anticipate that most people would think the work a piece of fiction.

What Does Sewer Fat Smell Like?

Although I am drawn to the idea of fatbergs much in the manner of Jane Bennett's (2010) vibrant matter, I do not actually know what sewer fat smells like, but I would conjecture pretty unpleasant. The 'sewer fat samples' in *Smell the City* were made with lard mixed with smells from an aroma laboratory; it was an attempt to copy something real, to make a simulation (Baudrillard, 1994, p. 3) (see Figure 2.2). I suggest that the sewers below Leicester Square in London, an area known for its proliferation of cinemas, do not smell like popcorn. Neither would I imagine the sewers below Brick Lane, known as an area of London with a cluster of curry restaurants, smell of fragrant spices. Nor do I believe that the upmarket hotel area Mayfair smells like soap and perfume beneath its roads. Does Caroline Street in Cardiff, known locally as Chippy Alley, really have sewers that smell like potato fries? Probably not, but through the work I inferred that these were the smells of these locations.

Creating an 'Other Space'

A white marquee was set up in the skittle alley of the pub. Within it, the installation was set up to look like a museum outreach project, complete with its own *Smell the City* logo and graphic signage. The sewer fat samples were held within scientific-looking beakers covered by plastic

FIGURE 2.2 Sewer fat exhibit with explanation label.

Image courtesy of Diana Oliveira, 2013.

cloches and set up on plinths. In order to reinforce the idea that the sewer fats were from real locations in London, explanation notes outlining the history of the location where the 'sample' had been taken were placed next to each one. The two Cardiff samples were mystery exhibits. Visitors were encouraged to guess which location in Cardiff they were from. Hanging from one wall was an information board outlining the 'project,' and from another the exhibition signage. In this way it became another kind of space separate from the pub. Foucault (2008) defined other spaces, or heterotopia, to be spaces that co-exist with our everyday spaces, but we consider them to be 'other' in some way. This is manifest by the modification of our behavior. We behave differently in a museum than we do at home or in a pub.

Who Were the Visitors?

There were two types of visitor to *Smell the City*. They were either people going to the pub for social purposes who then happened upon the installation. Or they were people following a series of exhibitions on an art trail set out by the *Made in Roath* organizers, each exhibition assigned a number and a dot on a map. How the visitors reacted to the installation might depend on how they came across the work. If you were on the art trail you might have treated finding the work quite differently than if you had gone to the pub to have a drink. One set of people actively seeking it out, the other inadvertently discovering it.

FIGURE 2.3 Victoria sniffs *Smell the City*.

Image courtesy of Diana Oliveira, 2013.

Navigation

> You can tell people that it was going to smell like this, but that's no good. You could tell people a million things . . . but until you can smell it and really relate . . . until you can smell it, then you can make your own decisions and thoughts on that . . . but you definitely have to sniff it.
>
> *exhibition visitor*

Visitors to the installation navigated the exhibits as they saw fit. They were encouraged to remove the cloches and smell the exhibits by the invigilator (see Figure 2.3). They were reassured that the fats were safe and sanitary. The visitors could spend as much or as little time with the exhibits as they saw fit. The invigilators repeated the information about the provenance of the sewer fats verbally, but gave no additional instruction. As the sewer fats were numbered from one to eight, many visitors interacted with the exhibits in numerical order. Some navigated the exhibits by taking reaction cues from other visitors. If another visitor reacted negatively or positively to an exhibit, then I observed that this reaction would discourage or encourage a person from sniffing that same exhibit.

> They [drinkers in the pub] would drag people in with their pints and stuff . . . and they would be, like, "Smell this, smell that" . . . and a lot of the people wouldn't smell it . . . because obviously they thought it was real.
>
> *exhibition invigilator*

> I was thinking of the Leicester Square huge cinema where all the premieres are. All the different [foods] . . . you've got Mexican, steak houses, burgers all those sorts of things . . . and also because I am a vegetarian I think that I consciously or subconsciously thought ah that is all very meaty . . . . I recoiled from it thinking it might be meaty.
>
> *exhibition visitor*

How Did the Visitors React to *Smell the City*?

Visitors were interviewed in retrospect. The semi-structured interviews were conversational in style. Interviewing in this manner relaxes the interviewee and creates a framework for the interaction (Leech, 2002; Fetterman, 2010, p. 44). This manner of exchange was two-way. The participants were able to ask about aspects of the artwork that they were curious about, as well as being asked about their experience. The participants were asked about their first impressions of the work, how they navigated the space, to describe the smells, and how they understood and shared the narrative.

Memory and Smell

> It was probably a good smell of Cardiff city centre . . . and the old smell, which I think you don't get any more, of Brains Brewery . . . as a student coming into Cardiff on the train coming back to the city from home when I first lived here.
>
> *exhibition visitor*

For some, sniffing the exhibits triggered memories. Press and Minta (2000) argue that smells can produce "high emotional fidelity" (p. 175) when instigating memory recall. The emotional connection between memory and smell is commonly known as the Proustian phenomenon. This is the "ability of odors spontaneously to cue autobiographical memories which are highly vivid, affectively toned and old" (Chu & Downes, 2000, p. 111). It is named after a passage in a book called *Swann's Way* by Marcel Proust. In the passage (Proust, 2002, pp. 47–50), the protagonist describes how a memory of his aunt is triggered whilst drinking tea from a spoon containing a piece of madeleine cake. Through the taste and smell of the cake and tea, he is transported to a place, time and people of his past.

> It reminds me of my dad, because he always has sneaky pasties . . . and I can smell them on him.
>
> *exhibition visitor*

This comment echoes the sentiments expressed by Proust's protagonist. Evidently the visitor is experiencing the Proustian phenomenon. Some neuroscientists opine that the olfactory system is linked to the memory and emotion parts of the brain (Reineke, 2008). The pyriform cortex (the smell area of the brain) and the hippocampus (the memory part of the brain) are sited next to each other, and many neuroscientists consider that they share various fibers (Jacob, 2011). When describing the neurological workings of olfactory stimuli, Freeman (1991) describes that:

> The result is a meaning-laden perception, a gestalt, that is unique to each individual. For a dog, the recognition of the scent of a fox may carry the memory of food and expectation of a meal. For a rabbit, the same scent may arouse memories of chase and fear of attack.
>
> *p. 78*

Freeman is alluding to how we process smell memory according to our past experiences and context. Visitors to the installation would bear this out. Smells that were disgusting to some people were deemed palatable to others. The actual smells were facsimiles of: popcorn, a fish market, steak and chips, cardamom and curry, soap and perfume, chips and vinegar, hops and whisky, and greasy savory pastries."Yeah it was just like, ah, you kind of sniffed it and it just smelt awful. No real chain of thought . . . that is how I imagine a sewer to smell like . . . you know . . . of shit," commented one exhibition visitor.

Smell and Place

Through *Smell the City*, I attempted to explore how humans make sense of place. Sense of place is understood as the emotional resonance humans give to locations. Making place is a part of human identity. Without people, Relph (1976) argues there would be no places. Places are made by humans, and having an attachment to a place is part of being human: "to be human is to have and to know your place" (p. 1). Having been a long-term resident in both London and Cardiff, both cities are known to me.

One of the exhibition visitors expressed a seeming familiarity with a location through sniffing one of the smells: "It must be Cathays [an area of Cardiff], because it smells like students." Classen, Howes and Synnott (1994) argue that smell provides an immediate and direct connection with our surroundings, which is difficult to have control over, whilst Tuan (1977) suggests that smell helps us to "identify and remember" (p. 11) places more easily because smell is adding to our experience of them. For Lefebvre (1991) smell and place are inextricably intertwined. The reactions of many installation visitors bore this out:

> I can't remember the others in detail, but when I got round to that one that I really didn't like it was like "Ooh, this is a really dirty place" . . . it reminded me of dirty Central London.
>
> The potatoes, that was a big one . . . convinced that that one was from Chippy Alley.
>
> It would be like you smell it, you smell it was soap and you think of, like you know it came from Mayfair . . . but I was thinking, "It's rich people bathing . . . rich people bathing in thousands of hotels in Mayfair."

Pass It On . . .

The installation was intended as a story to be shared. Bauman (1986) suggests that we make sense of events we experience by constructing and sharing narratives around them. Sharing experiences through narrative is part of the human condition (Stevanovic & Peräkylä, 2015). The interviewed visitors were asked whether they had shared the *Smell the City* story or brought others back to experience the installation:

> Absolutely, I've told everybody I know [laughs].
>
> I was in work the next day . . . and told my work colleagues . . . I met with three of my friends on Sunday afternoon and told them about it . . . to see their reactions, really, before I said "She made it up . . ."; so yeah, because it's a really good story.

The invigilator also witnessed the practice of participants bringing their friends back to experience the installation: "Some of them were returning visitors . . . they've been in before they'd seen it and they brought other people back to see it because they couldn't believe what was there . . . ."

Unintended Consequences

Although some of the visitors understood the installation was a piece of fiction, one of the surprising outcomes was that a large proportion thought that the exhibits were real. Despite the clues that it was a fiction (the signage, the comedy smells, its location), the pub landlord reported: "I couldn't believe how they believed it . . . ." One of the visitors also recalled being told by a curator that it was a fiction: "She said, 'It's all a lie.' I said, 'What is all a lie?' 'The smell . . . the smells are all a lie . . .' I was a bit shocked . . . ." As the exhibition invigilator observed: "I think most people did think it was real, and they were fascinated by it . . . ."

But some visitors did realize it was a fiction. As one commented: "The fact that they were so specific and so well-defined smells you just think I wish that was true but it can't possibly be . . . ." Another explained how they used this realization to catch out others: "I dragged my mate Sue along the following day . . . but didn't let on to her until a while later . . . she was completely taken in by it."

Some of the participants who thought that the exhibits were real, and then were subsequently were told it was fiction, were keen to maintain the story. These people constructed plausible reasons why the fiction of *Smell the City* could really be feasible:

> Yeah, I can imagine that . . . I can imagine that . . . I mean . . . it wouldn't be as . . . so, yeah . . . the smell wouldn't be that apparent. I think that there would be a degree of excrement in there as well, but there would be a slight tinge to it. It would all smell similar, but there would be tinges of . . . you know . . . different smells.
>
> *exhibition visitor*

The social psychologist Leon Festinger (1957) coined the phrase "cognitive dissonance" (p. 3) to describe having two opposing views at the same time. This may explain why some visitors came to know the sewer fats were not real, but could not let go of the plausibility of the idea that certain sewer systems could hold location-specific smells. One visitor created a scenario to explain one of the smells:

> Knowing what Theatreland is really like, I am thinking about there is a huge variety of food shops there . . . a variety of different sort of foods and drinks which could be washed into the system. And so many people, it kind of did fit with the idea of what a sewer might smell like.

Smell the City as a Research Opportunity for Designers

As an artist, I am interested in the ways people connect with one another through common, universal experiences. My installation artworks are not fully realized unless people participate with them. The past two decades has seen a rise in artworks that provide a context for conversations and encounters between people. As with *Smell the City*, these artworks do not conform to traditional notions of object-based art. This type of artwork has variously been termed "relational" (Bourriaud, 2002, p. 14), "dialogical" (Kester, 2004, p. 10) and "participatory" (Bishop, 2012, p. 11). The potential for gaining insights from the audiences of participatory artworks to inform design projects has been relatively untapped. Projects which cross over the two disciplines are unusual, and yet artists are creating participatory works which seek to understand the human condition. This may create observations helpful to designers for research and commercial purposes.

As intended, the smells in *Smell the City* seemed to have evoked for the visitors memories of other places and people. The smells provoked different kinds of responses dependent on the person's past experiences and the people they shared the installation experience with. The visitors shared their own observations of the artwork with other people and retold the fictional narrative. The device of embedding smells within the artwork, encouraged visitors to invent their own narratives around the smells.

Smell the City is an example of an artwork that could be used by designers seeking to illustrate how smell is intrinsic to our sensory experiences. The smells were integral in supporting the fiction of a sewer fat museum. Without the smells the work would not have been the same. It would have simply been eight lumps of lard in the back room of a pub. I suggest that the curation of smells offers designers an opportunity to create more than environment masking or scenting of a location. There may be circumstances where designers may seek to actively engage users with a location and not want them to pass through it. The deliberate act within *Smell the City*, of drawing the attention of the visitors to the smells, created an opportunity for engagement with the artwork through discussion and exchange of memories and ideas.

References

Atkinson, W. (2014, February). Clogged arteries. Retrieved March 1, 2017 from http://digital.bnpmedia.com/article/Clogged+Arteries/1612466/192647/article.html

Baudrillard, J. (1994). *Simulacra and simulation* (S. Glaser, Trans.). Ann Arbor, MI: The University of Michigan Press. (Original work published 1981)

Bauman, R. (1986). *Story, performance, and event: Contextual studies of oral narrative* (Vol. 10). Cambridge: Cambridge University Press.

Bennett, R. J. (2010). *Vibrant matter: A political ecology of things*. Durham, NC: Duke University Press.

Bestor, C. (2003). Installation art: Image and reality. *ACM SIGGRAPH Computer Graphics*, *37*(1), 16–18.

Bishop, C. (2005). *Installation art: A critical history*. London: Tate Gallery Publishing.

Bishop, C. (2012). *Artificial hells: Participatory art and the politics of spectatorship*. London: Verso Books.

Bourriaud, N. (2002). *Relational aesthetics*. Dijon: Les presses du réel. (Original work published 1998)

Carson, C. E., Zimmer, B., & Solomon, J. (2014). Among the new words. *American Speech*, *89*(1), 89–110.

Chu, S., & Downes, J. J. (2000). Odour-evoked autobiographical memories: Psychological investigations of Proustian phenomena. *Chemical Senses*, *25*(1), 111–116.

Classen, C., Howes, D., & Synnott, A. (1994). *Aroma: The cultural history of smell*. London: Routledge.

Cresswell, T. (2004). *Place: A short introduction*. Oxford: Blackwell Publishing.

Festinger, L. (1957). *A theory of cognitive dissonance*. Stanford, CA: Stanford University Press.

Fetterman, D. M. (2010). *Ethnography: Step-by-step* (3rd ed.). Los Angeles, CA: SAGE Publications.

Foucault, M. (2008). Of other places. In M. Dehaene & L. D. Cauter (Eds.), *Heterotopia and the city: Public space in a postcivil society* (pp. 13–29). London: Routledge. (Original work published 1967)

Freeman, W. J. (1991). The physiology of perception. *Scientific American*, *264*(2), 78–85.

Jacob, T. (2011). The science of taste and smell. In F. Bacci & D. Melcher (Eds.), *Art and the senses* (pp. 183–206). Oxford: Oxford University Press.

Kester, G. (2004). *Conversation pieces: Community and communication in modern art* (revised ed.). Oakland, CA: University of California Press.

Leech, B. L. (2002). Asking questions: Techniques for semistructured interviews. *Political Science & Politics*, *35*(04), 665–668.

Lefebvre, H. (1991). *The production of space*. Oxford: Blackwell.

Press, D., & Minta, S. C. (2000). The smell of nature: Olfaction, knowledge and the environment. *Ethics, Place & Environment*, *3*(2), 173–186.

Proust, M. (2002). *Swann's way* (L. Davis, Trans.). London: Penguin Press.

Reineke, R. (2008, August 1). It's right under our noses: The importance of smell to science and our lives. Retrieved March 1, 2017 from http://serendip.brynmawr.edu/exchange/node/1751

Relph, E. (1976). *Place and placelessness*. London: Pion.

Smith, B. (2007). The state of the art in narrative inquiry: Some reflections. *Narrative Inquiry*, *17*(2), 391–398.

Stevanovic, M., & Peräkylä, A. (2015). Experience sharing, emotional reciprocity, and turn-taking. *Frontiers in Psychology*. doi:10.3389/978-2-88919-825-2

Thames Water. (2014). *Annual performance report 2013/14*. Retrieved May 19, 2017 from https://www.thameswater.co.uk/-/media/969e8efbdf454c418499a075bda1565f.ashx?bc=white&db=web&la=en&thn=1&ts=a5a26f63-eb67-4a12-8001-2131731e4eaa.pdf

Tuan, Y.-F. (1977). *Space and place: The perspective of experience*. Minneapolis, MN: University of Minnesota Press.

Urry, J. (2000). *Sociology beyond societies: Mobilities for the twenty-first century*. London: Routledge.

Vidal, J. (2013, August 6). Fatberg ahead! How London was saved from a 15-tonne ball of grease. *The Guardian*. Retrieved March 1, 2017 from www.theguardian.com/environment/2013/aug/06/fatberg-london-sewer-grease-blockage

Webb, S. (2013, August 6). I think we're gonna need a bigger plunger: "Fatberg" weighing 15 TONNES found blocking sewers under streets of London. *Mail Online*. Retrieved March 1, 2017 from www.dailymail.co.uk/news/article-2385168/London-Fatberg-weighing-15-TONNES-blocking-sewers.html

3

ON OLFACTORY SPACE

Gwenn-Aël Lynn

As an artist, I have always been interested in the ephemeral, the fragile and whatever defies the standard canon of what art can be. Though initially trained in sculpture, I quickly became interested in perception, and particularly in the sense of smell, because olfaction is transient, elusive, invisible, difficult to rationalize and nearly impossible to grasp. During my first trip to India, in the summer of 1998, I noticed that I was moving through an ever-changing smellscape, with no particular order between scents. I jotted down in my travel diary daily successions of smells which I would encounter while walking down the street, for instance: jasmine, incense, dog diarrhea, fried food, diesel, incense again, human sweat, coconut oil, cow dung . . . .

This experience made me realize that I was much more interested in the distribution in space of smells than in perfume composition. I was more interested in the horizontal deployment of unhierarchized smells than in the vertical arrangement of harmonious scents. From that point on, I began exploring olfactory installations. Though my initial inspiration came from the streets of India, for exhibition reasons all my subsequent efforts focused on indoor diffusion, as precise outdoor diffusion is much harder to achieve. Moreover, any outdoor artwork must be weather- and vandal-proof, and given my interest in fragility, outdoor locations are not ideal for my olfactory installations.

In this chapter, I am going to trace my evolving practice with olfactory installations, and how I built from one project to the next, learning by trial and error, on the job, through readings, workshops, and interaction with perfume professionals. I will also address the area of diffusion techniques. I found that techniques offering better solutions are interactive digital systems. I also reflect on how olfaction behaves in its own specific ways in relation to space and time, and therefore conditions the nature of each installation.

Diffusion Methods within an Artistic Setting

The first method is the simplest and the oldest – specifically, the use of strong odorant material, like plants, food and spices, where the olfactory molecules are volatile enough to be quickly perceptible. For the artist, these present a basic problem of the whole exhibition space becoming saturated with the scents. One solution is to use a single scented material, or an olfactory chord[1] that works well. One scent (provided it is not too intense or considered offensive) or a chord (the harmonious combination of two or more scents) do not overpower a space, because they operate as a single scent, as the scent of that given space. The simultaneous notes of coffee and wine in Artur Barrio's *Réflexion. . .(s). . .* comes to mind.[2] *Réflexion. . .(s). . .* was an installation made *in situ* at the Palais de Tokyo in Paris in December 2005. The artist alternately drank coffee and wine as he stayed awake for long hours while working on his project. The remnants of this 'wake' composed the installation, which included coffee grounds blanketing the floor, empty wine bottles (whose scent lingered, even towards the end of the month-long exhibition), bread, handwriting on the wall, and other minor objects.

In this particular work, the role of wine is important because it points to the fact that when in contact with air, it oxidizes. The oxidization process changes the taste, the aroma and the color of the wine, with significant differences depending on grape varieties. Oxidation, taken to its extreme temporal conclusion, turns wine into vinegar. Since ancient times, wine has been kept in sealed vessels such as amphorae, jugs and bottles to stabilize this process. Once the vessel is uncorked, the oxidation process begins. One can notice the changes in aromas only a few hours after sustained exposure to air. Perfumes also undergo oxidation. Like wine, they are best preserved in sealed airtight containers. The chemical compositions of wines and perfumes are inherently unstable. As we will see later in this chapter, time is in an important factor in olfactory composition and perception.

Getting back to the problem of a scent taking over an exhibition space, *Staple Cheese (A Race)* offers an instance of a single smell taking over a gallery space. In 1970, Dieter Roth exhibited 37 suitcases filled with different kinds of German cheese at the Eugenia Butler Gallery in Los Angeles. In the California heat, the odor wafted so strongly out of the gallery (while maggots became flies inside) that the neighbors called health inspectors, who served a summons on the gallery owner (Dobke, Roth, Vischer, Walter & Garrels, 2003, p. 130). In this case, the prolonged exposure of cheese clearly elicited negative reactions from the community. As with the previous example, time, along with the nature of very strong German cheese, had a major impact on the exhibition.

Artists who would like a different outcome for their exhibition than *Staple Cheese (A Race)*,[3] and who want to have more than one odor co-exist in the exhibition space, need to separate the scents and prevent the stronger ones from overpowering more subtle ones. In 2002, I conceived *Olfactory Path*, an installation composed of 14 odorant cells. The exhibition space was set in darkness (but not pitch black) to help visitors re-focus on their sense of smell. Each cell was cylindrical and made from muslin, enclosing a suspended shelf where I deposited some odorant material (curry, turpentine, dirty laundry, etc.). A string linked one cell to the next, helping visitors to navigate in the penumbra. In order to exit the installation, one had to travel along the same path in a reverse order, thus revisiting all the scents. There was some distance between each cell, to let the visitor's nose reset from the scent it had just experienced, and thus be ready for the next one (Figure 3.1). I looped an audio file with field recordings of urban sounds. I chose muslin and strings because I wanted a fragile, non-intrusive aesthetic. This fragility allowed visitors, less inclined to follow suggestions, to pass under the strings and chart their own paths (which happened occasionally). The scents were distributed along the path to emphasize contrasts (e.g. the warm, heavy scent of tobacco, next to the lighter and colder scent of basil). Though each cell was pretty effective at containing its scent, after a few days the scents started mixing up in the empty space between the cells, the stronger ones taking over. Hence an overall interstitial scent started developing outside of the cells. The visitors were still able to experience each scented cell independently, but this 'in-between' scent slowly built up. Time thus became, independently of my initial intentions, a strong element of this project.

Perfumers often speak of how essential time is to a perfume experience. They evaluate an olfactory composition in terms of its head, heart and base notes, which denotes the three main moments of smelling, although there can be more intermediary notes[4] (Roudnitska & Souriau, 1977). Latency

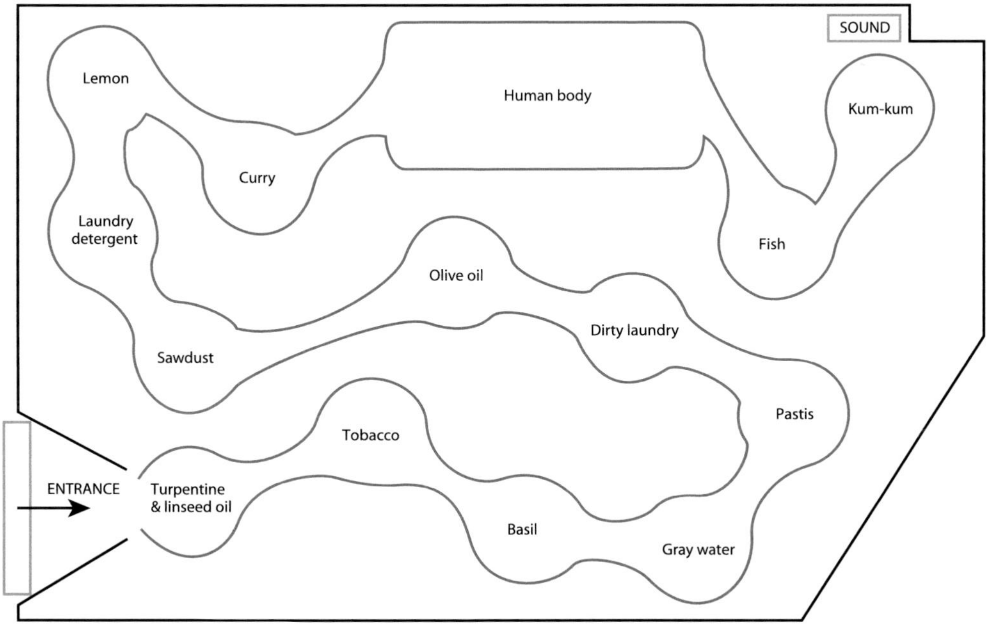

FIGURE 3.1 *Olfactory Path* installation layout, 2002.

Layout: Gwenn-Aël Lynn.

is also important: how long does a smell linger before completely dispersing? In my installation, although each smell had its specific time progression within its own cell, there was also the latency of the interstitial smell and its evolution over the days of the exhibition. As we saw at the beginning of this chapter, there cannot be olfactory space without olfactory time, and most professionals corroborate this (Roudnitska & Souriau, 1977). In astrophysics, and particularly cosmology, time and space form a continuum and have been renamed space-time. Thus we could speak of an olfactory space-time, as both space and time are intrinsic to the olfactory experience.

Olfactory Path left me with several questions: How could I create an installation without an interstitial scent, and retain each scent's singularity? Could I make a more flexible path, without strings to guide the visitor? Finally, now that I had experimented with scented material sourced from everyday places, could I work with scents that are not readily available (such as spices or aromatic plants), scents that are harder to reproduce, such as a sea fragrance or the smell of pressurized air? Interactive technology eventually answered the first two questions, but before solving the last one, I had to make a detour through a sonic performance with essential oils.

Performance Olfasonore

This performance was inspired by electrical perfume diffusers found in health food stores in Paris at that time (see Figure 3.2). To propel essential oil droplets into the air, these diffusers combine the Venturi effect with the Giffard system.[5] They utilize a modified aquarium air pump to compress the air that will project the essential oil onto the glass walls of a tube from which the odorant particles will nebulize out of the tube into the ambient air.

In the early 2000s, when I started this project, the air pump produced a drone.[6] This 'noise' would be considered offensive by many aromatherapists, because it is rather unpleasant and is not conducive to healing. I, on the other hand, found it very interesting, because by manipulating the

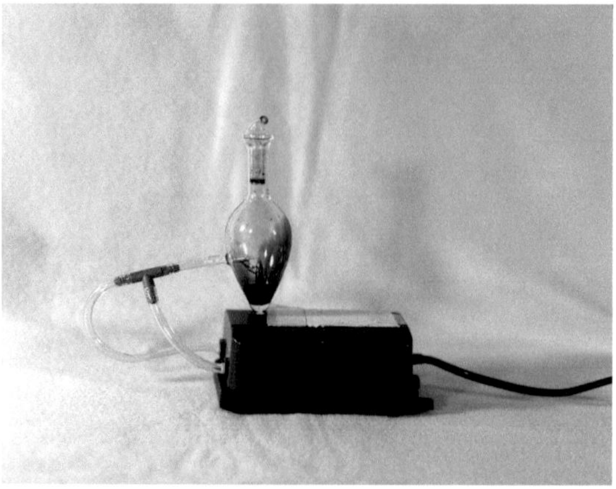

FIGURE 3.2 Scent diffuser from *Performance Olfasonore*, 2004.

Photo: Gwenn-Aël Lynn.

hoses of this little instrument, I could modulate the drone, and the glass parts of the device could be used as a tiny percussive instrument. I also found it amusing that a device considered to be 'relaxing' could produce such an apparently annoying noise. I used one manufactured by Florame, a well-known organic essential oil producer from France. Today, unfortunately, the design of these electrical diffusers has been modified to render them silent.

The following observations compelled me to associate sounds with smells:

- The auditory and olfactory cranial nerves perpendicularly cross paths in our skull.
- Sounds and smells propagate through air.[7]
- Both sound and smell are ephemeral phenomena (each having their own distinctive temporality).

The Florame scent diffuser was the instrument that could manifest this observation. As I was becoming aware of this association, the non-profit organization Cité-Cultures awarded me a three-month artist residency at the Cité Internationale Universitaire in Paris. It provided me with a studio and a small stipend to produce a *carte blanche* evening, when I was given full liberty to do anything. I used the stipend to purchase several electrical diffusers and essential oils. I enlisted the collaboration of Antony Maubert, an electroacoustic composer with whom I had worked on several projects.

We performed this piece, *Performance Olfasonore*, one evening at the Maison Internationale of the Cité, located in the 14th arrondissement, on March 1, 2004. For the title, I aggregated the words *olfactif* with *sonore*, its English equivalent being *olfactory* with *audio*, thus giving *audiolfactory*. I have used this expression subsequently on several occasions. Then we performed it two more times: on March 12, 2004 at the Forges de Belleville (now defunct) in Paris' 20th arrondissement, and on March 24th, 2005 at the Théâtre de la Villa Gilet, Lyon, by invitation from the Groupe Musique Vivante de Lyon. At both Paris-based events, we sat the audience in a circle around each diffuser (hence, there were several circles) to focus their attention away from the stage and towards the scents. We concealed the diffusers in a 'sculptural' container (in an attempt to prevent the audience from being distracted by the visuals of the apparatus). At the Théâtre de la Villa Gilet, because it had a stage facing rows of fixed seats, we had no option but to sit the audience facing the proscenium where we played. We also recorded one of our improvised rehearsal sessions. Our text for the promotional CD reads:

> . . . we propose a sound and smell event. An installation and a performance are combined. Several perfume diffusers are disseminated amidst the public while the musician electronically processes, in real time, a sound improvisation made by the visual artist with one perfume diffuser.

Antony used the software package Max/MSP to create a number of digital filters that would process and amplify the sounds I was making through my manipulations of the Florame electrical diffuser. At the same time, I had a switchboard to turn on and off the scent diffusers placed among the audience. I chose to deploy scents based on the sounds I was making. The relationship between the two was intuitive and very subjective, because there was no rationale that related any specific scent to any specific sound. As I was improvising the sounds, an intuitive feeling would compel me to choose one scent over another, but none of these decisions followed a particular logic.

The Florame electrical diffusers are interesting sonically, but as scent diffusers, they are actually quite limited. They have been designed for domestic use,[8] and after 12 minutes the essential oils have, for the most part, evaporated. With viscous oils, like juniper (*Juniperus oxycedrus*) or vetiver (*Vetiveria zizanioides*), the tiny glass nozzles get clogged. It becomes necessary to drop some alcohol in the glass pieces to clean them up. It can become quite inconvenient for long performances,[9] or for exhibitions, where the scents have to be diffused for longer hours. Something sturdier and more efficient was needed.

I also learned that essential oils, while readily available on the mass market, are somewhat limited. Standardized essential oils do not offer much beyond being pleasant raw notes: sage, juniper, ginger, lavender, orange and so on. What if one wants more urban smells – an 'after the rain on Parisian streets' scent, or the smell of urine and gasoline, or even fresh blood? As the market niche of essential oils is aromatherapy and well-being, such edgy and urban scents are simply not available. The obvious next step was to create my own olfactory compositions. I played with essential oils, and attempted combining them to create new olfactory mixtures, but the options were still limited. I figured the best solution would be to enlist the collaboration of a professional perfumer.

Interactive Odour and Sound Installation

That possibility became a reality when I was awarded another artist residency, in the Netherlands. I spent seven months in Enschede, from October 2005 to April 2006, researching and creating *Interactive Odour and Sound Installation* (see Figure 3.3).[10] This installation has been extensively documented (De Overslag, 2007, 2008; Lynn, Maubert, Roudnitska & Klomp, 2006; Picard, 2011; Pépinières européennes pour jeunes artistes, 2007, 2015) so I am only going to give a brief description here: it consisted of 16 distinct smells distributed in space. Each scent had an audio file attached to it. Both scents and sounds were triggered by the motions of visitors. In terms of the installation content, I wanted to address the fact that, at that time, despite years of immigrant labor and influx from the former colonies, identity in the Netherlands is still defined from the White Dutch majority, when in fact Dutch society has evolved into something more complex, precisely because of its colonial history (this problem is found in other parts of Western Europe as well).

From my perspective, this ethnocentric definition is historically and socially inaccurate, and very unfair to those who have toiled very hard in the metropolis. Parts of my in-residence work consisted in meeting and engaging with Dutch nationals who had some kind of affiliation with the former colonies. As expected, I met many different community members, some with very traumatic histories, because independence was not a peaceful process, others because when they came to the mainland, they had to deal with blatant racism. Hence, their experience reconsidered Dutch

FIGURE 3.3 *Interactive Odour and Sound Installation* (2006), De Overslag exhibition, Eindhoven, the Netherlands.

Photo: Maartin van Loosbroek.

culture from a postcolonial perspective. I asked them to relate this experience to scents, which ended up constituting the olfactory component of this installation. Antony Maubert, with whom I had worked on *Performance Olfasonore*, embedded some interview abstracts with his own sonic compositions based on the scents suggested by the interviewees. Thus the interviewees' olfactory and aural experience informed the installation. Given the fact that I had no formal training in perfumery, I enrolled the collaboration of Michel Roudnitska[11] to supply and create the olfactory compositions. I consulted with Paul Jansen Klomp,[12] from the Media Department of the Academie voor Beeldende Kunst & Vormgeving (AKI), for the interactive component of the installation. It was shown at Villa de Bank in Enschede (2006), De Overslag in Eindhoven (2007), at a group show, *Art4Lux*, at Casino Luxembourg, Forum for Contemporary Art, in Luxembourg (2007) and at the Abbaye de Noirlac in France (2015) for a symposium, *Arts et Territoires: Un Nouveau Défi Culturel.*

I still wanted to solve two main issues: namely, providing a flexible spatial experience, and preventing individual scents from contaminating the entire exhibition space. Regarding this latter problem, many olfactory artists (including myself) have resorted to containing smells. A diffuser that turns itself on only when a visitor is present avoids the need for such a container. Interactive technology allows the artist to localize the scent, and its short emission time rules out the eventuality of overpowering the space. In this manner, *Interactive Odour and Sound Installation* resolves some of the problems raised by *Olfactory Path* and *Performance Olfasonore*. Its diffusion system (which I describe below) is less noisy than the Florame diffuser. There is no interstitial scent, and there are no strings to direct the visitor along a pre-defined path. As noted above, the installation responds to a visitor's movement, his/her presence is picked up by infrared sensors, thus triggering the emission of a scent and its associated audio file. I spatially organized the scents in such a manner that the more overpowering ones are on the far end of the exhibition space while the lighter ones are closer to the entrance. Hence the visitor can chart his/her own path without the risk of the stronger scents eclipsing the more subtle ones. Rather than following the linear spatial progression of a path, the visitor moves within a network of scent stations, able to change directions whimsically. A multitude of scents can co-exist simultaneously. They become latent, waiting to be awakened by someone's presence.

Max/MSP, a software package initially conceived for musicians, operates the installation via a MIDI protocol. In computer sciences and telecommunications, a communication protocol is

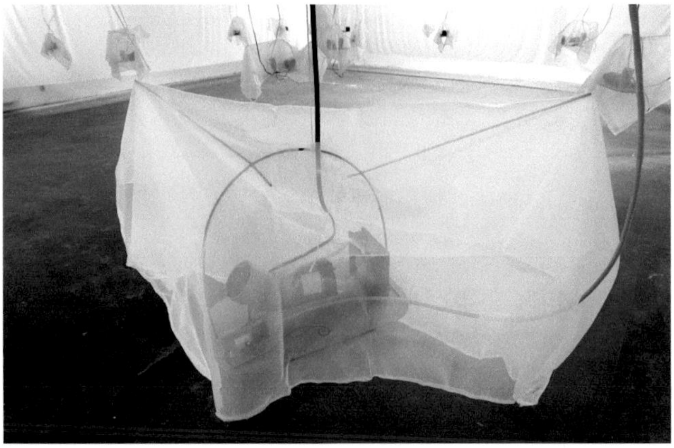

FIGURE 3.4 *Interactive Odour and Sound Installation* (2006), detail of kretek diffuser.
Photo: Maartin van Loosbroek.

a system of digital rules for data exchange within or between computers. Such systems must use well-defined formats to transmit messages. MIDI, short for Musical Instrument Digital Interface, is a technical standard that describes a protocol allowing multiple electronic musical instruments to communicate among one another, including with computers.[13]

A hardware interface relays the electronic signals of the infrared sensors, triggered by the motion of visitors, and transforms the abstract language of Max/MSP into the action of turning on a fan to actually propel olfactory molecules into the ambient air, and it also plays back the audio file attached to each specific diffuser. This interface is called the MIDIbox project, and consists of hardware and software originally conceived by Thornton Klose and licensed under GNU GPL. I built my own MIDIbox module while in residence, and as each scent diffuser was allocated its own specific MIDI channel, I was able to diffuse 16 scents and 16 audio files independently in space.

I made the diffusers, based on the advice of David Suissa, the CEO of Scentys, from a short piece of PVC pipe, a computer fan, and a Pebax patch (see Figure 3.4). Pebax stands for Poly Ether Block Volatile Amide Copolymer, a type of plastic. It can come in different forms, the one I used being granules. The fan blows through the Pebax patch soaked with the essence of each specific scent, thus releasing the olfactory molecules contained in each Pebax bead, and discharging them into the air. The website of Arkema,[14] one of the manufacturers of Pebax, describes the material thus: "Pebax . . . exhibits a high level of initial absorption (up to 50%) and a uniform and long lasting release of the fragrance compared to other supports"[15] (i.e. without the use of a fan, where the diffusion occurs through natural evaporation). It has a high retention rate for volatile substances (such as olfactory molecules), up to 60 days for certain scents, and does not affect the scent composition. Had I wanted to odorize a larger volume, Pebax's high retention rate would have required a stronger air draft to rip the olfactory molecules from it; a 12-volt turbine would have done the job, and we tested one, but it was very loud. However, since I only wanted the olfactory molecules to remain within the vicinity of their respective diffuser (an approximate radius of 20 cm), simple 12-volt computer fans proved to be adequate.

The paradox of scent diffusion is that, necessitating air as a medium of propagation, this propagation itself is likely to be noisy: the movement of air results in sound. Because this project was a sound as well as odor installation, audio files that played as the visitor inhaled the scents covered the low-level noise of my small fans. I used the same fabric, muslin, to have some visual continuity between the aesthetics of *Olfactory Path* and *Interactive Odour and Sound Installation*. Muslin, aside from being a beautiful, fragile material, forms a porous and semi-opaque veil, softening the appearance of the custom-made diffuser. Muslin behaves like scents. It moves when there is an air draft, is very soft to the touch, and only partially conceals what it contains. It is ambiguous.

Audiolfactory Creolization

In 2009, I began working on another interactive installation entitled *Audiolfactory Creolization* (Picard, 2011; Vargas, 2013; ISEA, 2014; Lynn, 2013, 2015). I investigated the 'mixed-experience'[16] in Chicago by engaging in conversations with community members: mixed-race, mixed-culture, transgender, bilingual or trilingual, and especially people who, through personal experience, have come up with their own definition of hybridity. I wanted to explore the question of hybridity for two main reasons. The first one is simply that I am myself mixed, though not racially, since I am both French and American (I have a parent from each country). Therefore I have always had a dual-cultural perspective on the world. The second reason is that I find the predominant mode of racialized identification in North America very constraining. It leaves very little room for people of 'mixed' identities who want to locate themselves in the interstice between racial, gender and linguistic categories. Hence, I asked fellow mixed folks to relate their experience to scents, which eventually constituted the olfactory component of this work. I chose to relate olfaction with identity because race is located on the epidermis. It is visible; it is first and foremost something that we see. However, there are many other invisible markers of differences: for instance, dialect and religion. I wanted to explore mixed identities through olfaction because it circumvents the limitations of race, or visual identities.

I located this project within Chicago, mostly because this is where I live, but also because it is a very interesting place to begin asking: why is there so little creolization happening in the so-called North American 'melting pot'? Chicago is often described as a 'city of neighborhoods,' but it is notoriously segregated and there is little interaction between neighborhoods. Yet the founder of Chicago was a French Haitian *métis*[17] named Jean Baptiste Point du Sable. This contradiction between the historical foundation of Chicago and its present demographic and cultural distribution seemed like a good location for such a project.

As with *Performance Olfasonore* and *Interactive Odour and Sound Installation*, I wanted sound, therefore I collaborated with sound artists to create audio files that supplemented the scents. My collaborators were: Christophe Laudamiel[18] and Michel Roudnitska[19] for the scents, and Christophe Gilmore (a.k.a. FluiD),[20] Spencer Hutchinson,[21] Rebecca Pavlatos and Travis[22] for the audio files (all of them coming from the mixed experience), and Ben Carney[23] who undertook the audio engineering in Pure Data.[24]

This time, the scent diffusers and speakers were placed inside modules based on the spatial deployment of equilateral triangles. These triangles have the advantage of being a geometrical form that can be articulated in infinite ways (Buckminster Fuller used them as the basic geometrical shape for his geodesic domes).[25] They also have the merit of presenting three equal sides that (to me) befit hybridity. Unlike the square, whose sides can be positioned with a 'top' and a 'bottom' and a 'left' and a 'right' (designations which carry strong hierarchical values, and operate in oppositional binaries), the equilateral triangle cannot be placed in such a spatial configuration, because its sides are not parallel. Its sides do not face each other, and therefore cannot be seen in relations of opposition, such as right and left, for instance. Hence, in my view, the equilateral triangle embodies the idea of 'third space' as theorized by Soja (2000). Its three intersecting sides cannot be grouped in pairs, or conceptualized in relations of oppositions, therefore it is difficult to conjure binaries (such as black and white, or male and female) when looking at an equilateral

FIGURE 3.5 *Audiolfactory Creolization* (2013), installation view, *Cultural Evolution, Migration, Hybridity, & Multiculturalism*, a group exhibition.

Photo: Sarah Baransky.

triangle. Any attempt at pairing, or forming binaries, gets disrupted by its third intersecting side, thus creating a third space away from any normative binaries. I made the modules from *hariko* (a Japanese rice paper maché technique), and I distributed them in three forms: hanging pods, wall pieces and corner pieces (Figure 3.5). I covered the modules with casts of noses and ears. The olfactory molecules came out through the casts of these orifices. Furthermore, paper responds very well to sound waves, and does not muffle the sounds or shift their pitch; rather, paper vibrates with the sounds (just like a speaker cone does).

This installation also utilized Pebax and computer fans as the diffusion system (Figure 3.6), but its interactive design differed. Instead of using Max/MSP, a proprietary software package, it entirely runs on open-source software. It is built with Processing[26] and Pure Data,[27] and instead of the MIDIbox, it uses an Arduino board[28] to communicate between the physical world and computer language. Instead of infrared sensors, I worked with the J-Myron video tracking library of Processing. Processing makes use of 'libraries,' a collection of code in a specified format. Developers other than the original authors make the libraries. They function as 'extensions' or 'plug-ins' to Processing. J-Myron is the cross-platform, cross-language, open source video capture and computer vision plug-in.[29] I used tiny CMOS (Complementary Metal-Oxide-Semiconductor)[30] video cameras to monitor the displacements of visitors in the installation. By using pixel coordinates, I can define very precisely the trigger range of each diffuser, but I can also modify it at will, and adapt it to any spatial configuration, something much harder to achieve with infrared sensors (whose range is set by default in the factory).

I allocated five sonic compositions to each scent. To manage the flow of information, thus ensuring a rich and diverse aural experience, Ben Carney created a random function in Pure Data that picks a different audio file from its pool of five each time a visitor activates a module.

FIGURE 3.6 *Audiolfactory Creolization* (2013), detail inside the rum scent diffuser.

Photo: Gwenn-Aël Lynn.

Conclusion

Over a period of more than 10 years, my work has moved from a somewhat constrained linear path – the strings from *Olfactory Path* (see Figure 3.1) – to scents freely distributed in space, triggered by the visitors who chart their own paths, at their own rhythms and whims (Figure 3.7). I started out by working from found smells, then I tinkered with essential oils, and I ended up collaborating with perfumers in order to either create or reproduce scents that are not ready-made. My diffusion systems have evolved from mere evaporation to interactive systems that respond to the visitor's presence.

The types of space these projects have generated are at once disseminated, networked, localized and simultaneous (see Figure 3.7). The scent stations are distributed throughout the exhibition

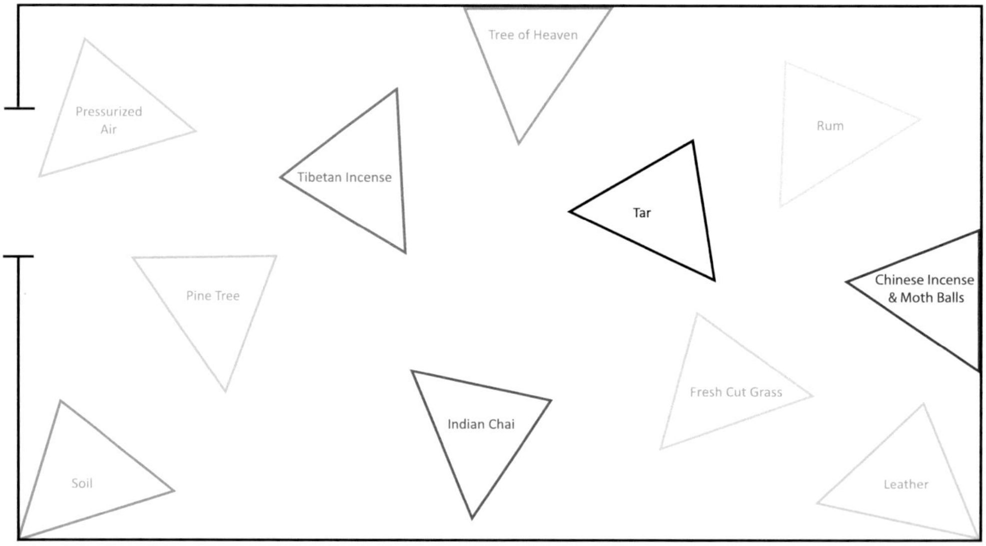

FIGURE 3.7 *Audiolfactory Creolization* installation layout, 2013.

Layout: Gwenn-Aël Lynn.

venue, in no easily recognizable pattern, with some empty space between each other to allow olfactory and aural silence, and prevent contamination. The installation is electronically networked, but it is also sonically networked, since the sonic compositions echo one another (especially in the latter two examples). The distribution of scents in space is organized according to their respective qualities, making sure that the visitor does not start with an overpowering smell (e.g. leather) that would prevent his/her nose from perceiving the more subtle ones (like pine trees). The scents are in dialogue with each other, sometimes harmonically, sometimes discordantly. It is the visitor who, in the end, after having charted a path, pieces together the various elements to make sense of the sensory and audio stimuli encountered during the visit (particularly in the latter two examples).

I say the scents are localized because they exist only in proximity to their respective stations. Hence, though the installation spreads across the gallery, each scent is localized. There are no airtight screens or barriers. In the latter two projects, interactive technology controls the dispersal of olfactory particles. In addition, since each scent results from the suggestions of participants, they are localized within the experience of these participants. Thus, the installation as a whole is disseminated, but it is localized in each singular scent. These works play with space and time, compressing dispersal with localities, and simultaneity with chronology. Their temporality is potentially simultaneous, yet experienced along a personal timeline. They are potentially simultaneous because their distribution in space privileges no particular path. The visitor charts a path along a self-assigned timeline (consciously or not). Hence, these works provide a meandering experience through olfactory space-time. The visitors reconsider their place in space-time by immersing themselves in an audiolfactory sensation, a much more intimate, and moving, experience than gauging space from the distance of their eyes.

Notes

1 An olfactory chord, just like a musical chord, is a combination of scent notes (usually at least three) that reinforce each other well, and create a 'new' scent. One of the most well-known examples is *fougère royale* (royal fern), composed by Paul Parquet in 1882. It combined coumarin with oakmoss, as its base notes, with geranium and bergamot as its top notes. In fact, it contained other notes, such as rose, lavender and sage, but the first four I listed are the main component of the 'fern' chord. Together they created a scent that actually does not exist in nature: ferns do not smell. A trained nose can recognize the components. In a way, though the chord is a new smell, its components can still be identified. This is what I mean by "they work well together." None of the components overtake the other ones.

2 Artur Barrio is a Portuguese artist, born in 1945. He lives and works in Rio de Janeiro, Brazil, and makes interactive installations.

3 Obviously Dieter Roth knew what he was doing. A pleasant olfactory experience is not what he had in mind; he wanted to provoke.

4 If we think again about the example of the royal fern chord example I gave in note 1, we can see that the base notes were coumarin and oakmoss, geranium was the heart note, and bergamot and lavender were the head notes.

5 These diffusers operate by combining fundamental principles of fluid dynamics known as the 'Venturi effect' and the 'Giffard system.' The Venturi effect describes a phenomenon in which a fluid increases its velocity. It is based on Bernoulli's theorem, which predicts that if a fluid flow is constant and the diameter of the pipe containing it decreases, the fluid's velocity necessarily increases. When a fluid flows through a constricted section of pipe, it causes a drop in its pressure. The drop in pressure results in an increase in velocity because of two fundamental principles of fluid dynamics: the principle of continuity and the principle of mechanical conservation of energy. In order to satisfy these principles, a fluid's pressure must decrease as its velocity increases. As a result, when a fluid flows through a narrowing pipe, it causes a higher pressure at the inlet than at the narrow end, thus resulting in a depression. Hence, the Venturi effect can be used to create a vacuum, thereby achieving suction. In 1858, Henri Giffard invented a pump that applies the Venturi effect. It is an injection system composed of two separate nozzles (in the form of cones) that face each other at a 90-degree angle, separated by a tiny free interval. The liquid, in this case essential oil particles, is situated at the base of the tube. The Venturi effect produces suction through the first nozzle while compressed air enters through the second nozzle, located on the side of the tube.

The Giffard pump propels the particles of essential oil through the tube and into the surrounding area, where (being naturally volatile) they diffuse. This is a form of nebulization: the projection of liquid into fine droplets.

6 The noun 'drone' is used here in a musical sense, and designates a sustained low tone.

7 This performance was conceived before Luca Turin published *The Secret of Scent: Adventures in Perfume and the Science of Smell* (2007). It is obvious that his vibrational theory (instead of the dominant one based on molecular shape recognition), would have highly interested us, had we known about it back then.

8 They are all one needs to odorize a room. Once the essential oil particles are airborne, they remain perceptible for quite a long time.

9 Our performance was only half an hour long, and since I was alternating between several scent diffusers, they never became clogged. In addition, we played in performance spaces that were modest in size. So it worked very well for our purposes. However, had we had more ambitious goals (much larger theaters, longer performance), we would have run into serious limitations.

10 This residency was supported by Pépinières européènnes pour jeunes artistes, and funded by Culture 2000, from the European Commission. It was facilitated in the Netherlands by Trans-artists.

11 Michel Roudnitska is a perfumer from Cabris, France.

12 Paul Jansen Klomp is an autonomous media artist (audio, video, installation), and a Professor of Media Art at the AKI in Enschede (now renamed Artez), the Netherlands. For 10 years he was also consultant for the Montevideo media lab in Amsterdam (which unfortunately closed in 2012).

13 A single MIDI link can carry up to 16 channels of information. The MIDI protocol is quite simple, and based on a 'yes' or 'no' type of communication: 'Note On' and 'Note Off.' In other words, if I press a key on a keyboard, it translates as 'Note On' (and thus opens one of the 16 communication channels). To reset this channel, I can press the same key again, which then becomes 'Note Off.'

14 www.extremematerials-arkema.com/en/ Retrieved March 1, 2017.

15 www.extremematerials-arkema.com/export/sites/technicalpolymers/.content/medias/downloads/brochures/pebax-brochures/pebax-product-range-brochure.pdf Retrieved March 1, 2017.

16 I am borrowing this expression from the *Critical Mixed Race Studies Conference*, a biennial conference examining mixed identities.

17 'Mixed race' in French.

18 Christophe Laudamiel is a scent composer and trained chemist who has created fragrances for houses such as Estée Lauder, Ralph Lauren, Burberry and Tom Ford. Christophe authored and managed the 30-plus scents for the first *Scent Opera* (Guggenheim Museums, New York and Bilbao, 2009).

19 See note 11.

20 FluiD is a Kansas City, Missouri-based experimental/multi-genre electronic musician and sound designer. He is Creole.

21 Spencer Hutchinson is a sound artist and painter based in Chicago. He is also of mixed identity.

22 Though they worked on *Audiolfactory Creolization* independently, Rebecca Pavlatos and Travis are both members of the experimental noise band oNo. Rebecca is Greek American, and Travis is of Black American and Chickasaw ancestry. He successfully petitioned the US government to have his last name dropped.

23 Ben F. Carney is a sound designer and new media artist based in Chicago.

24 Pure Data is a computer software package.

25 Buckminster Fuller (1895–1983) was an American engineer, architect and inventor. His design for geodesic domes remains one of his most significant achievements.

26 Processing is a computer software package. The developers' website describes the software thus: "Processing is a flexible software sketchbook and a language for learning how to code within the context of the visual arts. Since 2001 Processing has promoted software literacy within the visual arts and visual literacy within technology" (https://processing.org/ Retrieved May 30, 2017).

27 See note 24.

28 Arduino is an open source project and user community that designs and manufactures microcontroller kits to build interactive objects that can sense and control other objects in the physical world. Microcontrollers (or 'chips') are found everywhere, from phones to home appliances and cars. They are tiny computers running simple programs.

29 The goal of the J-Myron project is to keep computer vision free and easy for the new media education and arts community. Its main author is Josh Nimoy.

30 CMOS sensors are commonly used in cell phone cameras, Web cameras and digital pocket cameras. Just like any other image sensors, they convert light into electrons by way of an array of thousands or millions of tiny solar cells (depending on their quality), each of which transforms the light from one small portion

of the image into electrons. CMOS chips consume very little power, and are inexpensive, but their image is of lower resolution and lower sensitivity than other types of sensors, thus resulting in an overall lower quality. For my project, image quality was not really an issue since these cameras were used to monitor movement, not render it. There was no image displayed to the audience.

References

De Overslag. (2007). *Interactive Odour and Sound Installation.* Retrieved March 1, 2017 from http://home.iae.nl/users/overslag/gwennlynn.html

De Overslag (Ed.). (2008). *De Overslag 2006–2007.* Eindhoven: De Overslag.

Dobke, D., Roth, D., Vischer, T., Walter, B., & Garrels, G. (2003). *Roth time: A Dieter Roth retrospective.* New York: Museum of Modern Art.

ISEA. (2014). *ISEA 2014 location* [Exhibition catalog]. Dubai: ISEA, Zayed University.

Lynn G.-A. (2013). *Audiolfactory creolization.* Retrieved March 1, 2017 from www.gwennaellynn.com/audiolfactory-creolization/

Lynn, G.-A. (2015). Creolized olfactory spaces. In *ISEA 2014 Conference proceedings* (pp. 214–217). Dubai: Zayed University Books. Retrieved March 1, 2017 from www.isea2014.org/en/Submissions/ISEA2014_Proceedings.pdf

Lynn, G.-A., Maubert, A., Roudnitska, M., & Klomp, P. J. (2006). *Interactive Odour and Sound Installation.* Enschede: Villa de Bank. (Audio CD with booklet)

Pépinières européènnes pour jeunes artistes. (2007). Interactive Odour and Sound Installation. In *Parcour d'artistes/Artists' experience 2005–2008, map^{xxl}*. Marly-Le-Roi: Editions Pépinières.

Pépinières européènnes pour jeunes artistes. (2015). Interactive Odour and Sound Installation. In *Forum européen art et territoires.* Montreuil: Editions Pépinières.

Picard, C. (2011). Accents on the hyphen: Gwenn-Aël Lynn on hybridity. Retrieved March 1, 2017 from http://badatsports.com/2011/accents-on-the-hyphen-gwenn-ael-lynn-on-hyrbidity/

Roudnitska, E., & Souriau, É. (1977). *L'esthétique en question: Introduction à une esthétique de l'odorat.* Paris: Presses Universitaires de France.

Soja, E. W. (2000). Thirdspace: Expanding the scope of the geographical imagination. In R. Read (Ed.), *Architecturally speaking: Practices of art, architecture and the everyday* (pp. 13–30). London: Routledge.

Turin, L. (2007). *The secret of scent: Adventures in perfume and the science of smell.* London: Faber & Faber.

Vargas, C. (2013). *Cultural evolution? Migration, hybridity, & multiculturalism* [Exhibition catalog]. Exhibited at Arrowhead Gallery, Dickson Center, Waubonsee Community College, June 7–July 26, 2013.

4

THE SMELL OF SNOW

Experiencing Wonderland through Multisensory Art

Melissa Westbrook

Picture a white, open space in an historic Oxford building.

Now imagine it filled with vividly colored paintings and sculptures, flickering film installations, a cacophony of dissonant soundscapes and waves of disparate scent permeating from every corner.

Welcome to *Alice 3.0: An Alternative Wonderland.*

This multisensory art exhibition marked the 150th anniversary of Lewis Carroll's presentation of his original manuscript to the real Alice, Alice Liddell. The show explored Wonderland through the eyes of a 21st-Century Alice and brought to life some of the mystical characters that she might have encountered if she had entered Wonderland today. It was held in the summer of 2014 at the Museum of Oxford, which neighbors Christ Church College where Carroll taught.

Alice 3.0 showcased art from Oxfordshire school children, exhibiting it alongside, and given the same prominence as, over 20 international and UK artists to demonstrate my contention that artists of all ages can inspire us. The children's art was produced during sessions with young artists from Oxford High Junior School, Rose Hill Primary School and the Oxfordshire Deaf Children's Society.[1] Each group focused on a different scene from Lewis Carroll's writings. Oxford High Junior School examined the garden of talking flowers, Rose Hill Primary School explored the Mad Hatter's Tea Party, and the Oxfordshire Deaf Children's Society looked at rabbits and rabbit holes. Once created, the youngsters' artworks were organized into group installations for the exhibition itself. The *Alice 3.0* workshops, devised for children aged 6–10, promoted a multimodal approach to creative practice through specially designed games, which encouraged participants to use all of their senses in their artistic endeavors.

This chapter outlines the multi-disciplinary, multimodal art education methods that I devised for the *Alice 3.0* workshops, demonstrating how these games produced a stimulating atmosphere for art-making. Examination of two of the sessions, the *Mad Hatter's Feast* and the *Living Garden*, will demonstrate that olfactory exercises were particularly effective at helping young artists to gain an in-depth understanding of the physical aspects of their subject. Participants used scent as an artistic tool and as a vehicle for adding layers of meaning not always achievable through the use of a solely visual medium. The chapter concludes by analyzing how these practices impacted on this multisensory exhibition.

Six Impossible Things before Breakfast: Conceptualizing Alternative Wonderlands

Alice 3.0: An Alternative Wonderland set out to evoke a contemporary, alternative world: one that reflected the unbridled imagination and absurd humor that distinguishes Lewis Carroll's Alice adventures. The curatorial vision was of a disparate, chaotic reality laced with comforting doses of the familiar. *Alice 3.0* would be an exhibition unlike any other. It would combine children's and adults' work and intertwine the visual, the aural and the olfactory with the intention of engaging and bamboozling viewers in equal measure. Moreover, its accompanying workshops would enable young artists to fully explore their subject through multimodal exercises and give them a unique opportunity of adding scent to their work.

While I had recently created a contemporary interpretation of classic stories in my 2014 *Mirror Mirror: Heroes and Villains* show at the North Wall Arts Centre in Oxford, conceptualizing and staging *Alice 3.0* was truly a journey into the unknown. Like *Alice 3.0*, *Mirror Mirror* had celebrated creative difference by juxtaposing stylistically diverse works from artists of various ages and social backgrounds. This project also showcased my unique multi-disciplinary, multimodal approach to art-making and education. *Alice 3.0,* however, would take these elements in an entirely new direction through its use of olfactory tools during the creation and display of its artwork.

Integrating smell into the *Alice 3.0* exhibition and workshops provided a stimulating challenge. My training as a Byzantinist/curator, and my current work with Outsider artists, and the Neo-Outsider art movement, which I identified, provided little assistance. Spurred on by my childhood love of smell and awareness of how scent can encapsulate moods, places, people and objects, I trawled the Internet for artificial fragrances and essential oils. I soon located numerous scents that could be used to evoke Wonderland, and its cast of eccentrics. These included synthetic grass/dirt fragrances for rabbit holes and an angel cake scent to evoke High Tea. I amassed a kit of over 100 different bottled scents, from the environmental (petrol/dirt) and the gastronomic (roast beef/chocolate) to the downright odd (gym socks/soiled nappies). I planned to use these fragrances in educational games alongside tactile objects, thereby creating interplay between 'real' and synthetic scents. They would also be sprayed onto completed work where required. Moreover, scent would form part of a multimodal approach that linked sensory exercises with certain artistic techniques.[2] By teaching olfactory games alongside tactile exercises and role-play, I believed I could encourage children to approach creative endeavors using all of the sensory tools available to them.[3] Examination of the *Mad Hatter's Feast* and *Living Garden* will demonstrate how these ideas fared in the field.

Creating a Feast for the Senses

The *Alice 3.0* workshop inspired by the Mad Hatter's Tea Party was a feast for every sense. Held at Rose Hill Primary School, this session was devised to reflect modern celebrations rather than the traditions and social mores of an English afternoon tea. Its principal aim was to create a multi-ethnic feast, and by doing so, make the school's installation both contemporary and culturally relevant. Inspired by role-play, tactile and olfactory games, young artists were asked to create individual plates of three-dimensional scented food to reflect what they themselves would eat at a family party. To complete the installation, on a later date these pieces were organized on dining tables and paired with a simple live recording which captured the sounds of the children playing party games.

The *Mad Hatter's Feast* opened with two games that gave the young artists an opportunity to explore smell, touch and, to a lesser extent, smell.[4] Having split the class into two groups, the facilitators proceeded to blindfold each child. The first group was given foodstuff to identify. Food items were chosen for their contrasting textures and scent, and included bread, egg yolks/shells, vegetable oil, margarine, Demerara sugar, dried black-eye beans and oranges. While the first group were touching sticky, wet, crumbly and smooth objects, the second were presented with synthetic odors and essential oils. The children were asked to identify sweet and savory food smells such as apple pie, cola, meat and garlic.[5] When all the items in both groups had been identified, the teams were swapped.

These exercises were very successful at stimulating the young artists' imagination throughout the session. They provoked great discussion amongst the participants about their culinary likes and dislikes, and focused their thoughts on the nature of party food, which tended towards convenience food rather than dishes, which represented the ethnic mix of the participants. Having completed these preliminary sensory exercises, the children were provided with acrylic paint, foam, paper and other items to create a signature party dish. Young artists were given a choice between various plastic and china plates for their dishes as a way of encouraging interplay between the fake and the real.

The art-making concluded with each child choosing a synthetic or essential oil-based scent to match his/her work. Where no single scent was deemed appropriate, a specifically tailored blend was created on the spot – spaghetti with meatballs, for example, was evoked through a combination of roast meat and oregano. The essential oil and/or synthetic scent was then sprinkled on the mock food and prepared for the group installation.[6] The amount of scent used varied according to the strength of the fragrance itself and the intensity requested by the child.

Once the art was completed, children were given the opportunity of concluding the workshop with role-play.[7] The young artists were also invited to play a series of party games including musical chairs, which formed the soundscape that would accompany the school's installation and add further realism to the olfactory art. This aural element gave exhibition viewers the impression that there was a party in full swing in the next room. Moreover, food preparation sounds were added to the beginning and washing up noises to the end of the soundscape. This enhanced the olfactory art, actively blurring the lines between the real and the imaginary.

Analyzing the Sights and Smells of Wonderland

In the *Mad Hatter's Feast*, scent became as an artistic tool like tone and perspective. Young artists employed it to explore and convey the physical characteristics of their subject matter, such as the flavor of cupcakes or the scent of strange flowers. The *Living Garden* session demonstrates how olfactory tools can also be used to suggest context and express more abstract concepts, as described below.

One young artist scented her pink feather flower mask with a lily fragrance. She coupled this with fresh grass and soil to replicate the sense of being in a garden. Another used scent to convey a dystopian urban landscape; she felt a blend of curry and motor oil scents encapsulated the stale smell of the city where her flower was located.

One of the more unusual olfactory requests came from a young artist who asked me if I could help make her delicate blue flower smell of fresh air (see Figure 4.1). After some discussion, she wanted it to smell of snow instead. This was a true conceptual challenge. The artist did not want to evoke a sense of coolness, something which was achieved by a fellow pupil through the use of a menthol-heavy peppermint oil. Nor did she want to pair her work with scent associated with snow activities, such as the smell of a smoky log fire. None of the fragrances or combinations in my extensive kit were deemed suitable, so she chose to leave her work unscented. The smell of snow currently remains an interesting/curious challenge for her to capture and to explore.

FIGURE 4.1 *The Smell of Snow.*

Work created by I. G., an Oxford High Junior School student who participated in the 2014 *Living Garden* Workshop.

FIGURE 4.2 *The Bad Flower.*

Work created by M. W, an Oxford High Junior School student who participated in the 2014 *Living Garden* Workshop.

Two other participants in the *Living Garden* had similarly sophisticated requests. One young artist happily sprinkled her graffiti-style work with the odor of soiled nappies. Why? Because she said that the foul smell was her way of conveying the evil nature of her flower (see Figure 4.2). In contrast, her classmate asked me to help her decide on a smell that would convey the notion of something gothic. After several suggestions, she chose to use the smell of smoke to evoke the dark, mysterious and brooding nature of this genre.

The Dark Side of Wonderland

The examples examined above demonstrate that scent is an effective way of adding meaning to a piece of art. This sensory artistic tool was crucial in enabling young Oxfordshire artists to realize a truly contemporary and alternative Wonderland. Using scent as a medium, however, presented two main curatorial challenges: one regarding the longevity of the aromas and essential oils, and the other concerning the olfactory integrity of the children's art and possible aural, olfactory and visual overload experienced by exhibition visitors.

The first issue, which was easily resolved, was entirely foreseeable because all the art from the *Mad Hatter's Feast* and *Living Garden* workshops – with the exception of the 'snow smell' piece – was created with individual scents weeks before the exhibition opening. Synthetic fragrances and essential oils were applied neat on the outer edges of the canvases and mixed into soil of the potted flower sculptures. To ensure a consistent viewing experience over the three-month exhibition run, I organized bi-weekly 're-scenting' trips. These expeditions successfully maintained the olfactory element of the children's work so that viewers could fully appreciate every aspect of it.

Balancing the desire to protect viewers and museum staff from the sensory overload with the need to showcase the numerous olfactory voices in the exhibition was no easy task. As curator, I had intended for *Alice 3.0* to be confrontational, to mimic the sense of confusion that Alice might have felt upon entering Wonderland. Swirling lights, intermingling smells and dissonant sounds were used to evoke a sense of unreality, and any visual, aural and olfactory disorientation felt by those attending the show was entirely intentional. Great care, however, was taken to ensure that no work – or accompanying soundscape – was overly dominant. The Wonderland installation piece by American artists Bill Ross, Bob Scheadler and Sharon Butler (see Figure 4.3), which

FIGURE 4.3 Bill Ross' *Happy Birthday Alice 1.*

Original artwork: Bill Ross, *Happy Birthday Alice 1* (2014), acrylic on canvas.

comprised a film, 12 acrylic paintings, a china tea set and a tablecloth stained with animal prints and stale cake, was installed in a far corner, to minimize the unintended intermingling of individual smells from Rose Hill's feast.

This solution, however, was not suitable for the children's installations. The confines of the museum space made it impractical to juxtapose 40 individual pieces of scented art whilst maintaining their olfactory integrity. As a result, I decided to select a cross-section of 8–10 works per installation and to intersperse these with unscented pieces in the show. This approach made sure that all the participants exhibited in the museum and that the display itself was as varied as practically possible. Considering that a number of works were created with the same scents, there was minimal compromise on the olfactory front. Vanilla cookie, angel cake and chocolate, for example, were commonly used in the *Mad Hatter's Feast*, and a high proportion of flowers in the *Living Garden* smelt of rose.

"Begin at the Beginning . . . and Go on Till You Come to the End: Then Stop"[8]

The curatorial challenges outlined above are resolvable, and should not preclude the staging of similar events in the future. One can get around them by sourcing a longer-lasting range of scents, working with fewer young artists or choosing a larger exhibition space. The artistic understanding and creative freedom that come with using scent as an artistic medium far outweigh potential difficulties.

During the *Alice 3.0* workshops, I observed that olfactory exercises gave children greater insight into the physical attributes of their subject matter and allowed them to view it from newer and more unusual angles. It encouraged to them to think beyond shape, color, proportion and shading. By adding scent and sound to their work, Oxford High Junior School and Rose Hill Primary School children were able to imbue their pieces with a sense of naturalism. This technique enabled them to make bold creative statements and express subtle aspects of their subject matter – aspects not always easily conveyed through conventional artistic means. Most effective when combined with music, drama and tactile exercises, the sensory tool of olfaction not only changed the way in which the young artists conceptualized their art, but how they created it.

The experience of devising and teaching the workshops, combined with the enthusiastic response of both children and educators to the sessions, has encouraged me to continue to develop this style of workshop, as I believe that through this multimodal approach, children come to art creation with a renewed vigor and produce work of greater dimension and depth than would be achieved through conventional means.

The multimodal nature of the *Alice 3.0* workshops gave Oxfordshire school children the tools to create a personal slice of Wonderland. Firmly rooted in the real and the fantastical, this project gave young artists the opportunity to realize their own vision and place their own creative stamp on a classic Oxford narrative. It is my belief that this methodology will allow other children to create similarly unique pieces, and that one day, it may even be possible to create the smell of snow.

Notes

1 Drama specialists Nicola Pike and Samantha Michaels King devised and facilitated the drama games for the *Alice 3.0* workshops. Brian Gibson facilitated the art-making during the *Mad Hatter's Feast* and *Living Garden* sessions.

2 Tactile exercises provide concrete examples of actual food texture and consistency. They assist children with their exploration of a subject's physical characteristics and choice of medium.

3 The young artists from the Oxfordshire Deaf Children's Society and Oxford High Junior School were encouraged to literally become their subject during the *Rabbit Hole* and *Living Garden* workshops as a way of thinking about spatial relationships in art-making. Role-play encouraged this first group to hop around like rabbits and to think about the rooms inside the modern rabbit hole – most had modern technology, and one, more unusually, an interfaith prayer room. In contrast, participants in the *Living Garden* became flowers. They imagined the space around them, or rather the type of garden that they would inhabit, and the feelings of being rooted in one place. Examination of the symbiotic relationship between person and object served to give participants a greater insight into their subject matter. Young artists were encouraged to create work based on physical experiences rather than intellectual or observed concepts.

4 Potential food allergies were confirmed prior to the workshop, and the items presented to the children were altered accordingly.

5 A young artist in a similar Alice-themed workshop held at New Marston Primary School in Oxford in 2015 raised an interesting objection regarding the olfactory exercises. This Year 4 child wanted to know if the generic synthetic meat smell I had given her to test was halal. Given the symbiotic nature of smell and taste, this raises ethical issues regarding the use of bovine and porcine scent in art education and in public spaces. Islamic law is strict about interaction with any non-halal meat products, so great care will need to be taken about meat-related smells for future classes. Jewish law is less strict regarding kosher food, as the rules only prohibit the eating of certain food, and not obtaining other benefits from it.

6 Young artists were enthused by the taste memories evoked through the olfactory games. The synthetic scents of sugar cookies and chocolate truffles even had a physiological effect, as much of this class left the workshop craving sweets.

7 Through various rowdy drama games, the young artists explored how context affected their subject matter – e.g. how differently people behave in fine dining and fast-food restaurants, or what it feels like to be crawling through the desert craving food and water.

8 Carroll (2015), p. 163.

Reference

Carroll, L. (2015). *Alice's adventures in Wonderland*. London: Macmillan Children's Books.

PART II
Representing Smell
INTRODUCTION

Chris Perkins

Daffodils in our garden are blooming as I draft this introduction, their sweet and pungent aroma carrying on the wind. The aromatic qualities of the flower change over time. When picked and displayed in a vase, their smell permeates a room, becoming slightly sickly, and as their freshness declines, so the meaning attached to them alters – their smell comes to signify death and decay.

A scientist could analyze the particular chemical composition that comes together to make up their smell, and in so doing could distill and replicate the compounds that convey the aroma. The work of, for example, Wang, Yu, Qi, Gao and Yang (2013) and Koksal, Kafkas, Sadighazadi and Kulahlioglu (2015) on daffodils illustrates the potential of this kind of technological intervention. In that sense, smells exist 'out there,' as volatile but ephemeral molecular entities, and industries and fortunes depend on this science to capture essential oils. But the meanings that are attached to daffodils depend upon much more than replication.

Daffodils are also smelt by individuals, in richly diverse everyday practices, that reflect different ways of apprehending, conceiving and imagining what their smell might mean as part of human experience, and how it might relate to the striking and romantic visual image of the flower. In the wild, they have come to signify something uniquely English, emerging in early spring to evoke life after the depths of winter, a sweet harbinger of an emerging natural summer landscape, frequently planted as a central part of park and garden flora. Wordsworth's romantic and poetic evocation of the flowers growing near Ullswater in the Lake District (voted the nation's most popular poem) has played no small part in this process. But across the country, daffodils are largely no longer wild. Cultivars of *Narcissus pseudonarcissus* are grown as an industrial crop for the cut flowers market, or marketed as a plethora of carefully modified bulbs for planting out (see Tompsett, 2006). It is Wordsworth's *representation* of the wild plant that carries the cultural power to move us to cultivate versions that are crafted, and tame, with larger blooms and different aromatic qualities. The poet chose to try to capture qualities of daffodils, to *represent* things that are beyond science. These representations and the meanings associated with them vary from culture to culture, and across time. It is these cultural representations, beyond the plant itself, which change lives and landscapes.

This subjective experience, perception and representation of a smell make a significant difference to the culture and sociality of any smellscape. Part I focused on the creative potential of subjective work with smell, with evidence from artists engaging with the smellscape. By way of contrast, Part II focuses upon translation, and the challenges that face anyone trying to *represent* smells. As a strongly individual sensory modality, smell presents significant challenges for those who wish to communicate something of its essence to other people. Translation is necessary. Perfumes are advertised in visual media. Writers convert slippery smells into words to capture and fix their qualities. The changing spatialities of a smellscape may be mapped using a cartographic imagination. This part gets to grips with these challenges by exploring how other modalities might be deployed to share something about smell in other media, and in particular by artists, cartographers and people creating written narratives around smell.

Chapter 5 by Miriam Songster shows how artistic interventions can recreate aspects of a smell experience in order to facilitate a critical engagement with wider issues. Using her *GhostFood* exhibit, the artist describes the design of the public art project which incorporated food samples, and the creation of a highly targeted and controlled scentscape to invite participants to question taken-for-granted ideas about global warming and urban sustainability. Songster explores the selection, testing and evaluation of scents in the project, and documents the role of a serving van, servers, flavors and customer experiences. She charts the relations of taste to smell, and the roles that audience memories played, setting these against social and cultural contexts in which the project was enacted. Her chapter carefully marshals diverse evidence emerging from *GhostFood*, including rich evocations of how people felt about the experience and the issues that it raised. Together, this evidence hints at the cultural significance of smell and some of the ways in which olfactory design might exploit technological change, and in so doing throws light on issues from well beyond the apparent and literal smellscape.

By way of contrast, Chapter 6 by Julia Feuer-Cotter focuses on the potential of smell as a central part of storytelling, highlighting how aspects of life in a remote village community might be translated and brought to audiences far removed from their experiences. She deploys the insights of non-representational theory to argue for an embodied approach to smell. Using case evidence from the Kaktovik community in the North Alaskan oilfields on the shores of the Beaufort Sea, she explains the significance of her collective *Smells of Change* installation. In this installation, local people together with the researcher created and distilled a smellscape, to be set against other made objects evoking a sense of their place and smells. She enrolls these objects, smells and feelings into the construction of stories of place, and uses these multi-mediated experiences to exhibit changing and fleeting aspects of the past, present and future of the community, evoking embodied experiences of a threatened and remote way of life. Feuer-Cotter describes a project which reveals how smell can be used in narrative to contribute to an appreciation of often-complex issues, bringing a changing smellscape to distant strangers who would otherwise not have been able to experience it. Her arguments for a multisensory translation of smellscapes stress the importance of embodied cultural experience, and also hint at an important politics hidden in more literal distillation of smells.

In Chapter 7, artist and designer Kate McLean also explores the phenomenological possibilities of smell, but instead of narratives, she focuses on the creative potential of mapping a smellscape and exhibiting these results. She charts the differences between cartographic orthodoxies that see mapping as mimetic and entailing capturing objective qualities of the world, as against more artistic, creative and performative understandings of mapping practice which have recently come to interest theorists rethinking mapping, as well as artists remaking the medium. McLean draws on examples of smell mapping which focus on how we smell, exemplified in the work of Nicola Twilley. She also cites the influential explorations of relationships and lived experiences in the long-running series of interventions from Tolaas. But the majority of McLean's argument draws on her own diverse cartographic practice, investigating the challenges of mapping the temporalities of different smellscapes. This mapping entails a process of capturing of elements of the smellscape in walks across different smelly contexts. The size of a group, the numbers of participants, the route, the nature of artist intervention and the mode of recording alter what is apprehended, and hence completely change mapping practice. McLean also describes the design challenges of visualizing a smellscape, and justifies her changing use of graphic conventions, which at times have fixed smells by means of contours, dots and color, whilst at other times have sought to animate representations of the smellscape, or incorporate smells themselves into mappings. She considers the central importance of exposition in this mapping process, and the role of real and digital spaces and interfaces through which smell mapping might be deployed. McLean closes with a clarion call for why personal mapped representations of smells might matter: "Mapping, when viewed as a situated, collective process, rather than representative artefact with inherent meaning, has a creative potential to reveal the unseen, ephemeral and imagined" (p. 76).

These concluding words to my introduction highlight why the details of the translation matter. Tastes, stories, objects and maps have come together in this part to make something different about smell, in projects that enroll changing feelings about embodied sensory experience. These different projects move participants, both literally and metaphorically. They show the power of representing smells, and the importance of that work in conveying qualities of the smellscape that can make a difference to places. When you see the daffodils next spring, 'dancing in the breeze,' think about how their aroma contributes to the experience, the memories and feelings you associated with that moment, and ponder the challenges of communicating that moment in other media!

References

Koksal, N., Kafkas, E., Sadighazadi, S., & Kulahlioglu, I. (2015). Floral fragrances of daffodil under salinity stress. *Romanian Biotechnological Letters*, *20*(4), 10600–10610.

Tompsett, A. (2006). *Golden harvest: The story of daffodil growing in Cornwall and the Isles of Scilly*. Penzance: Alison Hodge Publishers.

Wang, J. Y., Yu, L. L., Qi, X. L., Gao, H. J., & Yang, J. X. (2013). Difference in aromatic components between Hyacinths and European Daffodil by GC-MS. *Journal of Beijing University of Agriculture*, *1*, 014.

5

THE *GHOSTFOOD* PROJECT

Enhancing Flavor through Personalized Smellscapes

Miriam Songster

> The sensorium is a social fact.
>
> *Howes, 2006, p. 127*

How do we experience climate change? Mostly, we don't. It's accessible to us through research and perhaps anecdote, but on a day-to-day basis, it's not part of our lived reality. Even for those of us who agree that it's a real phenomenon, it can seem so distant from our everyday lives that it's hard to imagine what it will mean for us down the road. It's as if we are 20-year-olds scoffing down double cheeseburgers: intellectually we know it's not good for us, but heart disease seems a long way off.

The *GhostFood* project was conceived as a way to create personal connections to climate change – that is, to move it from a distant, abstract concept to a relatable phenomenon that people could imagine impacting their lives. *GhostFood* accomplished this through exploiting our love of food, with all of the memories and hedonic dimensions that entails. By offering people 'taste experiences' that highlighted what they would be missing if certain species were to become extinct, *GhostFood* used eating (and smelling) as a means to experience climate change on the most personal of levels, and to encourage people to think both realistically and creatively about food and eating in a future marked by global warming and species loss. In the process, *GhostFood* also provided an opportunity for people to more closely examine the relationship between flavor and smell, and the impact each has on the way they experience food.

The individual experiences reported by people while eating and smelling *GhostFood* samples, when examined in light of the techniques and technologies used in the project, can provide some insight into possible futures for olfaction, food preparation and neighborhood scentscapes. These are discussed in the final "Learnings and Extrapolations" section of this chapter, which begins with a few observations about the effects of the project on participants, and then goes on to ask, "What scenarios might be foreseen if these simulations become popular, or if a direct olfactory stimulation device becomes a common accessory?" Individual responses to *GhostFood* have been consistent with research that shows context, memory and familiarity as determinative in our interpretation of scent and flavor. In this last section, I ask where these features of olfaction and taste might lead us, given certain social and technological trends that are already underway.

Context

GhostFood started as a proposal for *Marfa Dialogues/NY*, the second instantiation of this month-long series of artworks, lectures and events on the topic of climate change.[1] *Marfa Dialogues/NY* was organized by Ballroom Marfa, funded by the Robert Rauschenberg Foundation, and realized in the fall of 2013. The *GhostFood* project was performed in two locations: at the Rauschenberg Foundation's Chelsea, New York gallery during the opening of the show *Quiet Earth*, and on Main Street in Newark, New Jersey during the Newark *Open Doors Arts Festival*. The project has continued with performances in other locations; however, it was with these initial two audiences in mind that the decisions guiding the project's development were made, and it is their responses that inform the discussion below.

What Is *GhostFood*?

GhostFood is a participatory performance based around a food truck (see Figure 5.1) that serves 'taste experiences.' These taste experiences are based around three foods that normally are made using a species that is threatened with extinction due to the impacts of climate change. We chose species from three different ecologies: chocolate (rainforest), cod (ocean) and peanuts (grasslands). Each taste experience combines a synthetic scent[2] and a partially simulated version of a food that requires one of the threatened species.

For the edible component of each GhostFood, the potentially unavailable ingredient is replaced by an edible textural substitute made from climate change-resilient foods. The flavor of the missing ingredient is simulated by a lab-made scent delivered via *GhostFood*'s Direct Olfactory Stimulation Device (DOSD) (see Figure 5.2), which is worn on the face and positions a pod with a scented

FIGURE 5.1 *GhostFood* truck with servers.

Image courtesy of Miriam Simun and Dan Scofield.

FIGURE 5.2 The Direct Olfactory Stimulation Device and 'chocolate' milk.

Image courtesy of author and Miriam Simun.

wick under the wearer's nose (see Figure 5.3). This simulation is possible because of the role that scent plays in perceiving flavor.[3] When ordering a taste experience from the *GhostFood* menu, you are getting something that is partially what you think you ordered and partially a simulation via the DOSD.

The experience was facilitated by trained servers: actors who knew how the three species are threatened by climate change, and how the DOSD was designed to affect flavor perception. These servers were a combination of waiter and guide, helping participants use the DOSD, and shaping their experience through the use of scripted lines and context-sensitive dialog.

GhostFood's menu included beer-battered deep-fried Atlantic cod, Arizona peanut butter with grape jelly sandwich, and chocolate milk with West African cocoa. There were no actual cod, peanuts or cocoa in these foods, but the geographical sources of each were specified in the menu so as to call attention to place-specific effects of climate change and to echo the descriptive style used in menus of upscale restaurants that tout locally or organically sourced foods.

Ingredients from non-threatened species (e.g. batter, bread, jelly, milk) were used in the GhostFoods just as they would be in the 'natural' version; so the foods were mostly what they purported to be, and looked respectively like a fried morsel, a quarter of a sandwich and a small container of milk (shown in Figure 5.2). The normal-looking food combined with the menu descriptions to set expectations and influence perceived taste.

How Does It Work?

GhostFood works through two main physiological routes. First, in retaining the texture of the original food, a serving of GhostFood has the mouth-feel of the original version. The textural cue is consistent with expectations (Shepherd, 2011, p. 123). Second, a strong orthonasal and some

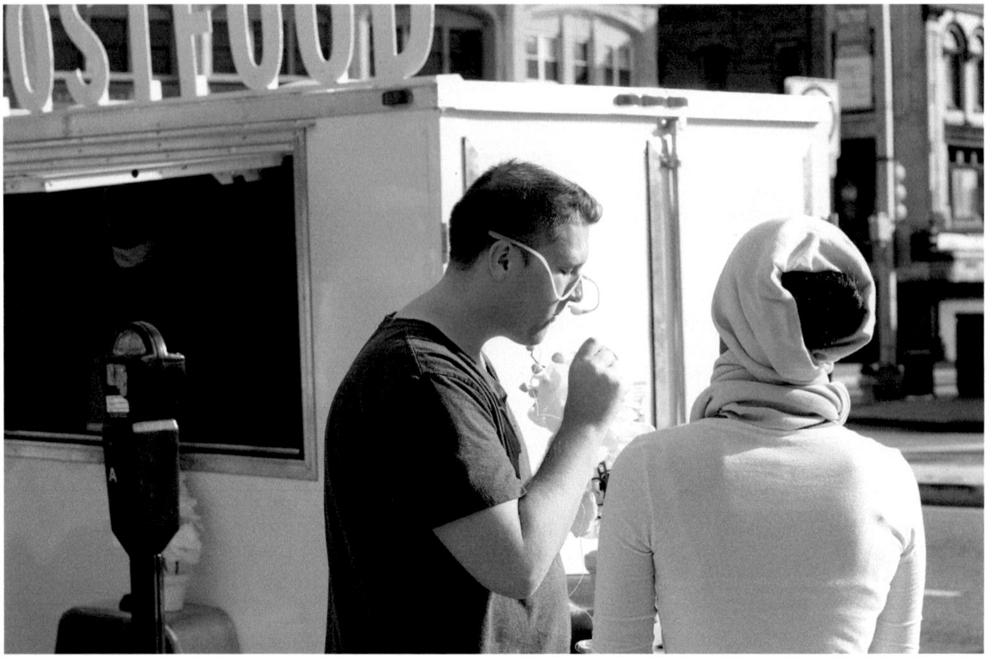

FIGURE 5.3 A customer savors a *GhostFood* sample.

Image courtesy of author and Miriam Simun.

retronasal olfaction can be achieved by synthetically boosting the smell that normally comes from the missing ingredient, and positioning the DOSD pod between the nose and the mouth (see Figure 5.3). During performances, the *GhostFood* servers encouraged participants to eat with their mouths slightly open, so that the retronasal effect – and therefore the illusion of eating the 'missing' ingredient – was enhanced.[4]

In order for these physiological cues to add up and create the illusion of eating the normal version of the food, that food must be familiar to the participants and their expectations must be managed. These requirements guided both the choice of foods and their presentation, as will be discussed later.

Over hundreds of participants, the effectiveness of the *GhostFood* simulation differed from person to person. For each of the three scent/food combinations, some people reported that their sensorial experience was similar to eating the original. Meanwhile, under the same conditions, others reported that the combination of the scent and the food did not cohere into a recognizable taste experience. (Sample responses appear in Table 5.1.)

One way to explain the difference in taste experience is to look for physical causes. For example, how much odor entered the nose and mouth of each participant? This would depend on the placement of the scent pod, the amount of scent left on the wick, the relative timing of the person's in-breath and chewing, and whether he/she was breathing in through the mouth or not. Another physical explanation might be differences in taste and smell acuity, although it's hard to say whether more sensitive perceptive apparatus helps or hinders the experience.

However, the physiological and mechanical aspects of eating and smelling GhostFood are only part of the story. Research shows that "the brain actively creates the sensation of flavor" (Shepherd, 2011, p. 4), and that while "simple tastes are hardwired from birth . . . retronasal smells are learned and thus open to individual differences" (p. 3). It appears that exposure to scent creates odor images

TABLE 5.1 Quotes from the participants

"So is this not peanut butter in here? It's good. It tastes just like peanut butter and jelly. I'd be really curious what's in it. It's really wild. It tastes just like peanut butter and jelly. That's very cool. Where is it coming from?"
"I get peanut butter and jelly. Oh my god, you're right, peanut butter and jelly! That's what's crazy. That's amazing, because it gives you the other piece, the food that's missing."
"I don't taste peanut butter. I think I'd like to have the real thing."
"I'm aware that I'm smelling cod and eating something tasteless and smelling fish at the same time. When you smell your food, it makes you really taste something. I can differentiate between the taste of what I'm eating and the smell of the cod. So I guess there was a moment when my brain was tricked into thinking I was eating cod."
"It's really interesting. Because it doesn't come naturally. I have to think of the process: chew, and breathe. I didn't get it to work at first. The smell is pretty cool, very like cod. But the soy part not so much. Maybe I wasn't being adept enough at eating and smelling it together?"
"It smells totally like cod fish, like bacalao. So this is not codfish? Wow it tastes just like fish. That's what they are talking about: in the future, we will just be smelling this, and eating whatever, grass or something. But I got the chocolate milk more. I really got the flavor of chocolate milk. This was a different experience."
"I tasted milk, but I just smelled the chocolate."
"Wow. This milk is unflavored? It tastes just like chocolate milk. It was really rich. It's interesting how much scent influences the taste of food. It's shocking. Quite dramatic. Deceptive. Great. Interesting."
"This is a legitimate taste experience. The correspondence between senses makes it seem real. I'm glad you'll be here, because it's inevitable that one day we'll have to experience the taste of chocolate this way."

FIGURE 5.4 Bye Bye Baby Cod.

Image courtesy of author and Miriam Simun.

in our brain, which we combine with tastes and other senses to realize the perception of flavor (p. 5). In other words, our past eating and olfactory experiences, along with their associations, play a significant role in our interpretation of the scents and flavors we experience later on (Damian & Damian, 2006; Herz, in Nguyen, 2015). Moreover, "there appears to be a strong tendency to judge unfamiliar smells as unpleasant; this finding relates to the concept of habituation" (Porteous, 2006, p. 90). Therefore, whether *GhostFood* worked as a recognizable taste experience, and how the experience was interpreted, would depend in part on the personal history, associations and expectations of each participant.[5]

Moreover, because *GhostFood* incorporated not just the food and scents, but also the truck, signage, menu, serving tray, 3D-printed DOSD headset, food packaging, napkins (Figure 5.4) and server outfits and interactions, a participant's reaction to *GhostFood* could be affected by his/her past experience of factors as seemingly far-flung as contemporary and performance art, modern design, and street food. The full range of influences is impossible to know, but it is certainly indicative of how designing with smell is contingent on both the other designed features of the experience and much that is unknown and largely unpredictable.

Design Decisions and Social Context

In designing *GhostFood* – from the food/scent combinations to the visual branding and the server interactions – the social contexts surrounding the performances were a major consideration. We asked ourselves questions such as, "What foods might be meaningful to the participants?", "What expectations will they come to the performance with?" and "What should the server's persona be like, and what should they say?" These were asked with reference to local demographics and the social fabric of each performance location. All *Marfa Dialogues/NY* performances were open to the general public, with the *Quiet Earth* exhibition expected to draw a fine arts and ecology-oriented crowd, and the Newark shows expected to attract a broad local audience.

The Truck

GhostFood was served out of a food truck to take advantage of the familiarity of this food delivery model, and thereby position *GhostFood* as part of an everyday eating experience. Food trucks are ubiquitous in many US cities, including New York and Newark, and all kinds of foods are now served from them. Therefore, it's not surprising that a few customers did not realize at first that the truck was part of an art performance. On close examination, the disparities between the *GhostFood* truck and other food trucks stand out: the *GhostFood* truck didn't emit any cooking smells, and its design was clean and futuristic. The intent behind this visual aesthetic was to gesture toward the future, but not to venture into science fiction cliché or suggest a specific future date. The hoped-for impression was: 'This could be your neighborhood food truck at some point.'

The Servers

Similarly, the outfits of the servers also fit this clean, minimalist and ambiguously futuristic aesthetic (see Figures 5.1, 5.3 and 5.5). The servers maintained a calm, encouraging, friendly aspect so as to create a non-threatening environment that allowed participants to savor their taste experience in a relaxed manner. Key phrases consistently used by the servers included: "*GhostFood* serves simulated flavor experiences of species threatened with extinction by climate change," "*GhostFood* serves 'taste experiences' of soon-to-be unavailable flavors," and "This is the *GhostFood* device, it

FIGURE 5.5 Serving *GhostFood* at the *Newark Open Doors Festival.*

Image courtesy of author and Miriam Simun.

inserts direct olfactory stimulation into the eating experience." If someone eating fried cod asked what was in it, he/she would be told it was an edible textural substitute made from climate change-resilient foodstuffs.

This unfamiliar language, combined with ambiguities around the truck – Was it a regular commercial food truck? Was it set in the future? – destabilized participants' interpretive framework. People didn't know what to think, which was an intentional effect of the project design – to create imaginative openings where participants could explore their own thoughts about the future of food, and become more aware of their sensory experiences of tasting, smelling, listening and seeing.

The Flavors

Potential menu items were restricted by the production process. Not all foods threatened by climate change can be effectively simulated or have their odors recreated in a lab. Just as important was choosing foods that would be both familiar and meaningful to local audiences. In order to experience how a food might be different in the future, you have to know what it tastes like now, in its contemporary real form – 'real' in the sense of using natural ingredients, and also because the food has an associated odor image that makes it seem real or right to us (Shepherd, 2006, p. 317). Without having actually eaten that real version, there can be no moment of recognition when experiencing the *GhostFood* version, nor any sense of what has been lost in translation. Moreover, for the purposes of *GhostFood*, we wanted participants to have an emotional response if their eating experience fell short of what they were used to. These responses are related to smell images and food attractions (or addictions). If there was a strong pre-existing relationship with the food a participant chose to eat, that person might care more about its potentially being unavailable in the future.

Given the performance locations and the practical considerations of scent and food preparation, we chose codfish, peanut butter and chocolate as the 'ghost' foods. These seemed likely to engage our specific audiences on an emotional level. Peanut butter and jelly sandwiches are a staple childhood meal throughout the United States, so almost anyone who is not allergic to peanuts will have had years of associations with this particular food combination, often spanning from their own childhood to those of their children and grandchildren. Fried fish is also common, and cod in particular would be well known in Newark, which has a large Portuguese population. Chocolate evokes such strong feelings among its fans that some call themselves chocoholics.

The Customers' Experience

Participants viewed the *GhostFood* menu, placed an order at the truck's window, and upon receiving their tray (see Figure 5.5), went to a table where one of the *GhostFood* servers helped them put on the Direct Olfactory Stimulation Device. It was important that participants chose which GhostFood they wanted, so they could have the taste and smell experience that was most likely to resonate. It is these resonances, along with the emotions brought forth by the gap between the real and the simulation, that form the key moments of the *GhostFood* project.

Were the simulations perfect, there would be no sense of loss and therefore little chance of creating a personal connection to the threatened species and to the issue of climate change. This moment of difference – where the simulation is experienced as both familiar and yet not – is a moment when the reality of climate change can feel personal. It is also a moment that can lead to new thoughts about scent and flavor.

Learnings and Extrapolations

GhostFood explored one scenario about how we might eat in the future. It did so through a combination of chemistry (synthetic scent), technology (via a 3D-printed Direct Olfactory Stimulation Device) and food preparation (familiar foods prepared with substitutes for environmentally unsustainable ingredients). The resulting 'taste experiences' were consumed by the public, and many people found them to be a satisfying facsimile of the original food (see Table 5.1 for representative responses).

Evaluating the project in terms of one of its primary aims – affecting attitudes about climate change – it's worth noting that quite a few participants asked why a species was threatened, and some discussed what it would be like for themselves and for their children or grandchildren to live in a world where these species are extinct. This interpretation led many participants to express an expectation of personal loss, in that they preferred the 'real' food, and did not think that replacing it with a scent/food combination would be as enjoyable. Not infrequently, participants also expressed a need to do something to reduce climate change and/or species extinction. Measured against the intentions of the project, these lines of thinking were productive:

> [This makes me feel] sort of helpless. The problem is huge to try to take on. It would be a replacement, but not like the real thing. I guess as long as we're getting nutrition.
>
> It really tasted like cod. This was very tasty. Shockingly similar, but I wouldn't be okay with it, because I'd like cod to live. I'd like the species to endure.
>
> I tell you, I never thought peanuts would be endangered. So this could be what my granddaughter's granddaughter could be eating? So why is this soon to be unavailable?

> If there won't be peanut butter in the future, then she will only know this experience. Someone like me, in my elderly years, would be able to compare the two and see the similarities. And I wonder how much it will cost if those foods are in danger? Will it cost more than peanut butter does today?

The *GhostFood* project was also intended to encourage people to become more aware of the role of smell. Most people were either unaware of the role of odor in creating flavor or had not given it much thought, and they were very surprised to find themselves experiencing a familiar flavor via the combination of a scented pod and an edible snack that was missing a key ingredient. Sometimes this led to a new appreciation for the role of smell:

> Oh it's so good [the smell]. The smell's delicious. Oh gosh, it tastes like chocolate milk. Nice. Very good. I realize that I have to smell my food before I eat, because that changes the palate.

GhostFood demonstrated that combining a synthetic scent with an imperfect version of a known food can be an effective way of creating an eating experience that people like; and existing research suggests that this is the case because we can use both our senses and our memories to fill in the missing components of a familiar food. The success of the simulations, when considered alongside the science of olfaction and certain cultural trends, has implications for what we might eat and how food could be prepared and consumed in the future. These in turn have implications for the environment – both the natural environment as well as local scentscapes that are impacted by human activities. Moreover, because our taste – both in terms of what we like to eat and smell, and what we think is stylish or cool – is dependent on what we are exposed to, it's worth considering some longer-term scenarios that might occur in a world accustomed to *GhostFood*-like eating experiences and the use of head accessories to transmit hyper-local personalized scent.

To see where we might be headed, let's start by considering some of the comments from participants who emphasized the benefits of scent/food pairings:

> So this way we will have a good time with chocolate without needing to have the chocolate, and you don't gain weight.

> It's actually quite good. It has good mouth feel, a neutral taste, like a fish of some kind. I could eat a couple more of them. So these are foods, when the earth is all covered with water, we will be eating. I guess I'm all for climate change, then.

> Wow, it works. I'll be damned. It really tastes like chocolate. It's amazing. I guess climate change is not so bad. I guess I'm prepared now.

For these people, the future scenario presented by *GhostFood* was not perceived as negative, but as a means for new, and positive, opportunities. If it's true that 'the way to a man's heart is through his stomach,' then it's unclear what the effect will be on threatened species if consumers experience a *GhostFood*-like simulation as an effective substitute. Will more species be able to thrive because we will no longer over-fish, as in the case of cod? Or will we start to see more species as dispensable, and do less to prevent the rising temperatures that threaten peanut and cocoa crops? In a parallel observation regarding natural scents, Alan R. Hirsch says:

> If nostalgia for the odors of the natural world experienced in childhood is an important impetus behind our concern to preserve the environment, that concern may fare poorly in

> another generation when the children of today become adults who can only experience olfactory-evoked recall in response to manufactured chemical odors.
>
> *Hirsch, 2006, p. 188*

Looking beyond issues of species unavailability, how might olfaction's connection to flavor be utilized in different contexts, and how might olfactory environments be impacted? The positive responses to the scent/food pairings of *GhostFood* suggest untapped potential, especially when considered in light of how malleable are our hedonic responses to flavor and odor.[6] What scenarios might be foreseen if these simulations become popular, or if a direct olfactory stimulation device becomes a common accessory?

Already there is one device that introduces synthetic food odors into the ritual of eating. Targeted towards the elderly (who often have a diminished sensitivity to retronasal smell (Shepherd, 2006, p. 320), and particularly to those with dementia, the Ode device diffuses food-related scents into a room as a way to stimulate appetite before meals. In the future, perhaps restaurants will apply a similar technique, diffusing a series of scents to enhance each course in a diner's meal. This could be through a family-style diffuser for groups, or through individual scent pods, similar to the DOSD used in *GhostFood*.

A DOSD delivering personal scentscapes might also become a weapon in 'the battle of the bulge.' By delivering a scent of the wearer's choice, it could block tempting odors wafting from fast food chains and bakeries; or it might dispense odors that are designed specifically to encourage weight loss. This idea is already being used in products such as the Aroma Patch.[7] With continued research into the neurobiology of food cravings and advances in molecular gastronomy, the replacement of enticingly odorous but unhealthy cooked foods with combinations of scent and synthesized food could become an enjoyable and healthy way to eat. Encountering fewer food smells would also mean that people will not have the opportunity to form olfactory images around some of the 'unhealthy' food odors, so they will not experience the same food cravings that haunt many of us today (Shepherd, 2011, pp. 190–199).

The role of olfactory images and the way our hedonic response to particular foods and smells is determined through exposure suggests that over time, we might not only become accustomed to synthesized versions of 'real' foods,[8] but we might also learn to love creations that we have so far not imagined. Similar to the fantasy flavors[9] created by the food flavoring industry, these new sensory experiences need not mimic the natural world. Perhaps there will develop a whole new realm of sensory experience makers, with mixologists, curators and DJs (or 'SJs,' smell jockeys?) catering to many niche olfactory and flavor palettes, including some who prefer dissonance, some who want to experience extinct flavors, and some who are synesthetes who want to share the scent of the color red. Possibilities may even reside in the rifts between smelling and tasting, not just in the synchronicity of the two. Could we learn completely new mappings of scent/food combinations, so that what we now experience as an inadequacy (in cases where a combination does not match the flavor of any food we currently eat) becomes a synchronicity through repeated exposure to previously unavailable combinations? Where are the limits to what we can learn to enjoy?

But there is a down side to losing touch with cooking smells. As I write this, the neighboring restaurant's fan is pumping out fried canola oil exhaust, an odor I dislike intensely and would gladly never smell again. Others may feel this way about the smells from their local fast food or fish and chips shop. However, research shows that even when we may not enjoy a particular food smell, we may still appreciate its role in defining place (Henshaw, 2014a, p. 200[10]). Local scentscapes play a significant role in our sense of belonging and in our enjoyment of both the places we know and new places that we visit. Whether at home, on our street, or throughout a city or even a country, place-specific associations with food and other smells influence how we characterize a locale and understand our relationship to it (Porteous, 2006, p. 91[11]).

The importance of these smells is expressed by a patient of Oliver Sacks (2006):

> Life lost a good deal of its savor – one doesn't realize how much 'savor' is smell. You smell people, you smell books, you smell the city, you smell the spring – maybe not consciously, but as a rich unconscious background to everything else. My whole world was suddenly radically poorer . . . .
>
> *p. 185*

This observation – by an anosmic person who could no longer smell anything at all – emphasizes how greatly we are affected by the specific odors that we smell in the course of daily living, and suggests the impoverishment we feel when they are no longer available to us.

Place-specific smells are already waning, part of an ongoing transformation of local scentscapes due to zoning regulations, gentrification, air conditioning and the movement of industry out of city centers: '[T]he smellscapes of city streets are turning into sterilized clones of one another' (Henshaw, 2014b, p. 29). A formerly practical concern with urban cleanliness and disease reduction has evolved into an increasing intolerance for any smell we deem unpleasant or inconvenient.[12] In a future with fewer competing food odors, plus the use of a DOSD to serve individualized smellscapes, will this intolerance become even more pronounced? Will any odor that we do not intentionally choose seem as annoying as the music we are forced to hear at the mall or the gym, which interferes with the customized content we stream through our ear buds? If we have the technology, will we try to control more and more of our sensory environment, insulating ourselves with controlled 'content' and sharing less of our immediate surroundings with the people around us? Will we still be asking our travel companions, "What *is* that smell?" or bonding over the familiar smells of our hometown? Would a disconnection of smell and place matter?

I would argue that certain shared identities and experiences are contingent upon place, and that the homogenization of place through any means, including smell, does threaten these shared identities and limit the meaning we can derive from time spent in specific locales. Humans have the capacity to distinguish very specific odors, and we equate those subconsciously with an experience, and often a place. Proust did not smell a baked good, he smelled a madeleine, and it reminded him of his aunt and her home at Combray. The memories we recall when faced with a familiar odor are bound up in our own personal histories – of the when and where of smelling. If a food always smells exactly the same no matter its source, or all places smell alike, the memories associated with those foods and places become non-specific and lose their emotional impact. Therefore, the departure of odors that are produced by the people living or working in a neighborhood and their replacement with non-location-dependent odors will perforce change the meaning of those neighborhoods. These non-location-dependent odors might come in the form of a DOSD or in a more familiar form, for instance interventional odors such as piped-in bakery smells at the supermarket and custom branded scents in retail stores and hotels, but in both cases the specifics of locality – the daily activities of home or industry that would traditionally create a local scentscape – are irrelevant. Traditional, unplanned and heterogeneous smells are out, while new interventional, proprietary and uniform smells are in.

While the *uniform* nature of interventional smells has been emphasized above, our accommodation to their *interventional* nature also bears examination. What if our personal olfactory stimulation devices were controlled in large-scale public or corporate interventions? These interventions are not a new idea. In one example, research at Duke University, initially conducted to explore how scents could help patients, has been repurposed for insights into controlling aggression in subway and prison environments (Damian & Damian, 2006, p. 155). The purposeful control of scentscapes by technocrats, politicians and corporate interests might be facilitated by a general acceptance of interventional approaches to environmental scenting, even if individual DOSDs could be hacked to circumvent state control. At issue is a similar concern to that which Jim Drobnick expressed regarding ventilation systems: "The question that continually arises regarding controlled climates is: Who will be in control – engineers or users?" (Drobnick, 2010, p. 148). Drobnick references the history of ventilation systems, in which "an emerging climate sector . . . aimed to instrumentalize the atmosphere," initially for industrial purposes. This intervention was based on an engineering ideal that "ignored more complex, contextual factors such as individual preferences, cultural circumstances and adaptability" (p. 148) Similar contextual factors are at risk of being ignored in scentscaping.

Whether it's controlling temperature and humidity or controlling scent, the possibility of unitary manipulation or control of a sensory environment requires additional attention to both intended and unintended consequences. Whose interests will be served, and how might residents or any persons with less social and economic power exercise control over their sensory environment? As Victoria Henshaw (2014a) has expressed, "Given that many odours in urban environments are experienced differently by different people, it is important to think about wider ethical complexities when dealing with odour in the environment" (p. 163). With the current scenting of arenas, hotels, retail stores and offices,[13] Henshaw's call is timely.

Perhaps scentscapes will remain a free-for-all, but that is not where trends are pointing. While it is possible that our local pizza joint will formulate their own 'grandma's pepperoni' scent wick and that we will all be designing our own 3D-printed olfactory accessories, it is probably more likely that we will be buying our olfactory devices from Coach and our wicks from Pizza Hut. But, for now, we still have cod, peanuts and cocoa, and perhaps we won't ever need or want the DOSD, the scent wicks or the partially simulated foods. Maybe we will find a way to preserve the natural environment and the species that we depend on for our favorite taste experiences. Either way, it seems likely that we will adapt, continue to create sensory experiences that we enjoy, and turn up our nose at those we do not.

Notes

1 The request for proposal stated: "Driven by the collective impulse to broaden public discussion, evaluation and understanding of climate change and artistic practice, RRF [Robert Rauschenberg Foundation] is announcing a request for proposals from organizations interested in presenting arts-related projects on the topic of climate change." Myself and fellow artist Miriam Simun submitted our proposal in April 2013, and in mid-June we learned that our project would be funded. *GhostFood* was presented in the fall of that year. The project was also chosen for the Human Impact Institute's 2014 Creative Climate Change Awards.

2 The scents were provided by Takasago International Corporation.

3 Flavor is detected when those volatile molecules reach the olfactory receptors. Research from 2014 indicates that with our approximately 400 different types of scent receptors, humans are capable of detecting at least 1 trillion different odors (Bushdid, Magnasco, Vosshall & Keller, 2014). In contrast, our tastebuds can detect only five basic tastes: salty, sour, bitter, sweet and umami.

4 Orthonasal stimulation refers to sniffing in through the external nares of the nose to activate the sensory cells in the olfactory epithelium. This is the route used to sense odors in the environment. Retronasal stimulation occurs during food ingestion, when volatile molecules released from the food in the mouth are pumped, by movements of the mouth, from the back of the oral cavity up through the nasopharynx to the olfactory epithelium (Shepherd, 2006, p. 317). "It is important to realize that flavor does not reside in a flavorful food any more than color resides in colorful object. Color arises as differences in wavelengths of light given off by an object; our brains transform those wavelengths into color to give it meaning for our behavior. Similarly, the smells that dominate the sense of flavor arise as differences between molecules; our brains represent those differences as patterns and combine them with tastes and other senses to create smells and flavors that have meaning for our perceptions of food" (Shepherd, 2011, p. 5).

5 For example, "In my work in English cities and further afield I have found the combination of smell information with insights gained from other sensory modes, and prior experiences of specific, or similar places, related associations and resulting expectations, to be critical in creating an ambiance of place and perception and producing resulting behaviors" (Henshaw, 2014b, p. 202).

6 "The most striking characteristic of human odor perception is its plasticity. This plasticity is expressed in many ways: qualitatively, in that learning and experience can modify the way odors are reported to smell, their similarity to each other, and their discriminability and hedonics, in that motivational state and past experience can alter whether an odor is liked or disliked" (Wilson & Stevenson, 2006, p. 175).

7 "Specific aromas can deprogram overweight people whose normal response to the smell of rich, unhealthy foods like chocolate, doughnuts and pizza was to become hungry and overeat. Scientists tested the benefits of food odors to suppress appetite rather than stimulate appetite, and found that there seemed to be certain smells that caused overweight individuals to reduce their cravings, and therefore eat less. . . . By forming an association between a specific smell and a desire to manage your food intake, you can learn to eat healthy foods and regain control of your weight" (Aroma Patch, 2003).

8 "People raised on strawberry Kool-Aid will prefer the Kool-Aid-type berry flavor over a completely natural berry. Actually, if you squished fresh berries and blindfolded these tasters, they would say it wasn't fruit" (Dr. James A. Steinke, quoted in Damian & Damian, 2006, p. 156).

9 On fantasy flavors: ". . . if you look at a drink like Red Bull, it's kind of powdery and medicinev and grainy and cherry. It's all over the place. But it sells. In beverages, it's one of the few flavors that has no relation to anything in nature. It's purely fantasy. I find it very cool that it's become so beloved" (Hedy Kulka, flavorist for Givaudan, in La Gorce, 2011). See also http://gastropod.com/savor-flavor/

10 ". . . smellmarks, even things as low-key as ventilation emissions released from one of Doncaster's basement restaurants, can be argued to contribute towards a sense of place. They do so by forming a part of people's everyday memories and associations with the street in which they are experienced, regardless of whether those same people would like or dislike the same odours if they were sniffing them from a test-tube. These odours instead become biographical, forming part of people's individual and social memories and their sense of 'insideness' or belonging" (Henshaw, 2014a, p. 200).

11 "In combination with vision and tactility, smell and the other apparently 'non-spatial' senses provide considerable enrichment of our sense of space and the character of place. We are all familiar with the fact that places may be characterized individually, or even typed, by smell, from the smell of India, of Mexico, of London from a generation ago during a 'pea-souper,' of Los Angeles today during a smog alert, to hospitals and the smoking section at the rear of a passenger airplane" (Porteous, 2006, p. 91).

12 It is arguable that even practical motivations for urban cleanliness campaigns have been tainted by intolerance. See, for example, the contributions of Gale Largey and Rod Watson, Martin Manalansan and Alan Hyde in *The Smell Culture Reader* (Drobnick, 2006).

13 Starting in the late 1980s, the Japanese construction company Shimizu began incorporating an 'Aroma Generation System' into offices and hospitals, accompanied by 'scent plans' designed by Takasago to serve the needs of clients who wished to encourage calmness, invigoration or other emotional states among their staff or customers (Damian & Damian, 2006, p. 148). Many other companies are now in the environmental scenting arena, including Aromasys, ScentAir and Prolitec, and both signature scents and subliminal scenting are increasingly being used in retail and hotel environments. The Barclays Center arena is scented, as are the arenas for the Dallas Cowboys and the Atlanta Hawks (Albrecht, 2013). The author notes, however, that the scent was not noticeable on a recent January 2016 visit to Barclays. Perhaps it departed when Jay-Z sold his stake in the Nets?

References

Albrecht, L. (2013, May 20). Barclays Center's "signature scent" tickles noses, curiosity. Retrieved March 1, 2017 from https://www.dnainfo.com/new-york/20130520/prospect-heights/barclays-centers-signature-scent-tickles-noses-curiosity

Aroma Patch. (2003). Retrieved March 1, 2017 from www.aromapatch.org/

Bushdid, C., Magnasco, M. O., Vosshall, L. B., & Keller, A. (2014). Humans can discriminate more than 1 trillion olfactory stimuli. *Science*, *343*(6177), 1370–1372.

Damian, P., & Damian, K. (2006). Environmental fragrancing. In J. Drobnick (Ed.), *The smell culture reader* (pp. 148–160). Oxford: Berg.

Drobnick, J. (Ed.). (2006). *The smell culture reader.* Oxford: Berg.

Drobnick, J. (2010). Architecture: Guarded breaths and the [cough] art of ventilation. In P. di Bello & G. Koureas (Eds.), *Art, history and the senses: 1830 to the present* (pp. 147–166). London: Ashgate Publishing.

Henshaw, V. (2014a). *Urban smellscapes: Understanding and designing city smell environments* (Kindle ed.). London: Routledge.

Henshaw, V. (2014b). Welcome to the smellscape. *New Scientist, 222*(2972), 28–29.

Hirsch, A. R. (2006). Nostalgia, the odors of childhood and society. In J. Drobnick (Ed.), *The smell culture reader* (pp. 187–189). Oxford: Berg.

Howes, D. (2006). Charting the sensorial revolution. *The Senses and Society, 1*(1), 113–128.

La Gorce, T. (January 17, 2011). The tastemakers. *New Jersey Monthly*. Retrieved March 1, 2017 from http://njmonthly.com/articles/eat-drink/the-tastemakers/

Nguyen, K. (2015) *The smell of skunk*. Retrieved March 1, 2017 from https://vimeo.com/135299665

Porteous, J. D. (2006). Smellscape. In J. Drobnick (Ed.), *The smell culture reader* (pp. 89–106). Oxford: Berg.

Sacks, O. (2006). The dog beneath the skin. In J. Drobnick (Ed.), *The smell culture reader* (pp. 184–186). Oxford: Berg.

Shepherd, G. M. (2006). Smell images and the flavour system in the human brain. *Nature, 444*(7117), 316–321.

Shepherd, G. M. (2011). *Neurogastronomy: How the brain creates flavor and why it matters* (Kindle ed.). New York: Columbia University Press.

Wilson, D. A., & Stevenson, R. J. (2006). *Learning to smell: Olfactory perception from neurobiology to behavior*. Baltimore, MD: Johns Hopkins University Press.

6

SMELLSCAPE NARRATIVES

Designing Olfactory Spaces as Infrastructure for Embodied Storytelling

Julia Feuer-Cotter

Thinking through smells and seemingly mundane objects can help frame considerations about connections between people and places. For example, sometimes I open a little cardboard box that I keep on my desk, filled with brass bookbinding screws, just to catch a whiff of dull patina. The little screws, which always slightly stain the book pages, hold different kinds of narratives in place while still allowing parts of the narration to be added, replaced or reordered. There is the possibility that a text can remain an open and never-finished project; there is always a chance to come back later and change the story and its sequence. Experience, practice and storytelling are not linear, and research that tells stories about and through memories, people and landscapes should not be, either.

This chapter presents the story of designing a narrative that is told through and framed by smells about experiencing Arctic change. The project was carried out in collaboration with a women's group in Kaktovik located in Northern Alaska, and UK-based visitors to various Arctic smellscape events in public everyday locations.[1] This collective design process allows various voices and interpretations to take place in the present moment of practice to explore the implications of displacing smells through the creation of alternative geographies. With my argu-ment grounded in non-representational theory and by reflecting on the collective process of designing immersive olfactory spaces and exploring the link of tacit fragmented stories to everyday lived space, I argue for collective and open design practices utilizing smells as infrastructure to tell stories without recourse to semiotics. The concept of smellscape narratives draws attention to the relationship between experiences and expressions as well as the way these connections are performed. Smells can tell stories and stories can be told through smells, and in this process, places and people are entangled through the embodied practices explored in this chapter.

Embodied Storytelling as Non-representational Practice

Through stories, the relationship we have with the natural and made environment is collective and shared between people. Narrative structures are inclusive; they engage with a participant view of processes and stories (Giovannelli, 2008). Stories take place in various ways, and narratives intersect with places to "accumulate layers of history organizing the physical and imagined changes in the landscape" (Potteiger, 1998, p. 136). Feminist scholars have pointed out that reflexive engagement with stories and identities can help to personify the storytelling process and thereby destabilize the authoritative and disembodied practices that reinforce common power structures in everyday life and academia (Beatty, 2005, 2010; Bondi, Davidson & Smith, 2007; Curti, Aitken, Bosco & Goerisch, 2011; Haraway, 1988; Urry, 2007).

The emerging tradition of non-representational thought values engagement, practice and experience over detached observation, and picks up on the concept of everyday life as a field of inquiry (Lorimer, 2003; Thrift & Dewsbury, 2000; Thrift, 2004). While the 'non-' in the name might suggest otherwise, non-representational theory is interested in representation, but through the practice of engagement (Simpson, 2010) – the theory is therefore sometimes referred to as more-than-representational (Lorimer, 2005). Broadly speaking, "non-representational theory is an umbrella term for diverse work that seeks to better cope with our self-evidently more-than-human, more-than-textual, multisensual worlds" (Lorimer, 2005, p. 83). Vannini (2015) explains the approach's broad foundation:

> With roots in the fine and performing arts, solid foundations in human geography, and expansions across cultural studies, the humanities, and the social sciences, non-representational theory is a mosaic of theoretical ideas borrowed from fields as different as performance studies, material culture studies, science and technological studies, contemporary continental philosophy, political ecology, cultural geographies, ecological anthropology, biological philosophy, cultural studies, the sociology of the body and emotions, and the sociology and anthropology of the senses – only to name a few.
>
> *p. 3*

Because this approach pushes the boundaries of conventional research practice, it is influential and at the same time controversial (Vannini, 2015). In "Dead geographies – and how to make them live," Thrift and Dewsbury (2000) criticize the apparent need to strip lived experience of its lived quality. They argue for interlinking momentary performativity and affect in what is termed "the present moment of practice" (Jones, 2011, p. 876). Non-representational thought is based on accepting that "we cannot extract a representation of the world because we are slap bang in the middle of it" (Thrift, 1999, p. 296). Shared and individual practices and sensuous dispositions are important in themselves, and not simply for what is produced through these practices. Therefore, a non-representational inquiry can add dimensions of lived experience and question power structures within various design processes. To ignore the complexity of being in the world and the messiness it involves, and instead remain focused on identifying and understanding patterns, is "to exclude a key set of relations through which lives are lived" (Anderson & Smith, 2001, p. 7).

Arctic Cultural Geographies in Practice

The Arctic is a place in motion, a geographically non-specific field that stretches in all directions with a northern focal point (Bravo & Sörlin, 2002). In attempting to explain human interaction with the environment, the significant factors are not the physical features of the landscape, but rather the idiosyncratic experiences of people living in or visiting a particular place. Two people can stand next to each other in the same place, at the same time, and yet their frames of mind can lead them to perceive the environment in different ways (Steel, 2015). This is critical to an understanding of the Arctic, a place where different understandings have had various implications in the past and will likely continue to do so in the future. The most powerful players are often located outside of the space that is defined as the Arctic. A contemporary and prominent example concerns not just the relationships between the oil industry and local indigenous and non-indigenous peoples (Catton, 1997; Coates, 1991; Keeling & Sandlos, 2009; Wilson, 2015), but also science itself (Wright et al., 2012). Centuries of funneling resources into knowledge centers south of the Arctic means that researchers engage with Northern people, their voices and their knowledge in only a limited capacity. The process of knowledge production about the North, which usually involves only briefly touching the ground to conduct projects that are often of little importance to the people who inhabit that place and who are often obstructed from even accessing the results of these studies, challenges the rights and agendas of Northern indigenous and nonindigenous peoples (Bravo, 2000; Cruikshank, 1992, 2005).

The Northern artist and researcher Chetwynd (Moffitt, Chetwynd & Todd, 2016, "Tacit Knowledge," para. 3) explains that "[i]n the North, knowledge is acquired through pragmatic engagement based upon directly integrating experiences grounded within . . . relations" between people and places. She stresses the importance of relational "information derived through stories, cultural teachings, and connection with place, people, and materials, through the acts of 'doing'" (para. 3). Although seemingly mundane practices, everyday performances link Northern people to their land (Bravo, 2000; Collignon, 2006; Cruikshank, 2005). Because of its tacit dimension that is gained through practices, Northern knowledge remains largely excluded from academia, which consequently ends up producing knowledge about the region without its inhabitants (Collignon, 2006; Cruikshank, 2005). It is important to consider this infrastructure of knowledge, evident through regulation of what valuable data is and how it is transmitted, when designing projects that engage with the Arctic (Catton, 1997; Nuttall, 1998).

A significant recent change in the outside understanding of the Arctic is the recognition of the capacity of Arctic communities to represent themselves (McConnell & Toopetlook, 2009). In the past, most Northern people were represented by the teachings of adventurers and anthropologists. From the outside, the popular perception mostly frames Alaska as untouched wilderness with ideas of the Last Frontier, expansion and natural conservation (Borneman, 2003; Catton, 1997; Cole, 2008; Kollin, 2001). These powerful concepts respond to imagined geographies where metropolitan cultures can reimagine a destiny in an unbound prospect (Kollin, 2001). Visual storytelling about Alaska is, as Duffy (2009) observes, particularly problematic. Wide-angle shots of the iconographic landscape that make the traces of the oil industry appear insignificant while the surrounding area seemingly remains undisturbed are often produced to appeal to the industry that sponsors everyday life in Alaska. Similar impacts have been observed in different Northern countries (Bravo, 2000; Cruikshank, 1992; Powell, 2005; Wilson, 2015). Duffy argues that this kind of art has impacted Alaska not just by creating structures in the landscape, but also by fostering artwork that uncritically embraces these physical traces by funding the creation of more art that does the same. This makes other non-visual and non-text-based media attractive for approaching stories from Alaska, as these can attend to other voices. The complex power structure surrounding the voice that is telling the story can be partially counteracted by attending to performance and relational, as well as tacit, knowledge.

Embodied Sensing and Knowing

Sensing, which is how we perceive the world around us, helps us recognize our place and situation in this world (Rodaway, 1994). The practice of sensing can be understood in terms of experiences through material encounters (Kearnes, 2010). Data, according to Kurgan (2013), "means nothing more or less than representations, . . . not presentations of the things themselves, but representations, figures, mediations – subject . . . to all the conventions and aesthetics and rhetorics that we have come to expect of . . . narratives" (p. 13). Collected data, then, is not about the world, but rather exists as part of our world and therefore alongside people who function as sensors for this data. Rather than negating representational elements to continue embracing the authority of data, Turan (2015) grants data an understanding that encompasses both. The senses are not just organs of perception, but also a powerful source of material data. The collected data is material in the sense that it is "registered in our consciousness, or in the case of repression, the unconscious knowledge, of our physical experience" (Stewart, 2005, p. 59). Often this embodied practice centers on data that cannot be adequately expressed with words; the result of this practice is not cognitive, but rather tacit knowledge that is as meaningful as data collected and communicated in different formats. Thrift (1996) argues that analyzing "the ways in which subjects know the world without knowing it" moves beyond interpreting texts and images, and include embodied information gained through practice (p. 10). Practice-led design processes rely on performance in the way they engage tacit information (Smith & Dean, 2009). This format for collecting data produces diverse information through different frames of reference that can shed light on different perspectives, which is fundamentally important when discussing such complex issues as Arctic change.

Yet it is equally important to the advocacy of practices to make space in the regulations that frame the applied inquiries and their results. Nelson (2013) sees institutional hindrances to engaging tacit knowledge. These constraints, Nelson writes, "range from strong academic traditions which privilege theory, to divisions between theory and practice in the very structures of education . . ., and regulatory frameworks which in some instances effectively exclude PaR [Practice as Research] by inscribing the scientific method" (p. 5). Peripheral participation of communities and practitioners challenges information exchange infrastructures that are oriented towards disembodied learning (Nelson, 2013). It is necessary to question these barriers alongside the process of designing, to allow different voices and frameworks for lived experiences. In order to substantiate this process in the case of designing with smell, the tacit knowledge produced by perceiving smells needs to be situated.

Placing Olfactory Knowledge

Smells are not neatly defined objects; they can travel large distances and appear seemingly unattached from their origin. Smellscapes can vary greatly in scale, ranging from continents to drawers (Drobnick, 2010). At the same time, however, smells can only partially detach themselves from the objects from which they originate, and thus they occupy a place somewhere in between object and idea (Drobnick, 2010). As Henshaw (2014) points out, places do not have a single identity or understanding and therefore do not have a single authentic or universally associated smellscape. Instead, smellscapes are always in process and always in the making (Quercia, Aiello, McLean & Schifanella, 2015). Despite this invisible ephemerality, olfactory perception makes a substantial contribution to our understanding of the world around us, through the connections formed between smell, memory, emotion and place. The perception of smells consists not only of the sensation of the odors themselves, but also of the experiences, culturally primed associations, memories and feelings that are evoked. Smells challenge the problematic power structure surrounding the voice of the storyteller (or researcher) that claims to provide a single valid interpretation; rather, all smell interpretations are always as valid as they are inheritably personal and associative. Smellscapes cannot be isolated from the social constructions of place that produce them, or from the body or other senses that may draw awareness to them. Rather, smellscapes are concerned with an immediate localized and immersed experience that involves emotions (Porteous, 1985).

The explanation for the emotion–memory–smell connection lies in the anatomy of the olfactory system. Olfactory signals travel from the exposed olfactory bulb along olfactory neurons into the brain's limbic system, the control center for emotions and memory (Gilbert, 2008). The intense stimulation activates the hypothalamus, which activates the endocrine system, which in turn can release hormones to stimulate the autonomic nervous system that foregoes logical reasoning.

The processing of smell perceptions is therefore similar to unconscious full-bodied activities (Gilbert, 2008).

That smells are complex full-body experiences makes it challenging to communicate them as such. It often appears easier to describe a situation where one encountered a particular smell than to name the smell itself. To perceive a smellscape narrative means to engage and experience the story about a place and other people's reality. The effectiveness of such a smellscape, according to Rodaway (1994), depends on the smeller's "own experience and ability to imagine the [original] olfactory experience" (p. 72). Every narration asks the reader to do some of the work, "to fill in a whole series of gaps" (Eco, 1996, p. 3). How the gaps are filled depends on the smeller's particular frames of reference, memories and experience.

Collectively Designing Smellscape Narratives

Smell of Change, a collective smellscape narrative, tells the Kaktovik community's experience of perceived changes as they want to emphasize them for others located outside of Kaktovik. All smells consist of accords, which develop according to the exposure to air. These accords can construct an alternative geography in a bottle, an embodied experience that tells a story and forgoes semiotics (see Vignette 6.1). The material that was used to create *Smell of Change* comes from a network of Alaskans, as part of an exchange involving an Inupiaq women's group telling of their home, a once important bartering place on the coast that is located east of the North Alaskan oil fields.

VIGNETTE 6.1: AN OVERVIEW OF THE PROCESS OF MAKING *SMELL OF CHANGE*

Kaktovik's remoteness means that almost all exchange happens through email, Skype and postal services. Compensation for time and other expenses, such as offers the women make to the land in exchange for what they take, were sent to the village. The group collected and sent me pieces of blown-out tires, tools, mud, fur, rocks and plants over several months, so that the odor would capture the seasonal change in the Arctic. One of the women even became involved in the smell extraction and made cloudberry oil, which is part of the base note. Usually, the smell extraction from solid objects happens in my kitchen or in the laboratory through distillation, Soxhlet extraction and cold pressing methods. Once the smell is in a liquid form, a small amount is applied in a glass vial, which is heated so that the liquid dries out and gas forms. The gas is analyzed in a gas spectrometer that captures the chemical fingerprint of the odor. One smell can have tens of thousands of components, and only a small number of these are analyzed. This process compares the relative concentration of the atomic masses in a given spectrum, and can help identify these characteristic chemical components which are marked through high values in the fingerprinting analysis. Once identified, the molecular masses of these components are compared in order to understand how the molecules will evaporate. As a last step, the smell accords are layered, blended and reconstructed according to the design that was agreed upon by the group, and test vials are sent to Alaska for feedback.

Top note: muddy, dirty laundry, wet dog, puddle, compost, hint of red berries

Middle note: heavy oil note, locker room, coins, wet leather

Base note: rotting blueberries, cloudberry, sandstone dust, iron

FIGURE 6.1 Accords of *Smell of Change*.

Image: author's own.

The top note is perceived immediately; it consists of small, light molecules that evaporate quickly (Figure 6.1). These form a person's first impression of a smell. The more rounded middle note masks the initial unpleasant impression of the strong top note. The base note develops last; it has heavy molecules that evaporate slowly and are typically detectable for the longest time.

Physics dictates that odors will expand freely, and therefore it is necessary to contain smells in a vessel. The form of this vessel is associative, and one might imagine a medicine bottle or a window cleaner spray bottle – the container always provides a preconception of what to expect. To avoid such a powerful interception through the form of the container, the *Smell of Change* vessel is given a form that was collectively designed with people who experienced the smell in different UK cities. The smellers were encouraged to respond to the smell and create shapes, and these were recorded and transformed into a low poly shape (see Vignette 6.2 and Figure 6.2). Low poly is an art style that refers to a relatively small number of polygons in designs. Low poly sculptures are angular, abstract and crude, but remain clearly reminiscent of the shape of the modeled object. Today's sophisticated modeling techniques allow mimicking reality, and low poly works against this similar to paper craft or origami, where the details are added in the mind of the observer to identify the object. It has been argued that the historical precedent for low poly can be found in the expressionist movement: during the advent of photography, which could capture the exact likeness of an object, expressionists started depicting abstract work focused on evoked emotions rather than accurate representations (Becker, 2014). Today's technology and image rendering allows seemingly accurately representations. Low poly takes a step back and offers the opportunity for the spectator to fill some of the gaps in the object's presentation. When people engage with *Smell of Change*, their intuitive responses explore the connection between the perception trapped in the mind and the body, similar to holding a paintbrush and transferring a mental image to a page. The completed result remains abstract, and the next visitor to the smellscape has to add his/her mind to the surface.

An interactive engagement between people and materials reframes key questions around interactions between the two, and identifies objects and things as reciprocally co-constructive agents to define form and meaning (Malafouris, 2013). The smell as material is then an infrastructure of storytelling that manifests itself in a shape through the active engagement of the body. The multisensory experience is the result of a collective practice that connects Kaktovik to people located in the UK. By establishing relationships between people in different places, between objects and experiences, new geographies to address these relationships are imagined. Geographical imagination is vital to help transcend the differences between nature and culture or urban and wild spaces, or concerning the Arctic, between here and there or us and them (Turan, 2011). Once the vessel is opened, the smellscape develops. An open window or a person wandering in can redefine the movement of the smell. Thinking through smells as storytelling devices means accepting this active involvement of the material in the final form-giving process. Just like a story held together by bookbinding screws, the story can be reordered and exchanged, or a page can be replaced by someone's body odor, movement or perfume.

VIGNETTE 6.2: EXPLORING THE FORM OF *SMELL OF CHANGE*

Initially, the smell is offered as diffused steam in a large vase that allows visitors to immerse their face in the warm, yet moist, developing smell. After this experience, which is accompanied by visual and tactile, yet predominantly abstract, low poly shapes, visitors are invited to manipulate a surface to explore the shape of the smell. During this process, photographs are taken in rapid succession to document the interaction with the surface. The process is framed as open, playful and exploratory. It is not the result of the movements, but the movement itself that is important. In Figure 6.2, one can see the shape emerging from a one-dimensional sheet to a three-dimensional object, and finally, the object that is the vessel for *Smell of Change.*

FIGURE 6.2 Exploring the form of *Smell of Change.*

Image: author's own.

Smellscape Narratives in Lived Spaces

Arriving at the limits of expressing experiences and failing to accurately represent them leaves a sense of epistemological insufficiency. Some information does not translate into our logocentric world; information communicated through words is invariably partial, which leads to excluding those who differ in their perception of the data (McBride & Nolan, Chapter 19 in this volume). Smell remains systematically suppressed from our knowledge system by the same power structures that enforce everyday relations of ruling, such as sexism and racism (Classen, Howes & Synnott, 1994; McBride & Nolan, Chapter 19 in this volume; Tuan, 1977). Telling stories through smells does not provide an answer to this epistemological dilemma, but perhaps it is an argument for the acceptance of practices themselves as valuable information, and therefore as a tool for better connecting with different realities and forms of data by removing some of the restricting power structures.

Smellscape narratives are a lived space within the environment through which the visitor and the smells themselves trace their own stories. A small part of the Arctic has become "a meaningful location" (Cresswell, 2004, p. 7), in the sense that it renegotiates the relationships between the artefacts people interact with. The smell's design acknowledges multiple relationships that may exist between site perception and visitor (Morris & Cant, 2006). Non-representational theory asks how we can approach and situate relational experiences and urges us "to encounter the world in whatever form it happens to turn up" (Lorimer, 2007, p. 97). *Smell of Change* has a chance to become part of the smeller's experience, and the visitor becomes part of the story by giving it a form. The performance, practice and activity of smelling the Arctic is then not about uncovering meaning, but rather about creating alternative trajectories to question how "life takes shape and gains expression in shared experiences, everyday routines, fleeting encounters, embodied movements . . . and sensuous dispositions" (Lorimer, 2005, p. 84). Non-representational thought, then, can add to the process of designing with smell by unsettling social science obsessions with authority, and instead instating different frames of reference that embrace practice and foster inclusion of diverse voices (Vannini, 2015).

Note

1 *Smell of Change* was exhibited as part of the following events: University of Nottingham May Fest (Nottingham, May 2015); Embodied Arctic (Arctic Futures, London, July 2015); Beyond Perception 15 (Scottish Agricultural College, Aberdeen, September 2015); British Science Festival (Bradford Brewery, Bradford, September 2015); Arctic Change (University College London, London, October 2015); Being Human Festival (Henderson Gallery, Nottingham, November 2015).

References

Anderson, K., & Smith, S. (2001). Editorial: Emotional geographies. *Transactions of the Institute of British Geographers, 26*(1), 7–10.

Beatty, A. (2005). Emotions in the field: What are we talking about? *Journal of the Royal Anthropological Institute, 11*(1), 17–37.

Beatty, A. (2010). How did it feel for you? Emotion, narrative, and the limits of ethnography. *American Anthropologist, 112*, 430–443.

Becker, M. (2014). Configurations of intensity. In *ACADIA 14: Proceedings of the 34th Annual Conference of the Association for Computer Aided Design in Architecture* (pp. 589–596). Los Angeles, CA: ACADIA/Riverside Architectural Press.

Bondi, L., Davidson, J., & Smith, M. (2007). Introduction: Geography's "emotional turn." In J. Davidson, L. Bondi, & M. Smith (Eds.), *Emotional geographies* (pp. 1–18). London: Routledge.

Borneman, W. R. (2003). *Alaska: Saga of a bold land.* New York: Perennial.

Bravo, M. (2000). Cultural geographies in practice: The rhetoric of scientific practice in Nunavut. *Ecumene, 7*(4), 468–474.
Bravo, M., & Sörlin, S. (2002). Narrative and practice: An introduction. In M. Bravo & S. Sörlin (Eds.), *Narrating the Arctic: A cultural history of Nordic scientific practices* (pp. 3–32). Canton, MA: Science History Publications.
Catton, T. (1997). *Inhabited wilderness. Indians, Eskimo, and National Parks in Alaska.* Albuquerque, NM: University of New Mexico Press.
Classen, C., Howes, D., & Synnott, A. (1994). *Aroma: The cultural history of smell.* London: Routledge.
Coates, P. (1991). *The Trans-Alaska Pipeline controversy.* London: Associated University Press.
Cole, D. (2008). *North to the future: The Alaska story 1959–2009.* Fairbanks, AK: Epicenter Press.
Collignon, B. (2006). *Knowing places.* St. Lambert, Quebec: CCI Press.
Cresswell, T. (2004). *Place: A short introduction.* Malden, MA: Blackwell.
Cruikshank, J. (1992). Oral tradition and material culture: Multiplying meanings of words and things. *Anthropology Today, 8*(3), 5–9.
Cruikshank, J. (2005). *Do glaciers listen? Local knowledge, colonial encounters, and social imagination.* Seattle, WA: University of Washington Press.
Curti, G. H., Aitken, S. C., Bosco, F. J., & Goerisch, D. D. (2011). For not limiting emotional and affectual geographies: A collective critique of Steve Pile's "Emotions and affect in recent human geography." *Transactions of the Institute of British Geographers, 36*, 590–594.
Drobnick, J. (2010). Toposmia: Art, scent, and interrogations of spatiality. *Angelaki: Journal of the Theoretical Humanities*, 7(1), 31–47.
Duffy, A. (2009). The natural resource industry visual impact paradox: A perspective on Alaskan imagery and art. In S. Jakobsson (Ed.), *Images of the North: Histories – identities – ideas* (pp. 139–148). Amsterdam: Rodopi.
Eco, U. (1996). *Six walks in the fictional woods.* Cambridge, MA: Harvard University Press.
Gilbert, A. N. (2008). *What the nose knows.* New York: Crown.
Giovannelli, A. (2008). In and out: The dynamics of imagination in the engagement with narratives. *The Journal of Aesthetics and Art Criticism, 66*(1), 11–24.
Haraway, D. J. (1988). Situated knowledges: The science question in feminism and the privilege of partial perspective. *Feminist Studies, 14*(3), 575–599.
Henshaw, V. (2014). *Urban smellscapes: Understanding and designing city smell environments.* London: Routledge.
Jones, O. (2011). Geography, memory and non-representational geographies. *Geography Compass, 5*, 875–885.
Kearnes, M. (2010). Geographies that matter – the rhetorical deployment of physicality? *Social and Cultural Geography, 4*(2), 139–152.
Keeling, A., & Sandlos, J. (2009). Environmental justice goes underground? Historical notes from Canada's northern mining frontier. *Environmental Justice, 2*(3), 117–125.
Kollin, S. (2001). *Nature's state: Imagining Alaska as the last frontier.* Chapel Hill, NC: University of North Carolina Press.
Kurgan, L. (2013). *Close up at a distance: Mapping, technology, and politics.* New York: Zone Books.
Lorimer, H. (2003). Telling small stories: Spaces of knowledge and the practice of geography. *Transactions of the Institute of British Geographers, 28*, 197–217.
Lorimer, H. (2005). Cultural geography: The busyness of being "more-than-representational." *Progress in Human Geography, 1*, 83–94.
Lorimer, H. (2007). Cultural geography: Worldly shapes, differently arranged. *Progress in Human Geography, 31*(1), 89–100.
Malafouris, L. (2013). *How things shape the mind.* Cambridge, MA: MIT Press.
McConnell, S. S., & Toopetlook, D. M. (2009). Welcome to Anaktuvuk Pass! In S. Jakobson (Ed.), *Images of the North: Histories – identities – ideas.* Amsterdam: Rodopi.
Moffitt, M., Chetwynd, C., & Todd, Z. (2016). Interrupting the northern research industry: Why northern research should be in northern hands. *Northern Public Affairs, 4*(1).
Morris, N. J., & Cant, S. G. (2006). Engaging with place: Artists, site-specificity and the Hebden Bridge Sculpture Trail. *Social & Cultural Geography*, 7(6), 863–888.
Nelson, R. (2013). *Practice as research in the arts: Principles, pedagogies, resistances.* Basingstoke: Palgrave Macmillan.

Nuttall, M. (1998). *Protecting the Arctic: Indigenous peoples and cultural survival*. Amsterdam: Harwood Academic Publishers.

Porteous, J. D. (1985). Smellscape. *Progress in Human Geography, 9*(3), 356–377.

Potteiger, M. (1998). *Landscape narratives*. New York: John Wiley.

Powell, R. (2005). Northern cultures: Myths, geographies and representational practices. *Cultural Geographies, 12*, 371–378.

Quercia, D., Aiello, L. M., McLean, K., & Schifanella, R. (2015). Smelly maps: The digital life of urban smellscapes. *Proceedings of the 9th International AAAI Conference on Web and Social Media (ICWSM)* (pp. 327–366). Oxford: ICWSM.

Rodaway, P. (1994). *Sensuous geographies: Body, sense, and place*. London: Routledge.

Simpson, P. (2010). *Ecologies of street performance: Bodies, affects, politics*. PhD thesis, University of Bristol.

Smith, H., & Dean, R. T. (2009). *Practice-led research, research-led practice in the creative arts*. Edinburgh: Edinburgh University Press.

Steel, D. (2015). Polar, social, science: A reflection on the characteristics and benefits of the Polar social sciences. *The Polar Journal, 5*(1), 1–7.

Stewart, S. (2005). Remembering the senses. In D. Howes (Ed.), *Empire of the senses* (pp. 3–4). Oxford: Berg.

Thrift, N. (1996). *Spatial formations*. London: SAGE.

Thrift, N. (1999). Steps to an ecology of place. In Massey, D. B., Allen, J., & Sarre, P (Eds.), *Human geography today* (pp. 295–322). Cambridge: Polity Press.

Thrift, N. (2004). Summoning life. In P. Cloke, P. Crang, & M. Goodwin (Eds.), *Envisioning human geographies* (pp. 81–103). London: Arnold.

Thrift, N., & Dewsbury, J. (2000). Dead geographies – and how to make them live. *Environment and Planning D: Society and Space, 18*, 411–432.

Tuan, Y.-F. (1977). *Space and place: The perspective of experience*. Minneapolis, MN: University of Minneapolis Press.

Turan, N. (2011). New geographies. In K. Wooler (Ed.), *20/20: Editorial takes on architectural discourse* (pp. 219–227). London: AA Publications.

Turan, N. (2015). How do geographic objects perform? *ARPA Journal, 3*.

Urry, J. (2007). The place of emotions within place. In J. Davidson, L. Bondi, & M. Smith (Eds.), *Emotional geographies* (pp. 77–86). London: Routledge.

Vannini, P. (2015). *Non-representational methodologies. Re-envisioning research*. New York: Routledge.

Wilson, E. (2015). *Energy and minerals in Greenland: Governance, corporate responsibility and social resilience*. London: IDED.

Wright, S., Lloyd, K., Suchet-Pearson, S., Burarrwanga, L., Tofa, M., & Bawaka Country. (2012). Telling stories in, through and with country: Engaging with indigenous and more-than-human methodologies at Bawaka, NE Australia. *Journal of Cultural Geography, 29*(1), 39–60.

7

COMMUNICATING AND MEDIATING SMELLSCAPES

The Design and Exposition of Olfactory Mappings

Kate McLean

Exploring and sharing individual and collective olfactory urban experience through mapping is problematic. In this chapter, I introduce how contemporary artists and designers have approached the topic of olfactory mappings, and then continue to consider some practical concerns in researching urban smellscapes and presenting such works.

Smellscape mapping, as a creative practice, is a flight through the imagination and an engagement with individual perceptions and memories. It kindles olfactory experience, forming associations between an invisible sensory input (odor), a visual image (symbolization of a smell experience) and a place. Smellscape mapping foregrounds the sense of smell over other sensory information, and is initiated by the whimsical and ephemeral world of olfactory perception.

Smell is fundamentally attached to place (Feld, 2005; Hoover, 2009; Rodaway, 1994; Tuan, 1977). Smellscape mapping, as a creative artwork, is a processual series of events designed to communicate an evanescent world that is "spatially ordered or place-related" (Porteous, 1990, p. 25), but simultaneously one that is "non-continuous, fragmentary in space and episodic in time . . ." (p. 25). But which mapping paradigms serve best to chart and suggest the rich complexity of such invisible spatial information?

The role of the map in contemporary society bifurcates. Traditional maps attempt to render legible our world, seducing us with symbolic reality, and digital mapping satisfies a "mania for accuracy, exactitude, concordance with reality" (Breebaart, 2015). To this end, smells that need to be measured, the nuisance and noxious smells of industry, can be defined quantitatively using remote sensors and bound to place through the exactitudes of geolocation. But when the myth of the map as a depiction of reality is questioned, we can perform mappings of ephemeral and sensed invisibles. Instead of seeking replications, reproductions and accuracy, we can use performative and artistic mappings to explore human relationships with smell in space and in place.

There is an inherent contradiction between the top-down view of mapping and the lived experience of smelling (Margioles, 2001). Smells resist being objectively mapped because of their incoherent and imperceptible structure: "specific flows of odours cannot be represented according to the principles of physical-metrical space: they have no sides, no parts and no perspective" (Diaconu, 2011, p. 229).

However, when we regard the map as an abstract concept, and the smellscape as a moment in time, we join space and time in ways that best represent the dataset. The volatile ephemerality of smell is such that smellscapes alter rapidly, unlike their land-based counterpart, posing questions as to how to represent temporal factors. Approaching mapping as performative, and therefore no longer seeking objectivity or fixity, smell mapping may link individual users with their lived experiences of a particular place, and to consider how everyday immersion in a smellscape may *feel* instead of merely identifying 'what is where.'

Recent Smellscape Mappings

Urban smellscape mapping has a handful of practitioners, each contributing a unique approach to an emerging field.

Mapping the Characteristics of Smell and External Factors Affecting Its Perception

Wu's map *The Whiff of Uncertainty* (2007) shows us what the senses detect – in her case, the Dunkin' Donuts scent in New York City that she and her classmate encountered daily in 2007. She shows us that it is possible to map human smell perception and ascribe some criteria against which it is measured. The visualized 'odor canopy,' the space in which smell was perceived, is indicated on a sequence of small charts. She also introduces meteorological factors affecting smell perception, including temperature, air moisture, and wind speed and direction. Wu's choice of sequential day-by-day charting is labor- and space-intensive, but communicates the randomness of individual, situated, temporal smell perception.

Mapping How We Smell

Twilley considers smell as an internal construct as much as an external, physical reality. Based on research asking hundreds of New Yorkers from a range of backgrounds to describe 66 different scents, then comparing the answers against DNA, gave results that Vosshall describes as an "olfactory demography of New York." It revealed the preferred scent overall to be vanilla, and the least favored to be a sweaty gym/sock smell. The map, *Scratch'n'Sniff New York City*, includes micro-encapsulated smells on scratch-and-sniff stickers. In Twilley's (2010b) "amateur explorations of the olfactory universe," the manufacturing process is described in detail, and it is quickly apparent that access to specialist suppliers of synthetic odors and screen-printing equipment is a prerequisite to the creation of such a mapping. Twilley advises, "I'd have to recommend not trying it at home unless you are an organic chemist for whom terms like 'coacervation' and 'partition co-efficients' hold no fear" (Twilley, 2010a).

Mapping of Smell to Explore Relationships and Lived Experiences

Tolaas has been exploring urban smellscapes for over a decade; she researched the 20 *arrondissements* of Paris (Tolaas, 2003) through extensive engagement with the local population, and subsequently synthetically reproduced a summary scent for each neighborhood. This type of work invites participation, and requires collaboration with professional scientists to recreate the perceived city smells. In *Talking Nose – Mexico City* (Tolaas, 2009), she researched 210 neighborhoods over nine years, recreating the identified smells synthetically. The exhibit includes 2000 written responses labeled on hundreds of identically bottled odors arranged on a tabletop. The project tackles the issue of pollution, a key component of the Mexico City smellscape, and asks people to consider their own relationship with, and contribution to, a notoriously polluted urban environment.

A much smaller-scale project, *Smell It* (Marketou, 2008), encouraged a local population to create a do-it-yourself smell map by recording their perceptions and then making graffiti marks on an artist-created 3.3 m × 1.8 m map of Philadelphia. It dealt with both the transgressive behaviors of smelling in public spaces and the creation of individual notation systems to share smell perception. This hands-on participation has the benefit of immediacy and community involvement in locally generated public art.

Mapping to Explore Temporalities of Smellscapes

My own early work in the field of smell mapping (e.g. Edinburgh and Glasgow smell maps) dispensed with cartographic conventions like neat lines, scale, north arrow, boundaries and topography. I am attentive to the experience of smelling, what might happen when a smell disperses, and how clouds of smell molecules overlap and distribute themselves, combining together to form pockets of complexity across an urban landscape. To explore these possibilities, I have developed symbols to indicate speculative trajectories of scent, and keep the notional title of 'map' to convey a place, and a legend to indicate smell descriptors. Viewers with local knowledge can read the artwork as well as any conventional map. In my work in Amsterdam, Pamplona and Singapore, I deal more with the lived experience, individual smells, contestation and temporalities of urban smellscapes. The remainder of this chapter highlights some of the challenges and practical issues I encounter in my own practice.

My Practice of Mapping Smellscapes

My artistic practice maps smell from a human-centered perspective that investigates "relationships between urban smellscapes and city experiences and perceptions" (Henshaw, 2014, p. 5). Whilst technologies to analyze the olfactory profile of the smell are used by other practitioners to formulate synthetic odor molecules, I explore the descriptions as written by amateurs. This is a form of volunteered geographic information (Goodchild, 2007) in which people work as sensors. I work from the premise that the human nose is superior, in terms of sensitivity, to any technological devices, and that direct human olfactory experience results in an embodied sense of a place.

From a lived perspective, smellscape contains information from a local population, but finding people to take part in offbeat, transgressive behaviors is occasionally problematic. My earliest work, *Smellmap: Paris*, used data collected via personal contacts on social media and secondary sources such as journalists' reports (McLean, 2012). It is possible to collect an intriguing mix of city smell data in this albeit time-consuming fashion. To collect larger datasets of smells across entire cities, it is possible to harvest geo-referenced social media content using known words for smells (Quercia, Schifanella, Aiello & McLean, 2015). However, my preferred method is walking with local residents, as this invokes an intimate, personal and collective, creative experience of mapping olfactory information.

Collecting the Data on the Ground through Walking

Mapping as experience occurs in two very different forms: initial smellscape mapping takes place during the data collection smellwalks – this experience is wholly immersive and performative, following in the traditions of the Situationists, notably Debord (1967), and ambulatory artists such as Long (2016) and Cardiff and Miller (2014). A secondary smell mapping occurs after the initial information is recoded, reframed and designed, and a final visual/olfactory artwork is displayed.

The smellwalk is the source of all data for the smellscape mapping; names, locations, intensity, colors. Without a smellwalk, there is no smellscape mapping.

To recruit participants, social media, especially Twitter, can generate interest, but collaboration with local business and local voluntary and academic organizations has been my prime recruiting ground for recent smellscape mapping projects. Currently, the concept of a smellwalk is intriguing enough to attract media attention, both supportive and skeptical.

Time of Day, Time of Year

Urban smell is particularly responsive to warmth and humidity. I have run smellwalks in Paris in November and February, when frozen fingers curtailed our encounter with the smellscape. In the summer, walks take place outside; they shift inside in the winter. To the north of the Tropic of Cancer and to the south of the Tropic of Capricorn, smellscapes are seasonally affected, whereas equatorial environments such as Singapore remain hot and damp throughout the year, although growing seasons, and thus local produce, affect scents in the marketplace. Temperatures above 18 °C and below 25 °C make for pleasant walking conditions, and at this level, the stinkier aspects of city smells (drains, sewers, garbage) do not dominate.

Henshaw (2014) used individual smellwalks along an identical, repeated route to research the olfactory environment at two experiential levels – detection and cognition. She noted that temporal factors make a difference, and that individual differences in the detection of odors are also significant. There remains further work to be done in this area, examining and analyzing the temporal differences of urban smellscapes.

In 2012, Henshaw and I ran a group smellwalk in Manchester to ascertain its suitability as a research method. As a designer and artist, I saw the communication potential of a group smellwalk where individuals could share their 'smellervations' and also explore the differences between individual urban smell perceptions. The walk had potential to be a creative mapping in its own right.

Since then, I have run and led group walks in cities, towns and rural countryside in the UK, USA, Spain, Singapore and France, working with translators when necessary. What follow are some of the more practical considerations to ensure a positive and memorable experience.

Optimum Group Size for a Smellwalk

The intimacy, revelation and novelty of the smellwalk all affect participants. My recommendation for group size is based on the quality of *their* experience rather than on my need to acquire data. I have led smellwalks with a sole participant (Hagenaars, 2015) and with 20 people walking in pairs as smell buddies (McLean, 2014a). Walking with an individual involves a greater participation and dialogue between researcher and subject (Henshaw, Adams & Cox, 2009), whereas group walks enable conversation between walkers (Hecht, 2014). With groups of up to 12 people, the approach can be quite responsive; if a smellwalker notes an interesting odor, then the group can be called over to share and comment. I try to avoid walks with more than 30 participants, since at this scale the experience becomes akin to a guided tour, and therefore antithetical to my intent of sharing olfactory knowledge.

Route Design and Overall Planning of a Smellwalk

Smellwalking is tiring; foregrounding an alternate sense takes effort and concentration. Through experience, the optimum length is between 45 and 60 minutes over a distance of no more than

FIGURE 7.1 Select sniffing sites for their smell potential – a wooden park bench in Pamplona.

Image courtesy of Victoria Lydford.

3 km. It is advisable for the organizer to walk the route in advance and forewarn local vendors, especially if a large group of people require access to a business premises. Although I do not want to 'lead' the walks, I do usually single out a couple of interesting olfactory stopping points (florists, extractor fans, alleyways, bus stops, green spaces, trash bins) in advance, leaving the remainder to serendipity. On separate smellwalks, I have sniffed metal railings, grass, park benches (Figure 7.1), smoldering oud (a dark resinous wood from southeast Asia used for incense and perfumes), lobster bait and a bemused motorcyclist in full leathers.

Processual Immersion for Participants and Leader Interventions

Sensewalking is a method used increasingly across the social sciences to qualitatively explore the "under-researched aspect of environmental experience" (Henshaw et al., 2009) and immersion into the experience of walking 'nose-first' benefits from a series of levels. This entails a gradual process of acclimatization analogous to listening in 'soundwalking' (Westerkamp, 1974), starting from a focus on the self, and moving outwards to the immediate environment and more distant sources. Smellwalking, too, benefits from staged sections.

I start with a brief introduction with some smell 'facts,' and relate some of the incisive, lyrical and comic moments from previous walks. I then explain what I term 'smell catching,' which simply involves attending to every smell that enters their nostrils. I ask smellwalkers to detect and record four such smells in a 15-minute period.

For the second stage, I introduce 'smell hunting,' whereby participants use all their senses to seek potentially interesting places where smells might be hiding and to record a further four smells over a second 15-minute period. The final section I call 'free smelling,' enabling the smellwalkers to detect and record smells in their own way. During the breaks after each 15-minute session, I will draw attention to the smellnotes (described below), and ask walkers to count their good versus bad smells or to recount a personal association. This discussion is integral to my work, as it questions our relationships with smell in the city – generally regarded as something to be feared, inducing disgust, or related to a lack of cleanliness – and instead draws attention to the smaller smells of everyday life.

Recording Smell Data

A number of sensory notation strategies have been used to record phenomenological olfactory perception: mental maps and monitoring maps (Diaconu, 2011), sensory sliders (Malnar & Vodvarka, 2004) and a perceptual radar (Lucas & Mair, 2009). Since 2012, I have refined my own-design 'smellnote' as a tool to encourage deeper engagement and reflection. Completion of the 'smellnote' (Figure 7.2) changes incrementally from the perceptual (noticing the smell) to a cognitive appreciation (associations and memory) of up to 13 odors. Whilst I am currently using a paper form for the 'smellnote,' a digital version exists in the form of a prototype app called Smellscaper that will enable smellnote data to be entered into a central database by anyone with a smartphone, anywhere, in the world.

Smellscape Mapping Design

In researching and depicting urban smellscapes, I lean towards an anthropological approach to landscape, as outlined by Ingold (2000): ". . . it is a process that is not about 'representing' or capturing what is observed but . . . coupling the movement of the observer's attention (haptic, aural etc.) with currents of environmental activity" (p. 220). My mapping seeks to portray the experience of olfactory perception and information – to reframe the individual, embodied accounts as collective.

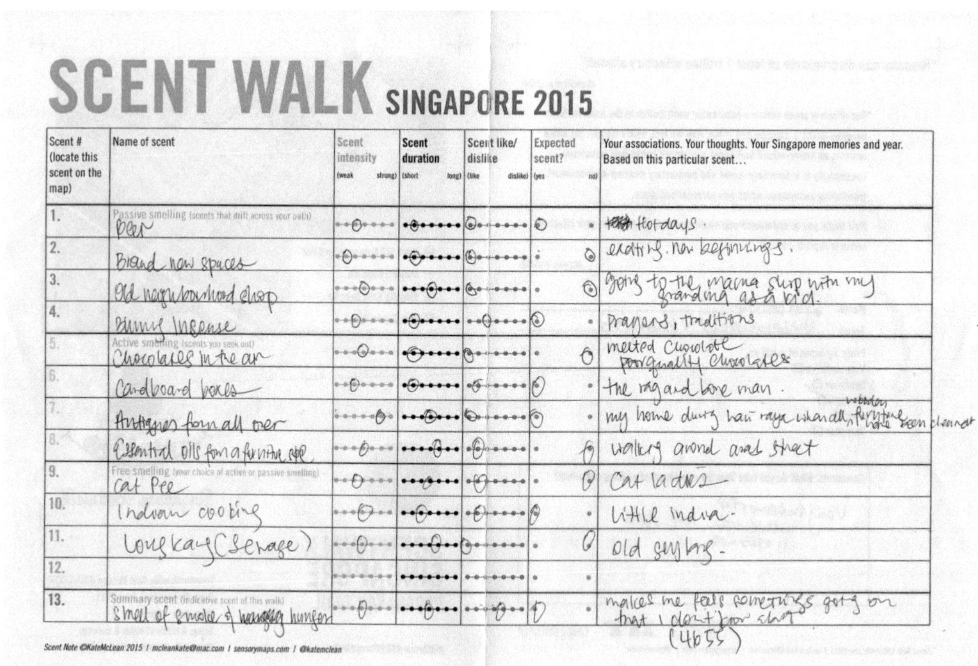

FIGURE 7.2 Moving from perceptual to cognitive appreciation of urban smell using purpose-designed 'smellnote' for *Scentscape 06, 2015 – Singapore.*

'Smellnote' design © Kate McLean 2015.

Components of a smellscape to be mapped derive directly from the smellnotes. I map individual geo-located smells by their source, range, intensity and duration, and shift these according to local wind strength and direction. The process is an assemblage of digital with hand-placed elements and manipulation. Despite the use of digital software, there is no computer-generated algorithm underlying this part of my creative process; all decisions as to what to include and where to place them are made one-by-one, by hand. In selecting colors, I use either samples from smell sources (McLean, 2013), select from the local environment (McLean, 2014b) or use colors chosen by the smellwalkers. As more elements derive from the walkers themselves, so the map more truly belongs to participants, and as the artist/designer, I am increasingly a mediator of experienced information.

So the artistic component is essential to my mapping practice, as Amoroso (2010) suggests when summarizing the guiding principles behind map-landscaping: "creativity can offer insight that straightforward tracing cannot. The connection between art and mapping information must never be downplayed . . ." (p. 157).

My decision to create Singapore's smellscape from tiny dots (Figure 7.3), in contrast to the contour lines I frequently use, was tacit, felt and embodied – an instinctive decision taken after a series of smellwalks talking with local people about the effects of humidity on Singapore's smellscape.

But What about the Smells?

Some of my maps are sensorially integrated with accompanying smelly artefacts. Odors can be made in-house using naturally occurring materials. Olivia Alice Clemence talks about this process of distillation, which I too occasionally deploy, in Chapter 24 in this volume. The advantages are

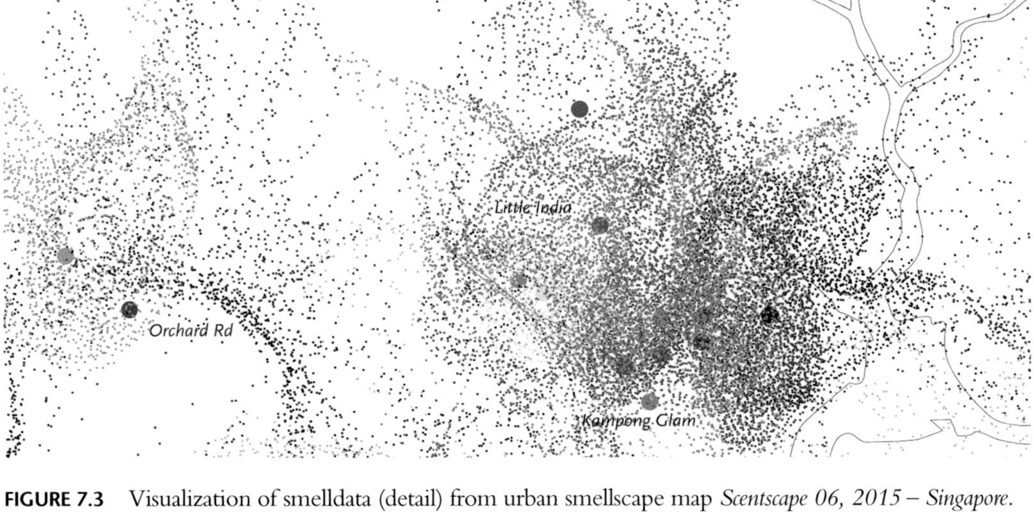

FIGURE 7.3 Visualization of smelldata (detail) from urban smellscape map *Scentscape 06, 2015 – Singapore.* Digital artwork © Kate McLean 2015.

that an essence of organic material can be extracted in this way; however, one problem with this method is that the shelf life varies depending on the purity of the distillation: it can range from one month to two years. If exactitude of reproducibility is required, it is essential to note that no two distillations will generate identical results. An even more basic approach than distilling is to bottle the raw ingredients themselves, or fabricate a close approximation; for *Smellmap: Edinburgh*, I thought to use the smell of penguin poo as the indicator of penguins in the zoo. But on the day I visited, I could not access enough guano to fill a bottle. A little lateral thinking, research into the penguins' diet (squid) and a willingness to leave some 'squid juices' to go off slightly in my kitchen was the recipe for one of the best smells I have ever created in terms of complexity and curiosity. During exhibition at the *Edinburgh International Science Festival* in 2011, the 'olifactors' (the sniffers) appreciated its qualities.

Tolaas, on the other hand, strives for such reproducibility, and works closely with International Flavors & Fragrances (IFF)[1] to generate a comprehensive synthetic scent library. I worked with IFF on the *Smellmap: Amsterdam* project, identifying and recreating the city's smells, and one important discovery was the difference in the qualities and sensations of IFF's initial set of urban smell creations, which were far more concentrated than smells are in the wild. Almost every synthetic urban smell required dilution, resulting in altered characteristics – not just weaker, but noticeably different, bearing out the theory that, "It is well-known, for example, that changing the concentration of an odorant, and thereby its intensity, may change its quality as well . . ." (Kruger, Feldzamen & Miles, 1955, p. 117). The benefit of reproducible synthetics, however, is their longevity; two years on (kept in the cool and the dark), the synthetically generated smells of Amsterdam retain their original quality.

Exposition

The exposition of an urban smellscape is a means to disseminate the experience to a wider audience. Tolaas chooses not to expose her work online, saying that the digital world remains one in which smell cannot yet permeate. My own smell maps can be seen online (http://sensorymaps.com) as an indication for the full-size and multisensory experience of an exhibition. Through their interaction with the maps, 'olifactors' have a chance to interpret their own reactions to a smellscape, sometimes an ocean distant from the original. When *Smellmap: Newport* exhibited as part of *Nostalgias: Visualising Longing* (Pie Factory, 2013), several connections were made between the site of the exhibition, Margate, UK, and the site of the map (Newport, Rhode Island, USA) through specific smells, and one viewer described an 'imagined Newport' through the whole collection of nine smells. Through performativity, the viewers bring an unvisited place into being, conjuring detail about construction materials, foliage and paint colors as a dimensional, lived environment – a true case of the "faraway nearby" (Solnit, 2014), where individual and place may be physically distant yet emotionally connected.

Signaling a Sniffing Experience

There are a variety of mechanisms by which the call to sniff can be invoked. In *Smellmap: Paris*, I used large text as a call to action, asking people to sniff and record the emotions and locations evoked by each scent. This call I repeated in several subsequent works. On a more prosaic level, fixing shelf height and small bottles at nose level with a graphic situated immediately behind (*Smellmaps: Newport, Edinburgh* and *Glasgow*) serves as a physical call to action (Figure 7.4).

The choice of smell vessel and dispersal method depends on the desired experience, the form of the scent molecules and the venue. There are many options from the realms of aromatherapy

FIGURE 7.4 Audience sniffing bottled (natural) 'smells' of Edinburgh at *Edinburgh International Science Festival*, John Hope Gateway, Botanical Gardens, 2011.

Image: author's own.

(essential oils, distillations, tinctures, candles), residential and commercial scent diffusion (synthetic single molecules in bead form or liquids with an alcohol base) as well as scratch-and-sniff and infused scent blocks. All of these may be activated using proximity sensors, and can also be fan-assisted (see Gwenn-Aël Lynn, Chapter 3 in this volume). Interestingly, many gallery exhibitions of olfactory art select an aesthetic of the invisible through their 'display' of scent on, or in, white cube spaces (Eichblatt, 2012; Yatzer, 2015).

Yet visual cues remain an important element in the smelling experience – from exhibiting my smell maps, I witnessed how the more curious viewers will smell the most rancid of odors (boys' toilets in primary school) knowingly after reading the smell map legend. Indeed, smell identification in relation to the legend of featured smells listed on the smell map often becomes a competitive game between individuals in a small group.

Towards the Multisensory

This chapter has described how the urban smellscape can be artistically explored in order to facilitate individual and shared interpretation of the olfactory in our understanding of place. Smellwalking can enable readers to create their own smellwalks independently or in larger groups. The artistic and designed representation of smell for shared interpretation will necessarily be highly subjective, but my sincere hope is that more and more people will become involved in the exploration and depiction of smell. The choices underpinning creative responses will facilitate shared and individual interpretations of proximal or distant urban smellscapes. Working from the premise that we gain olfactory knowledge through activity, I suggest that smellwalking as a practice is an effective agent for urban olfactory engagement, just as accompanying dialogues, descriptions and intersensory translation can open new avenues for exploration of the seemingly known in everyday surroundings.

Mapping, when viewed as a situated, collective process rather than a representative artefact with inherent meaning, has a creative potential to reveal the unseen, ephemeral and imagined. If the *Sensorium* show at Tate Britain in 2015 is an indication of how we can explore the existing anew, then we are in for further multisensory depictions of far more than existing artworks in the future as we make our reality through action, through the deliberate action of sniffing.

Note

1 www.iff.com/smell Retrieved March 1, 2017.

References

Amoroso, N. (2010). *The exposed city: Mapping the urban invisibles* (1st ed.). New York: Routledge.

Breebaart, R. (2015). *The humid canopy: A psychogeography of Brisbane* (eBook). Brisbane: SLF Publishing.

Cardiff, J., & Miller, G. B. (2014). *Walks*. Retrieved March 1, 2017, from www.cardiffmiller.com/artworks/walks/

Debord, G. (1967). *Society of the spectacle*. Orlando, FL: Bread and Circuses Publishing.

Diaconu, M. (2011). Mapping urban smellscapes. In M. Diaconu, E. Heuberger, R. Mateus-Berr, & L. M. Vosicky (Eds.), *Senses and the City: An interdisciplinary approach to urban sensescapes* (pp. 223–238). Vienna: LIT Verlag.

Eichblatt, S. (2012, November 30). Architects Diller Scofidio + Renfro design the "Art of Scent" show at MAD, NY. Retrieved March 1, 2017 from www.wallpaper.com/architecture/architects-diller-scofidio-renfro-design-the-art-of-scent-show-at-mad-ny/6218

Feld, S. (2005). Places sensed, senses placed: Toward a sensuous epistemology. In D. Howes (Ed.), *Empire of the senses: The sensual culture reader* (pp. 179–191). Oxford: Berg.

Goodchild, M. F. (2007). Citizens as sensors: The world of volunteered geography. *GeoJournal*, *69*(4), 211–221.

Hagenaars, M. (2015, July 16). *Smell and the city*. Retrieved March 1, 2017 from www.basenotes.net/features/3167-smell-and-the-city

Hecht, J. (2014, September 12). Put your nose first: Smellwalks for you and your dog. *Scientific American*. Retrieved March 1, 2017 from http://blogs.scientificamerican.com/dog-spies/put-your-nose-first-smellwalks-for-you-and-your-dog/

Henshaw, V. (2014). *Urban smellscapes: Understanding and designing city smell environments*. New York: Routledge.

Henshaw, V., Adams, M., & Cox, T. J. (2009). Researching urban olfactory environments and place through sensewalking. In *PhD Colloquium on "Understanding Places,"* University of Westminster, Marylebone Campus, July 6. Retrieved March 1, 2017 from www.manchester.ac.uk/escholar/uk-ac-man-scw:122854

Hoover, K. C. (2009). The geography of smell. *Cartographica: The International Journal for Geographic Information and Geovisualization*, *44*(4), 237–239.

Ingold, T. (2000). *The perception of the environment: Essays on livelihood, dwelling and skill* (reissue ed.). London: Routledge.

Kruger, L., Feldzamen, A. N., & Miles, W. R. (1955). A scale for measuring supra-threshold olfactory intensity. *American Journal of Psychology*, *68*(1), 117–123.

Long, R. (2016). Richard Long official website. Retrieved March 1, 2017 from: www.richardlong.org/

Lucas, R., & Mair, G. M. (Eds.) (2009). *Sensory urbanism proceedings 2008*. Edinburgh: The Flâneur Press.

Malnar, J. M., & Vodvarka, F. (2004). *Sensory design* (1st ed.). Minneapolis, MN: University of Minnesota Press.

Margioles, E. (2001). Vagueness gridlocked. *Performance Research: A Journal of the Performing Arts*, *6*(2), 88–97.

Marketou, J. (2008). *Smell it*. Retrieved March 1, 2017 from: www.jennymarketou.com/works_2008_7.html

McLean, K. (2012). Emotion, location and the senses: A virtual dérive smell map of Paris. In *Proceedings of the 8th International Conference on Design and Emotion: Out of Control*, September 11–14. London: Central Saint Martins College of Art & Design.

McLean, K. (2013). Smellmap: Glasgow. In *Final Programme of the 26th International Cartographic Conference*, August 25–30. Dresden: International Cartographic Association.

McLean, K. (2014a). *Smell colour sketch: Brooklyn*. Retrieved March 1, 2017 from http://sensorymaps.com/portfolio/smellsketch-brooklyn/

McLean, K. (2014b). Smellmap: Amsterdam – olfactory art & smell visualisation. In Forbes, A. G. & Chevalier, F. (Eds.), *Proceedings of the IEEE VIS 2014 Arts Program* (pp. 143–145). Retrieved March 1, 2017 from http://visap.uic.edu/2014/papers/18_McLean_Smellmap_VISAP2014.pdf

Pie Factory. (2013). Nostalgias: An exhibition coinciding with the Nostalgias: Visualising Longing conference taking place at the Pie Factory Gallery, Margate, UK. Retrieved March 1, 2017 from http://piefactorymargate.co.uk/exhibition/nostalgias-visualising-longing

Porteous, J. D. (1990). *Landscapes of the mind: Worlds of sense and metaphor*. Toronto: University of Toronto Press.

Quercia, D., Schifanella, R., Aiello, L. M., & McLean, K. (2015). Smelly maps: The digital life of urban smellscapes. In *Ninth International AAAI Conference on Web and Social Media* (pp. 327–336). Retrieved March 1, 2017 from www.aaai.org/ocs/index.php/ICWSM/ICWSM15/paper/view/10572

Rodaway, P. (1994). *Sensuous geographies: Body, sense and place*. London: Routledge.

Solnit, R. (2014). *The faraway nearby*. London: Granta.

Tolaas, S. (2003). *SIRAP mon amour* [Fragrances]. Retrieved March 1, 2017 from www.hkw.de/en/programm/projekte/veranstaltung/p_917.php

Tolaas, S. (2009). *Talking nose* [Fragrances, map, scratch and sniff map]. Retrieved March 1, 2017 from www.ediblegeography.com/talking-nose/

Tuan, Y.-F. (1977). *Space and place: The perspective of experience*. Minneapolis, MN: University of Minnesota Press.

Twilley, N. (2010a, October 6). How to: Make your own scratch-and-sniff map. *The Atlantic*. Retrieved May 23, 2017 from https://www.theatlantic.com/technology/archive/2010/10/how-to-make-your-own-scratch-and-sniff-map/64106/

Twilley, N. (2010b, September 22). You are here. Retrieved May 23, 2017 from www.ediblegeography.com/you-are-here/

Westerkamp, H. (1974). Soundwalking. *Sound Heritage*, *III*(4). Retrieved March 1, 2017 from www.sfu.ca/~westerka/writings%20page/articles%20pages/soundwalking.html

Wu, E. (2007). *The whiff of uncertainty* [Digital print]. Retrieved March 1, 2017 from http://estherwu.com/images/whiff_big.jpg

Yatzer. (2015, January 10). Smells like art: Olfactory art exhibition in Madrid explores over 100 years of perfume-making. Retrieved March 1, 2017 from https://www.yatzer.com/smells-art-olfactory-art-exhibition-madrid-explores-over-100-years-perfume-making

PART III
Smellscape Design and Monitoring

INTRODUCTION

Jean-Paul Thibaud

Is there such a thing as a 'smellscape,' and if so, is it possible to design and monitor it? By asking this question, we are required to specify what we understand by 'smellscape' and to reconsider the very idea of 'design.' The four chapters presented in this part offer diverse responses and describe numerous case studies related to both of these issues.

One could imagine that an olfactory environment is simple to decipher and monitor: on the one hand an environment that 'smells good' and needs to be promoted, and on the other hand one that 'smells bad' and should be prohibited. After reading the chapters presented here, the question is revealed to be more complex and certainly more interesting and subtle.

To start with, it is important to notice that odor alone is never sufficient to evaluate and characterize a smellscape. At least three additional aspects should be fully considered. First, we have to acknowledge the fact that sensibility to odors depends very much on individuals and groups. The olfactory sense is more or less developed depending on people, but an appreciation of an odor can also be very diverse. Hence, a smellscape involves not only odors, but also the experience of odors and the relationship we have with them. Secondly, one needs also to consider the role of context: as Michael Bull shows in Chapter 11, an odor can be pleasant or unpleasant depending on its concentration and combination with other odors, depending also on the place, the time and the activity during which it is experienced. Thirdly, as Jieling Xiao, Malcolm Tait and Jian Kang show in Chapter 8, smellscape cannot be dissociated from the other senses such as vision, sound, taste and touch. Hence, a smellscape involves an experienced situation as a whole; it has to be considered from an inter-sensorial perspective. Moreover, designing a smellscape brings stimulating issues and involves new questions related to the field of design. Let me mention and emphasize five of them.

First: Integrating Cultural Know-how

At first glance, smellscape design seems to be a new kind of activity, a contemporary project which stems from the growing interest in the senses. After reading these chapters, it becomes obvious that such an idea is incorrect. Not only do we learn that odorization has been more or less appreciated depending on cultures, but it also becomes clear that certain ancient traditions have developed unique sets of practices regarding the world of smells. Korean smellscapes are charted by EunSook Kwon in Chapter 10, and China and Japan also immediately spring to mind as demonstrating a rich and unique olfactory cultural tradition, but other traditions could also be mentioned. What is interesting here is that such cultures have also invented material devices and promoted specific know-how, be it, for example, burning incense described by EunSook Kwon, or using fragrances in gardening, evidenced in Victoria Lygum's Chapter 9. It is also worth noting that the use of smell is often closely related to rituals and religious ceremonies. Hence, smellscape is part of a cultural heritage that crosses the boundaries between materiality and spirituality.

Second: Paying Attention to the Background

Composing with odors demands careful consideration of the role of attention, and also that of background levels. Often taken for granted or supplanted by other sensory phenomena, the world of smell tends to be forgotten in everyday life. It is mostly noticed when it becomes a public problem and involves a form of nuisance or disturbance, as is demonstrated by Michael Bull in Chapter 11. But what happens when such is not the case? Smelling is not only passive – a mere reception of an olfactory environment – but also active. It relies on the act of breathing, inhaling and exhaling, and this activity can be enhanced, become intentional and be traced back to consciousness. One of the challenges for designers is to produce a smellscape that will be fully experienced. A complete vocabulary can be used to account for smell thresholds and sensitivity to odors. This smell environment is described and acted upon in terms of detection, concentration, dilution and dispersion. Hence, smellscape design highlights the importance of sensory background in the experience of an environment.

Third: Producing All Sorts of Effects

When designing a smellscape, one does not intend to only create a pleasant and comfortable environment, but may also aim at producing more specific effects. Such effects can be very diverse, tending to reduce stress or calm people, as described by Victoria Lygum (Chapter 9), to produce a sense of sacredness, to purify the air and the surroundings, to mark the space and foster urban identity, to attract visitors or passers-by, to trigger memories and past experiences (one could add to encourage shopping and consumption). Well-being is, of course, at the core of this activity, but odors may also concern the fields of health and therapy (see Chapter 9), architecture and urban planning (see Chapter 8) sociability and religion, or business and marketing (described in Part IV). Thus, the aesthetic dimension of smellscape has to be articulated in all its diverse pragmatic and praxiological dimensions. The need to design a smellscape, and the various consequences of such a project, can also raise political issues as well as ethical ones, since odors can be effective without a fully conscious perceiver.

Fourth: Composing with the Elements

Natural elements seem to be the primary means by which designers create and monitor smellscapes. Flowers, trees and all sorts of greenscapes diffuse specific odors and contribute to composing an olfactory environment. For instance, orange trees and other fragrant plants have the ability to pervade a place with odors, and their design potential is explored in Chapter 8 by Jieling Xiao, Malcolm Tait and Jian Kang, who also discuss the role of water in the diffusion and transmission of aromas. Waterscapes produce several effects and involve diverse modes of action that are important in order to monitor a smellscape: running water may accelerate ventilation, increase air humidity and bring a sense of freshness. Last but not least, air is the medium *par excellence* of the propagation of odors. This is maybe one of the most difficult problems encountered by designers:

how to master air movements and airflows that are very labile, unstable and transient by nature? We have just seen that water can be very useful, but the built environment is also to be considered. Urban forms rely on different types of boundaries or semi-closed spaces that can confine odors and limit their dispersion. In Chapter 10, EunSook Kwon considers this issue in the case of cultural festival site design. They also involve corridors and streetscapes that can orient and channel air movements. Thus, the monitoring of odors described by Michael Bull (Chapter 11) needs to understand the complex interaction between the materiality of a built space and the atmospheric dimension of a smellscape.

Fifth: Designing in Accordance with the Weather

Because smellscapes rely on air and elements, weather is a key feature of this kind of design. Humidity, temperature and wind, rain or dryness, affect the nature and the experience of an olfactory environment. One of the most stimulating lessons to learn from smellscape design is that it cannot be identical anywhere and any time. One will not design the same way in a place subject to a tropical climate, a continental, mild or polar one. Jieling Xiao, Malcolm Tait and Jian Kang's use of different cases in Chapter 8 illustrates this great diversity. The weather conditions are not the same, neither are the flora or the sensitivity to seasonal cycles. Geographical situations and seasonal change determine largely what it will be possible to do in terms of design and planning, and consequently three important implications emerge. First, designing a smellscape means designing space-time. These two dimensions – space and time – are inseparable, since the olfactory resources of a place depend very much on the weather and the season. Secondly, I am referring to a temporality that is cyclical and not linear. Smellscapes are not only subject to variation and transformation, but also to a periodic return, one season after the other. Thirdly, entering the era of the Anthropocene, smellscape design has an important role to play in order to sensitize inhabitants to climate change. Because it is so fundamentally embedded in natural conditions, smellscape design can help to raise awareness by creating new olfactory experiences significant of our time period, which were explored earlier by Miriam Songster in Chapter 5.

Finally, smellscape design is at the crossroads of numerous issues related to the contemporary environment. Situated between quality of life and sensory marketing, health concerns and ecological challenges, it is particularly effective perhaps because of this discrete but integrative intellectual and practical location. Moreover, dealing fundamentally with ambient and atmospheric phenomena, and with diffuse and invisible components, smellscape design can lead to a fresh questioning of what design is all about. The chapters in this part demonstrate this potential and show how design can be a way of immersing in life, pervading bodies and places, and enhancing everyday sensory experience.

8

THE DESIGN OF URBAN SMELLSCAPES WITH FRAGRANT PLANTS AND WATER FEATURES

Jieling Xiao, Malcolm Tait and Jian Kang

Designing urban smellscapes with aromas of natural elements can bring restorative effects to people's daily life in cities. People tend to prefer smells derived from natural elements, especially smells of plants and water (Henshaw, 2014; Moncrieff, 1966). In particular, these smells are found to have a significant positive influence on people's health, emotions, memories and perceptions of places (Bringslimark, Hartig & Patil, 2009; Chen, Chen & Yin, 2007; Henshaw, 2014; Porteous, 1985; Tuan, 1979). The cleansing and freshening effects brought about by water also can result in positive impacts on people's general experiences within urban spaces, so greenscapes and waterways are suggested as two main categories of restorative smell sources, which may aid in promoting higher-quality environments (Henshaw, 2014). In landscape designs, plants and water are also essential elements (Booth, 1989), allowing landscape architects to improve visual aesthetics, shape spaces and provide shading. Some plants, such as rose, jasmine and lavender, produce fragrant scents enjoyed by people, which can also mask negative smells. Water features have also been found to attract people visually, as well as cleaning the air and modifying the microclimate by increasing humidity, reducing temperature and changing wind speed, thus changing surrounding smellscapes (Booth, 1989; Kang, 2012; Henshaw, 2014). This chapter reviews principles and practices of using fragrant plants and waterscapes in urban spaces, particularly through four cases using these elements in the design of smellscapes. Within these discussions, it is argued that urban vegetation and water features significantly influence the general smell environment, and may play a more important role as climate change increases city temperatures, necessitating greening of streets and public spaces.

Design with Fragrant Plants and Waterscapes: Principles

Application of Fragrant Plants and Waterscapes in Landscape Design

Fragrant plants are widely found across many countries. For example, people in China and Japan have a long history of using fragrant plants in gardening (Liu, 2005). In Chinese landscape practice, fragrant trees, vines and herbs are deployed, with scents that may be floral, fruity, wooden or leaf-like, and with varying smell intensities and functions (Chen et al., 2013; Gu & Zhang, 2010). Smells from fragrant plants are suggested to offer therapeutic functions in some clinical practices of aromatherapy: for example, lavender may reduce stress and calm people (Buchbauer, Jirovetz & Jäger, 1991; Diego et al., 1998). Although some have argued that little scientific evidence supporting aromatherapy has been found in laboratory experiments (Cooke & Ernst, 2000), positive influences have been observed from physical, psychological and social studies through interactions between people and smells of plants in many other studies (Bringslimark et al., 2009; Classen, 2013; Elings, 2006). Introducing fragrant plants into urban spaces in this sense may bring benefits to people beyond visual aesthetics.

In many countries, theme parks and farms of fragrant plants are also successful in attracting tourists at blossom seasons, such as the UK's Mayfield Lavender Farm, and flower fields in Furano, Japan and Grasse, France. However, in most cases, fragrant plants are mostly planted in private gardens or indoors at small scales, rather than as basic urban elements at large scales. Recently, fragrant plants have been used as an alternative way of improving environmental qualities in urban spaces in many Chinese cities, like Kunming, Guangzhou, Xi'an and Xiamen. They have been largely planted along streets, and in community gardens, hospitals and theme parks. The actions of bringing fragrant plants into urban spaces in such cities has, according to its promoters, achieved positive effects on people's living qualities, creating unique urban identities and attracting tourists through smells (Gu & Zhang, 2010). Such efforts demonstrate that fragrant plants can create spaces where people want to visit and linger, with potential lessons for designers of the urban public realm.

Unlike fragrant plants, water normally does not produce a smell. However, people can sense freshness from water, as it increases air humidity and wind speed, thus affecting smellscapes (Henshaw, 2014). There are many cases of fountains serving as essential visual urban elements in historical European cities, like Rome and Paris, and modern water features in cities are designed and used for multiple purposes: recreation, irrigation, climate control and to mask traffic noise (Booth, 1989). However, their influence on smellscapes has largely been ignored. This chapter will discuss, using examples, how different types of waterscapes can inform smellscape design in public spaces.

Urban Smellscape Design Principles and the Challenges of Using Fragrant Plants and Waterscapes in Urban Spaces

Human perceptions of the smell environment are inevitably influenced by other sensory mediations (Porteous, 1985). Smellscape design cannot be separated from other sensory elements, such as vision and sound. Many designers concentrate on auditory, visual and tactile perceptions when bringing plans and water features into urban spaces. However, it is important to identify the full range of sensory impacts in urban spaces. Realizing the potential of fragrant plants and running water features to achieve restorative effects with smells is an important, but little understood, strategy for the landscape architect, and one that needs to be guided in a systematic way.

A number of general principles might guide the design of urban smellscapes with fragrant plants and running water. Moncrieff (1966) summarized four general principles of improving the smell environment: separation, deodorization, masking and dilution. Henshaw (2014) then developed these into sequential basic design and management principles: separating odors through planned activities or displacement; removing unpleasant odors, like smells of dirt and waste; masking any remaining unpleasant odors with pleasant smells to hide the original ones; and introducing an odor with a special quality to create new characteristics for the space. A number of environmental aspects are also suggested as important, including air movement and microclimates, activity density and concentration, and materials and topography (Henshaw, 2014).

Apart from these, it is emphasized that smellscape should be considered within the physical and social contexts of places (Classen, Howes & Synnott, 1994; Henshaw, 2014; Porteous, 1985). Selections of fragrant plants and water features need to consider local cultures and activities as well as wider environmental factors such as temperature and seasonal changes. This presents a challenge for designing an urban smellscape suitable for all seasons and a wide group of users in multicultural cities. In some countries, smells are highly appreciated and generally accepted in public spaces, such as Japan, while countries such as the US have been identified as generally against scenting in all types of public spaces (Damian & Damian, 2006). However, implementing large-scale fragrant planting in urban spaces carries the risk of causing difficulties for people who are allergic to certain vegetation smells and suffer from hay fever. Water features on a large scale in urban spaces may also be problematic for cities with cold winters.

It is important to define scales and types of spaces when designing urban smellscapes. Generally, they can be designed and studied at three scales (Henshaw, 2014):

1. micro level – a specific site-based scale;
2. midi level – neighborhood district;
3. macro level – citywide area.

Different characteristics and scales of spaces may result in intensity differences of smells and involve different interactions between people and environments. The following section discusses four cases selected using fragrant plants and running water features in urban landscape designs in China, the UK and across Europe.

Cases for Learning

In this section, the cases explore the design of smellscapes by examining fragrant plants and waterscapes at different scales, and with different purposes and elements. Generally, three types of spaces are included: streets, gardens and squares. Chuncheng Road, Kunming is selected for its multi-purpose urban street form and various types of fragrant plants, informing smellscape designs at a midi-level scale, whereas the case in Athens illustrates a much smaller and simpler streetscape, forming a distinct smellscape through interactions of both smells and taste at a micro level. A sensory garden in Sheffield, UK is used to illustrate how community entertainment and educational design purposes can be designed at the macro level, taking into account accessibility and surroundings. The case of waterscapes along the Gold Route in Sheffield is used to show how different types and scales of water features form smellscapes in urban spaces through micro level interactions with people. The Alhambra Palace in Granada provides a good overall example of combining fragrant plants and waterscapes in a sustainable fashion. Each case illustrates strategies and methods of designing smellscapes by using fragrant plants and waterscapes.

Street Plants on Chuncheng Road in Kunming, China

Kunming is located in the southwest of China, which is famous for its rich natural resources and mild climate. The urban area of Kunming city has much greenery and a tradition of landscape design that takes into account the local climate and plant species. Following the local government's intention of making the city green and aromatic, most of the trees selected for the city's landscape blossom and produce aromas delivering visual pleasure, reducing air pollution and providing attractive aromas and shade to residents and visitors during all seasons (Chen, 2005). These strategies have created in Kunming a particular identity derived from the vegetation smells and colors.

A clear example is Chuncheng Road, which provides a good case to learn how fragrant trees and plants can be used in urban districts to achieve an appropriate and satisfying smellscape. Chuncheng Road is one of the major roads in Kunming, with a two-way four-lane motorway, a bike lane and a sidewalk on each side (see Figure 8.1).

FIGURE 8.1 Fragrant trees and plants in summer on Chuncheng Road, Kunming, Yunnan, China.
Image: author's own, 2015.

Trees and plants are used as barriers to separate motorway, bike lane and sidewalk. Most are local fragrant species commonly used in urban spaces, where they can help clean the air by reducing dust and absorbing pollution. The trees and plants are mostly evergreen species, and blossom from February to November, with enjoyable scents. From a smellscape design perspective, in the case of Chuncheng Road, fragrant trees and plants are used in three ways to improve the smellscape: to separate car fumes, to dilute car fumes and to mask car fumes with floral scents (see Figure 8.2).

Bushes planted at about 1 m height help to block particles, car fumes and heat at low levels, improving the pedestrian environment and allowing visual interaction across the road. These

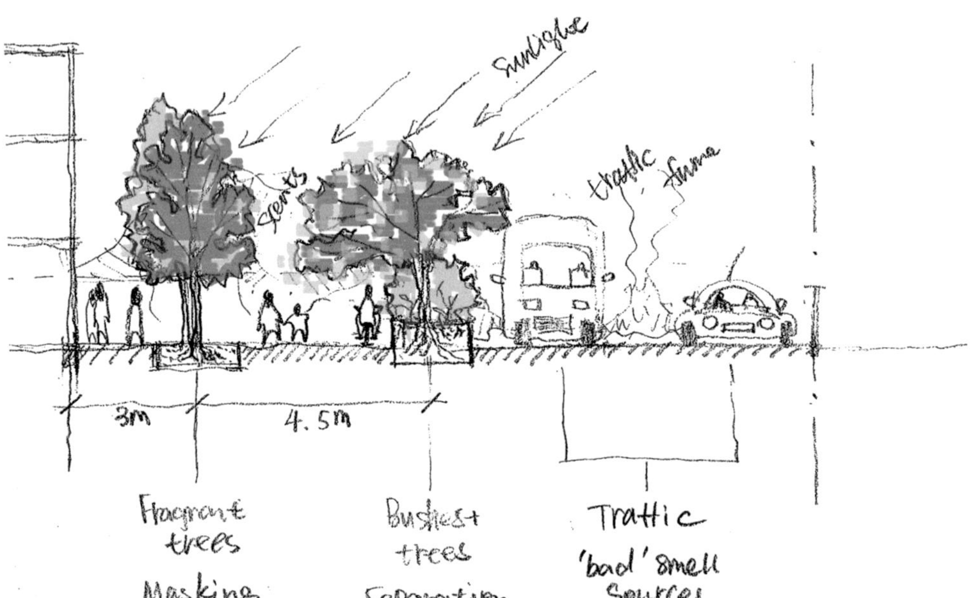

FIGURE 8.2 Interpretation of the smellscape on Chunchĕng Road, Kunming, China.
Image: author's own, 2015.

bushes are mainly *Ligustrum quihoui* (waxyleaf privet) and *Gardenia jasminoides* (gardenia), which are evergreen species and effectively reduce pollution of particulates such as sulfur dioxide and hydrogen fluoride. *Ligustrum quihoui* blossoms between July and August, and *Gardenia jasminoides* between May and July, both having white flowers with noticeable light scents, which can be enjoyed by cyclists and pedestrians. Other planted bushes include *Dendranthema indicum* (Indian chrysanthemum), which produces yellow flowers and a light scent from July to November. Above the bushes are trees, mostly cherry blossom trees and *Lagerstroemia indica* (crepe-myrtle); these are also planted with the intention of reducing the effects of harmful gases whilst also providing visual interest through their blossoms (Chen, 2005). Cherry tree blossoms of white and pale pink emerge between February and April, whilst *Lagerstroemia indica*'s bright pink flowers occur between July and September. Their flowers give off light but distinct scents which do not overwhelm the street smellscape. They have wide treetops, providing shade to the sidewalk, and also reducing temperature and the dispersion of traffic fumes across the bushes.

There are three main species of evergreen trees planted near to pedestrian pavements: *Osmanthus fragrans* (sweet olive), *Magnolia grandiflora* (laurel magnolia) and *Cinnamomum camphora* (camphor laurel). *Osmanthus fragrans* blossoms between August and October, with small white flowers with strong, pleasing scents, perceptible from long distances. *Magnolia grandiflora* produces noticeable white flowers from May to July, creating a light, pleasing scent. Both these species are planted every 5 m at the edge of the pedestrian pavement, connecting to the pavement and providing intense smells for pedestrians. By contrast, *Cinnamomum camphora* trees are usually planted non-continuously along the sidewalks, producing a light, fresh scent that keeps insects and mosquitoes away from pedestrians.

It can be concluded from this case that designing the urban smellscape with fragrant plants requires designers to undertake a preliminary study of local plant species, their growing conditions, scents, functions and blossom periods. A key priority in this location is to reduce pollution, with planting explicitly aimed at creating a higher-quality pedestrian environment. The plants selected may absorb harmful gases and reduce dust as well as producing pleasing scents. The selection of fragrant plants may also help to create combinations of different scents at the same time, such as *Osmanthus fragrans*, roses and lavenders. However, it may be more positive to have one distinguishable and dominant smell for pedestrians, to mask traffic, residential and sanitary smells, which also can easily be remembered and become a smell mark. The scale of planting should be carefully considered and tested to avoid being intrusive and allow enough lighting and visual interactions in the space. Another important aspect is to select species conducive to the human scale. For example, in this case, tree trunks are around 2–2.5 m high, which shortens the distances between perceivers and smells coming from leaves, flowers and fruits. Meanwhile, the color and shape of flowers and leaves offers better visual experiences, which may also indirectly add value to the general smellscape.

Bitter Orange Trees in Central Athens, Greece

> I am in Athens this spring and the city is drowning in the fragrance of the bitter orange trees that flood its sidewalks. The aroma is intoxicating, numbing the mind and the body's senses.
>
> *Frangouli-Argyris, 2013*

This section describes the particular landscape in the streets of Athens, with bitter orange trees (*Citrus aurantium*, or Seville orange) yielding scent for pedestrians (see Figure 8.3). These trees and their smells have become a defining feature for the city, certainly in the mind of this writer. Bitter orange is an evergreen species, which blossoms between April and May, with fruits harvested

FIGURE 8.3 Bitter orange trees in a neighborhood in summer, Central Athens, Greece.
Image: author's own, 2015.

between October and December. From late spring to winter, the bitter orange trees give sweet and fresh scents to the streets in central Athens. The wide-scale planting of bitter orange trees was initially for decoration (Varivaki, 2013), but gradually became part of the heritage of the city.

Bitter oranges are too sour to eat raw; however, local people use them in cooking and desserts. Historically, the harvest of bitter oranges was a chance for the local community to celebrate, but this tradition has died out and bitter oranges are now left unpicked throughout central Athens, forming a particular identity of the city. Tourists, who often have no idea that these are *bitter* oranges, frequently touch and taste them, potentially forming attachments between people and these trees based on the senses of smell and taste. Bitter oranges in this case give sensory meaning to the city, both historically and now.

In terms of designing smellscapes, it can be inferred from this case that the scale of street and human activities are important. The old streets in Athens are normally about 8 m wide, with pedestrian pavements 1.5 m wide, and with multi-story buildings on both sides (Figure 8.4). The street space is relatively small compared to contemporary building scales, which prevents significant dispersion of the smell from the bitter orange trees and concentrates its intensity. As they are planted in a confined streetscape, mobility remains a priority, which limits the design of planting involving human interactions. The bitter orange trees, as fruit trees, differ from regular fragrant trees and plants by involving the sense of taste as well as smell and vision. The bright color of the oranges hanging on the trees attracts pedestrians' attention, inviting people to smell and taste. This interaction, from seeing to thinking to smelling and eating, makes the smellscape in the street more active and interesting. Meanwhile, the light, fresh smell given by the bitter orange trees is pleasing at a background level, and free of pollen.

Fragrant Plants Applied in Community Space: Millhouses Sensory Garden, Sheffield, UK

A sensory garden has been designed and built by Sheffield City Council to provide a sensory experience of smells and sounds to local residents in this northern UK city (see Figure 8.5). It is located in the Millhouses Park, around 15 minutes by bus from the city center. Millhouses Park provides facilities for children, outdoor activities and sports, with a pond and playground. The

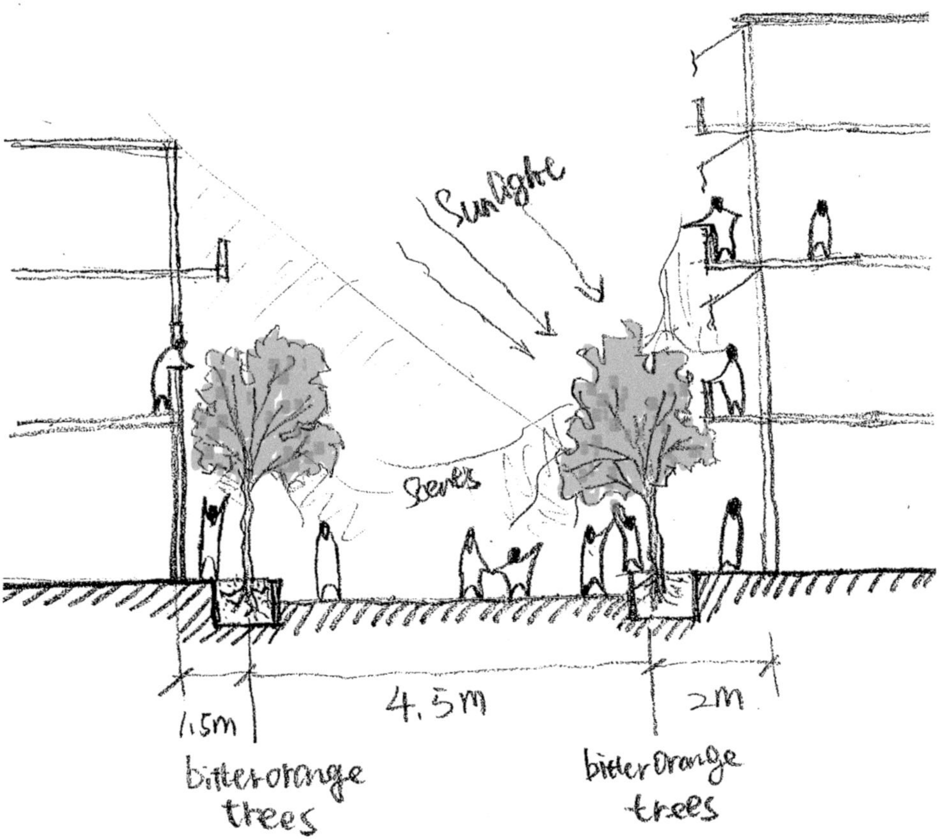

FIGURE 8.4 Interpretation of smellscape with bitter orange trees in Athens, Greece.

Image: author's own, 2015.

Millhouses Sensory Garden is between the tennis courts and a cricket pitch. Access into the garden is from both sides. There is an information board at the front entrance, saying:

> This tranquil garden is an area for contemplation and peace. Respond with your different senses to the lovely sounds, smell and feel of the garden. It provides an opportunity to watch and enjoy birds, bees, moths and butterflies. Smell the herbs and flowers which have been specially chosen for their fragrance . . . .

On the information board, there are pictures with short descriptions of selected herbs, including mint, oregano, basil, lemon balm, parsley, rosemary, thyme, sage, chamomile and lavender. These herbs are common cooking ingredients in the UK, and familiar to most visitors to the Sensory Garden. However, people normally take the experiences of smells for granted, and pay little attention to the smell environment (Henshaw, 2014). In this respect, the information board at the entrance to the Sensory Garden is particularly useful to make people aware of the distinct smell environment it offers.

The design aims to attract visitors by dispersing smells across quite large distances. Approaching the garden, the bright colors of various herbs stand out, and are designed to lead people in (see Figure 8.5). The core part of the garden is a semi-open space, surrounded by bushes and trees.

FIGURE 8.5 The Sensory Garden in summer, Millhouses Park, Sheffield, UK.
Image: author's own, 2015.

Roses have been planted to create an occasional strong aroma. In the center, a 1 m-high flowerbed with tall rose bushes in the middle emits intense fragrant smells. Two benches are placed facing towards the rose bushes, surrounded by herbs (Figure 8.6). The distance between the bench and the flowerbed is about 2.5 m, allowing a full view of the garden and, in summer, a scent of roses.

The garden is frequently used by residents of all ages and backgrounds, including young parents with babies, older people in wheelchairs, children, young couples and individuals who visit the garden on their way to other places in the park. The location of the garden allows good access, and its small scale enables people with little time to pass quickly through. Meanwhile, the garden has no physical boundaries, and the height of the outlying bushes and trees does not exceed the human scale, allowing people to see through the garden from outside and giving a more open feeling.

Waterscapes in the City Center: The Gold Route, Sheffield, UK

The Gold Route is a path designed to connect public spaces in the city center of Sheffield, from the railway station to Sheffield University, using a series of waterscapes along the route. These vary from silent streams to falling waters (see Figure 8.7). They are embedded in public squares, and attached to different functional spaces, such as the railway station, a university library, art gallery, shops and restaurants. These waterscapes provide multisensory experiences to visitors, particularly in terms of their soundscapes (Kang, 2012). The fresh smell and cooling feel from running water are also often described and enjoyed by visitors:

> I can see the water and it smells more fresh. Sometimes you can smell the chlorine in it. I quite like the smell, I find it quite fresh, to be fair, it is a clean smell. It is quite pleasant. I also like the sounds of the running water, I like to look at the running water, they have

different types of fountain in there, it is quite pleasing. If it is good weather, I would probably sit by the water with a coffee, reading, or people watching and time watching. It is a very relaxed environment.

a visitor at Sheaf Square

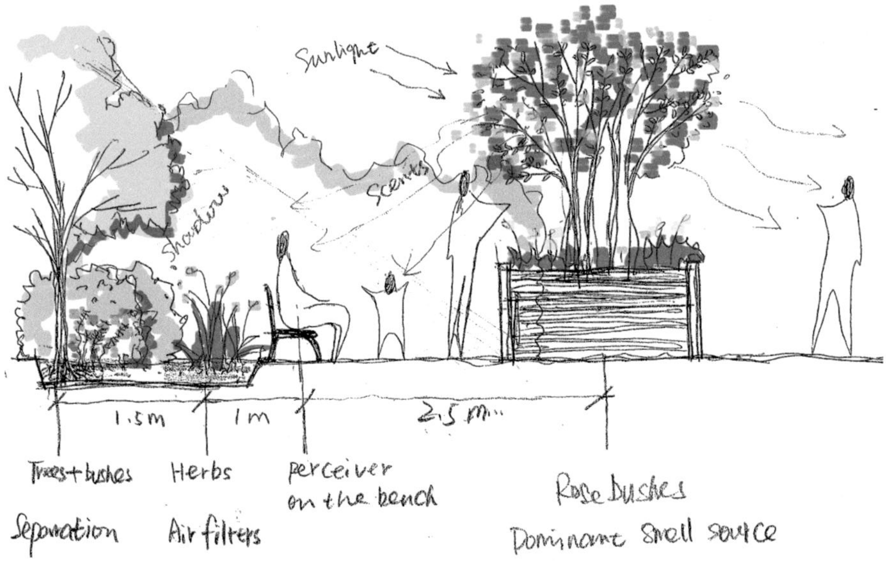

FIGURE 8.6 Interpretation of smellscape in the Sensory Garden, Sheffield, UK.

Image: author's own, 2015.

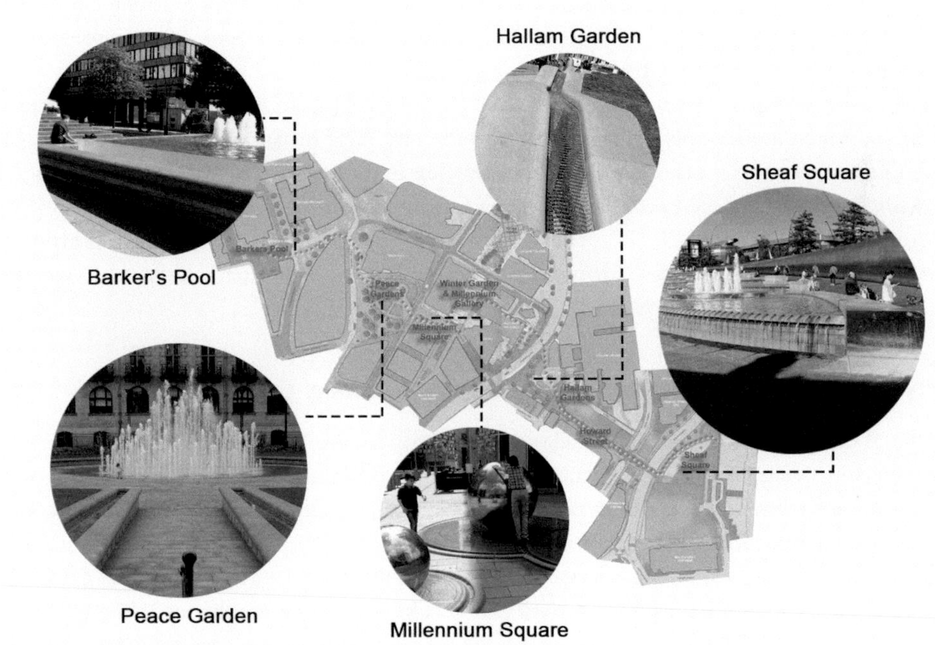

FIGURE 8.7 Waterscapes along the Gold Route, Sheffield, UK.

Image: author's own, 2015.

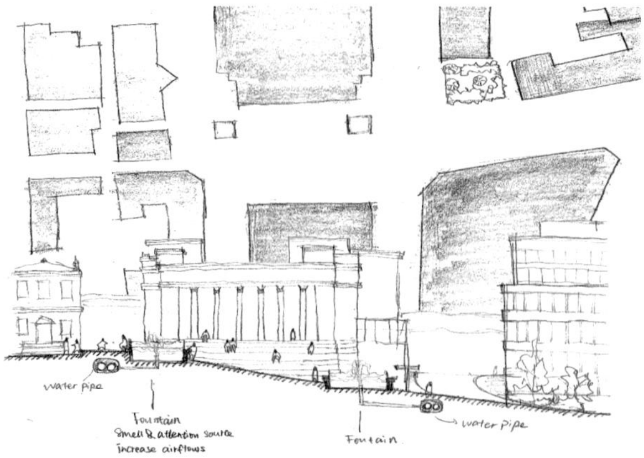

FIGURE 8.8 Urban context around Barker's Pool fountains.

Image: author's own, 2015.

Smellscapes along the Gold Route present a fresh and pleasing character to visitors. In the case of Sheaf Square and the Peace Gardens, there is high flow of traffic on one side of the space, producing fumes and noise. Waterscapes in both squares include large-scale running water, accelerating ventilation and removing and diluting traffic fumes. The running water helps clean the air and increases humidity, which offers a fresh smell to mask traffic fumes. Meanwhile, the sounds of running water also help mask traffic noise around these two squares, reducing general sensory nuisance. However, in the case of Millennium Square and Hallam Gardens, both pedestrian areas, the type of waterscape produces no splashes or sounds. They provide sensory experiences by attracting people to get close and play with the water (see Figure 8.7). People touch and smell the waterscape directly, and are distracted from bad smells in the environment, leading to an indirect improvement in people's perceptions of the city.

In Barker's Pool Square, there are two modern fountains built to reflect the historical meaning of the place, which used supply drinking water to the city. The urban fabric around Barker's Pool is quite different from the other four waterscapes (see Figure 8.8). The square is surrounded by multi-story buildings and connected to streets on four sides, with a wide opening towards Fargate, the main shopping street in the city center. Running water increases the ventilation around the square and forms wind corridors through the surrounding road and path network, providing a continuous refreshing effect for visitors.

Waterscape and Fragrant Plants: The Alhambra Palace, Granada, Spain

The Alhambra Palace is located in Granada, and was built by the last Muslim empire in Spain. It is famous not only because of its architectural beauty, but also because of its gardens, which combine fragrant planting with water features, creating a pleasant environment in hot weather. Though smell may not be the main reason for people to visit the Alhambra Palace, for many people it provides strong memories. The gardens also follow a long tradition in Islamic culture of including waterscape for aesthetic and agricultural purposes (Ansari, 2011). The smellscape is enhanced through reflected water increasing the surface temperature of the plants, which helps to accelerate the transmission of aromas (see Figure 8.9), whilst the high humidity caused by the water also improves perception of smells in the garden. The contrast to the surrounding buildings also enhances the general experience in the Alhambra Palace.

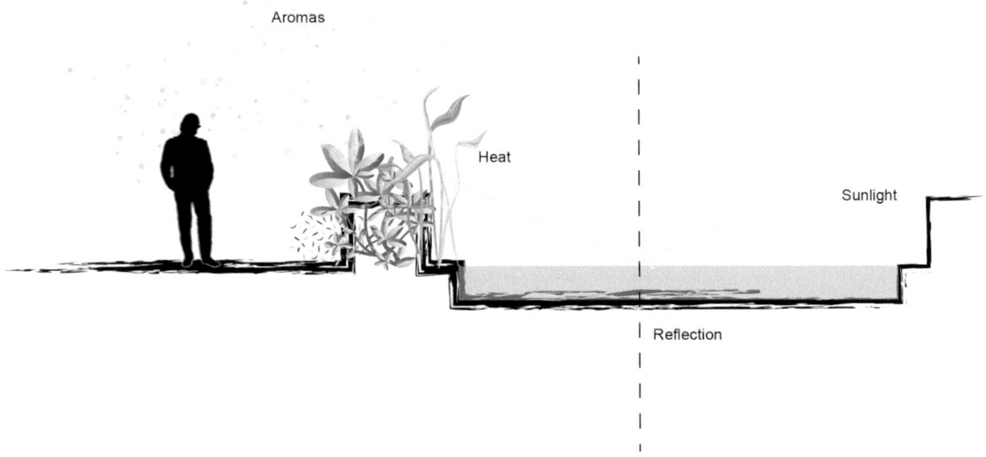

FIGURE 8.9 Interpretation of smellscape in the Alhambra Palace.

Image: author's own, 2015.

Advice for Designing with Fragrant Plants and Waterscapes

Select Appropriate Types of Fragrant Plants and Waterscapes

As Henshaw (2014) suggested, the initial step in the design process of a smellscape is to establish a clear objective. The four cases included in this chapter indicate general aims and relate planting to problem-solving, recreation and education. For example, the Chuncheng Road case particularly indicates how bringing fragrant plants into the urban environment may mitigate the effects of traffic pollution and noise whilst creating character and urban identity. The case of the Millhouses Sensory Garden provides a good example of designing for educational and recreational purposes. There are also places where fragrant plants and waterscapes have become part of a heritage that needs to be preserved to maintain historical smellscapes, as in the Alhambra Palace. However, selection of plants and types of waterscapes needs to be defined within local contexts and established objectives. A strong understanding and exploration of local fragrant plants, types of waterscapes and climate are a necessary prerequisite to designing smellscapes with these elements.

Design at a Human Scale: Distance + Intensity + Boundaries

Smells, which disperse and transmit with air movements, are temporary and unstable in an open urban environment (Henshaw, 2014). When designing urban smellscapes, the intensity of smells needs to be controlled above the threshold level for them to be detected and perceived. As suggested by the cases above, different types of boundaries are important in creating semi-closed spaces, which separate smells and concentrate scents. Fragrant plants and waterscapes can both serve as smell sources and act as boundaries, as shown in Chuncheng Road and Sheaf Square.

Distances between perceivers and smell sources also influence their detection. In theory, the closer the source, the stronger the smell. In the case of fragrant plants and waterscapes as smell sources, the scale of the space will determine distances and potential interactions. These interactions, like playing with water and picking oranges, also reduce the distance between these elements and perceivers. In a complex urban environment, integration of fragrant trees and waterscapes with people and spaces may need to reflect accessibility to these elements and an appropriate human scale.

Design with Activities + Other Sensory Stimuli + Context

Fragrant plants and waterscapes involve sensory perceptions other than just smell. Vision, sound and taste should be considered in smellscape design, making it essential to identify the characteristics of plants and water which imply other sensory features such as the sounds of running water, the taste of fruit and the color of flowers. Meanwhile, it is necessary to consider how people associate with these elements, like playing with water and cooking with oranges. Therefore, the introduction of fragrant plants and waterscapes into public spaces, as elements of designing smellscapes, can be more easily accepted and enjoyed by local communities. For instance, in the Chuncheng Road case, the plants selected are local species, which are familiar to local residents from history. Designers should engage with local residents to understand their preferences not only for an aesthetic, but also an 'aromatic' identity to a place. In many countries, like India, Malaysia, Iran and Mexico, where fragrant plants have been part of culture and life, promoting fragrant plants in public spaces can be a symbolic and natural feature, such as spice markets in Mumbai. Similarly, cities like Venice and Barcelona have smells associated with water and sea wrapped up with their branding. People interact with water through boating, swimming and surfing, bringing with them general sensory experiences, including smells. However, the feature of 'watery' space also needs to be designed within its local context.

Challenges: Seasonal Changes and Sustainability

Seasonal changes are essential issues to consider when designing with fragrant plants. In the cases outlined above, evergreen plant species seem preferable as they provide greenery throughout the year. Plants are selected with different blossom periods and scents. In places, where the climate is mild throughout the year, the influence of seasonal changes can be very small. The challenge of designing smellscapes with fragrant plants is to create smellscapes that respond to seasonal changes. However, in places where weather contrasts between seasons are large, it is a significant challenge for plants to produce fragrance throughout the year. It is similar in designing with waterscapes, which can be problematic in winter and rainy weather. Meanwhile, the artificial waterscape needs to consider the sustainability of piped water and recycling systems in areas with low rainfall and hot climates. It may be possible to learn from the Alhambra Palace, where water circulates in a sustainable way and contributes to the irrigation of surrounding plants. However, it can also be a major challenge to design such an eco-system with both plants and water within urban contexts.

References

Ansari, N. (2011). *The Islamic garden*. Ahmedabad: CEPT University.

Booth, N. K. (1989). *Basic elements of landscape architectural design*. Long Grove, IL: Waveland Press.

Bringslimark, T., Hartig, T., & Patil, G. G. (2009). The psychological benefits of indoor plants: A critical review of the experimental literature. *Journal of Environmental Psychology*, *29*(4), 422–433.

Buchbauer, G., Jirovetz, L., & Jäger, W. (1991). Aromatherapy: Evidence for sedative effects of the essential oil of lavender after inhalation. *Zeitschrift für Naturforschung C*, *46*(11–12), 1067–1072.

Chen, C., Chen, L., & Yin, L. (2007). Application of healthful plants in the garden. *Jiangxi Forestry Science and Technology*, *2*, 58–60.

Chen, D., Xian, Y., Ma, L., Zhang, Y., Xiao, C., & Ye, L. (2013). Species and application of ornamental aromatic plants in rehabilitation and reconstruction of Wenchuan seismic disastrous area: with Beichuan as an example. *Journal of Guangdong Landscape Architecture*, *35*(6), 69–74.

Chen, X. (2005). Tree species selections for urban greenery in Kunming City. *Forest Inventory and Planning*, *30*(5), 104–107.

Classen, C. (2013). Green pleasures. In B. L. Ong (Ed.), *Beyond environmental comfort* (pp. 175–183). London: Routledge.

Classen, C., Howes, D., & Synnott, A. (1994). *Aroma: The cultural history of smell*. London: Routledge.

Cooke, B., & Ernst, E. (2000). Aromatherapy: A systematic review. *British Journal of General Practice*, *50*(455), 493–496.

Damian, P., & Damian, K. (2006). Environmental fragrancing. In J. Drobnick (Ed.), *The smell culture reader* (pp. 148–160). Oxford: Berg.

Diego, M. A., Jones, N. A., Field, T., Hernandez-Reif, M., Schanberg, S., Kuhn, C. . . . & Galamaga, R. (1998). Aromatherapy positively affects mood, EEG patterns of alertness and math computations. *International Journal of Neuroscience*, *96*(3–4), 217–224.

Elings, M. (2006). People–plant interaction: The physiological, psychological and sociological effects of plants on people. In J. Hassink & M. V. Dijk (Eds.), *Farming for health* (pp. 43–55). Amsterdam: Springer Netherlands.

Frangouli-Argyris, J. (2013, June 23). Athens, the city of bitter orange trees [Blog post]. Retrieved March 1, 2017 from www.huffingtonpost.com/justine-frangouliargyris/athens-the-city-of-the-bi_b_7128710.html

Gu, Y., & Zhang, X. (2010). Application of fragrant plants in urban gardens. *Journal of Beijing Landscape Architecture*, *4*, 37–40.

Henshaw, V. (2014). *Urban smellscapes: Understanding and designing city smell environments*. New York: Routledge.

Kang, J. (2012). On the diversity of urban waterscape. Paper presented at *Acoustics 2012*, April, Nantes, France.

Liu, Z. (2005). The application of therapy by means of aromatics to gardens. *Forest Inventory and Planning*, *30*(6), 91–93.

Moncrieff, R. W. (1966). *Odour preferences*. London: Leonard Hill.

Porteous, J. D. (1985). Smellscape. *Progress in Human Geography*, *9*(3), 356–378.

Tuan, Y.-F. (1979). *Space and place: Humanistic perspectives*. Berlin: Springer.

Varivaki, M. (2013, July 30). *Bitter orange* [Blog post]. Retrieved March 1, 2017 from www.organicallycooked.com/2013/07/bitter-orange.html

9

STOP AND SMELL THE FLOWERS

Plant Scent in a Healing Forest Garden for Patients with Stress-related Illness

Victoria L. Lygum

The Healing Forest Garden, Nacadia

In the late 20th Century, researchers from various academic disciplines such as environmental psychology and landscape architecture started focusing on the relation between nature and health. Their work has entailed a significant amount of studies which indicate that contact with nature – for example, watching a beautiful tree from indoors, sitting on a bench in a garden or doing horticultural activities – can lead to stress reduction and restoration (Marcus & Barnes, 1995; Sherman, Varni, Ulrich & Malcarne, 2005; Ulrich, 1984; Verderber, 1986). The research has led to the overarching term 'healing gardens' to denote gardens designed for specific user groups so as to facilitate people–plant interactions through which users may develop well-being (Diehl, 2007).

Such a garden is in focus at the Section for Landscape Architecture and Planning, Department of Geosciences and Natural Resource Management, University of Copenhagen. Here, Professor Ulrika K. Stigsdotter and her research group investigate the connections between nature, health and design. Part of this research is a healing forest garden that facilitates nature-based therapy for patients with stress-related illnesses (Stigsdotter, 2014). 'Nacadia,' as the forest garden is called, serves as a treatment setting where patients strengthen their capacities to overcome illness, as well as a research and education setting with the goal to obtain and disseminate knowledge on healing garden design and nature-based therapy. The patients include war veterans suffering from post-traumatic stress disorder and individuals on long-term sick leave due to work-related stress. The treatment takes its starting point as experiences and activities in natural settings. It is led by therapists, and focuses on restoring health on a physical, mental and social level (Stigsdotter, 2014).

Nacadia is located north of Copenhagen in the Hoersholm Arboretum, which is Denmark's biggest collection of woody plants. The forest garden covers 2 acres (0.8 ha), and includes tall trees, bushes, shrubs and grassy meadows. Nacadia also has a building with offices and restrooms, a greenhouse with an outdoor kitchen, different types of water features, and a system of terraces and paths that connect the elements of the forest garden. Nacadia is designed as part of an evidence-based health design process. Here, research results and valid practical experience, knowledge about the patient group, as well as landscape and architectural skills have formed a solid basis for the decisions about design. A variety of garden rooms offer different types of experiences and activities which aim to match the needs and preferences of the patients at each stage of the treatment. At their own pace, and guided by therapists, the patients seek out garden rooms that are increasingly demanding as their health improves (Stigsdotter, 2014).

Sensory qualities such as color, texture and sound are carefully worked into the design of Nacadia. The forest garden offers vegetation that is rich in species as well as materials such as wood, stone and water. To make the sensory qualities accessible, there is an imaginative use of raised planting beds, pergolas, different types of seating, raised platforms, differently paved paths, as well as still and running water (Stigsdotter, 2014). In the nature-based therapy, the sensory qualities help bring the patients' attention to the present moment. They are also used as starting points for strengthening body awareness as well as metaphors in conversations with the therapists (Corazon, Stigsdotter, Moeller & Rasmussen, 2012).

Smell is an important part of these sensory qualities. Smell in Nacadia depends on various factors, such as the vegetation, water and soil conditions, building materials, the season, the weather, the time of day and the activities carried out in the forest garden. Smell can also come from the surrounding arboretum, or even from further away. All things considered, the selection of plants is especially important for the smell in Nacadia, as vegetation makes up the main building material of the forest garden.

The Use of Plant Scent in Nacadia

The following section focuses on the use of plant scent in Nacadia, and gives perspectives on how plant scent may support the treatment of patients with stress-related illness. It takes as a starting point a walk and talk interview with the Arboretum Supervisor, Ole Rasmus Byrgesen. Ole is a key person when it comes to knowledge about the plants in Nacadia, and he has been assisting the therapists in their work with the patients.

It is an early morning in August, and the air in Nacadia smells fresh. Ole and I start our tour on the wooden terrace connected to the building and overlooking large parts of the forest garden. Along the facade, there are several raised planting beds with built-in benches, so it is easy to get in close contact with the scented and edible plants, such as mint, Mexican marigold and basil (see Figure 9.1). Ole diligently hands me scented leaves and flowers, and says that the patients use them in both tea and cooking. In general, the scent of plants is stored as essential oils in cells in the outer parts of the plant, such as flowers, leaves and roots (Kaiser, 1993), or in resin, which is found in wood and bark (Frosell, 1982). Seeds and fruit may also contain essential oils (Lacey, 1995).

FIGURE 9.1 The raised planting bed gives easy access to scented plants such as lavender, thyme and sage.
Image: author's own, 2015.

Ole tells me that at the beginning of the treatment, the patients are often encouraged to touch and smell the plants. Therapists may hand out so-called sense cards – small pieces of paper describing an activity that the patients can perform if they want to. An activity could be: 'Take a walk in the Arboretum and experience the different smells. Go slowly. Go a bit faster. Stop. Take your time. You may write in your logbook. Be as specific as possible.' It might take some time, but gradually the patients open up to their surroundings and begin to seek out the qualities of Nacadia on their own initiative. I reflect upon how scent may help the patients become familiar with the forest garden by tempting them to interact with the plants. At the same time, scent may motivate the patients to strengthen their body awareness by reaching out, bending down and stretching in order to smell the plants.

From where we stand, it is possible to overlook the roof of a pergola that connects the wooden terrace with a lower-lying meadow. Ole says that the honeysuckle that scrambles over the pergola smells wonderful in the evening (see Figure 9.2). In fact, the release of scent from flowers is controlled by the lifecycle of the plant and an internal biological clock that is specific for each species. Light intensity, day and night cycles and temperature changes can promote or inhibit the release of scent, depending on the plant species (Kaiser, 1993). In the end, the function of the flower scent is to attract pollinating animals with the aim of ensuring the maintenance of the species (Lacey, 1995).

We continue our tour, and walk towards the greenhouse, which we enter. Inside, the air feels warm and heavy with sweet-scented plants such as passion flower and star jasmine. The greenhouse is quite large, and offers several smaller separate rooms with different amounts and types of vegetation, which makes it possible for patients to seek out or avoid powerfully scented plants. This is a good thing, because the patients' relation to smells can be influenced by the mental and physical imbalance and by the intake of medication. Depression symptoms can reduce the ability to smell (Pollatos et al., 2007), while a tendency to headache can make some patients vulnerable to scent (Kelman, 2007).

Ole picks a leaf from a eucalyptus and rubs his neck with it. He tells me that the sharp scent scares the mosquitoes away. Essential oils in leaves, wood and bark have a different function from flower scent. Insects and grazing animals are repelled by the sharp smell and taste, which can prevent an animal bite or insect attack. In addition, the essential oils can be antiseptic and protect the plant against diseases (Rawlings, 1998). In leaves, the scent is released by a light touch, or by

FIGURE 9.2 In the evening, the honeysuckle fills the air with fruity sweetness.

Image: author's own, 2015.

FIGURE 9.3 'Sensitive plant' delights the senses.
Image: author's own, 2015.

rubbing or crushing it. In some cases, no contact is required (Lacey, 1995). Scent is also released by pressure and damage to wood and bark (Raven, Evert & Eichhorn, 1995).

Near a stepping-stone in the greenhouse grows a very peculiar plant. It is a 'sensitive plant' (*Mimosa pudica*), the pink flowers of which look like magic light balls. Its name relates to its leaves, which fold up in pairs when touched. Besides being beautiful and fun to touch, the flowers release a delicate, sweet scent. In Nacadia, patients have a tendency to direct their attention inwards at the beginning of the treatment. I wonder if scent may be used to awaken the patients' attention for the surroundings, as smell has a special ability to provoke feelings that are not conditioned by the overall mood of a person (Engberg-Pedersen & Meyhoff, 2004). In order to keep the patients interested, it would be an advantage if the source of scent were connected to other kinds of appealing sense impressions, such as bright colors or unusual textures. The 'sensitive plant' is a great example of a plant with such qualities (see Figure 9.3).

Ole and I head out of the greenhouse, across the meadow and into the forest. The forest floor is covered with needles from the conifers that surround us, and there is a refreshing scent that is markedly different from those in the greenhouse (see Figure 9.4). A number of research studies actually show positive pharmacological effects of forest scents, such as pine (Jo, Fujii & Cho, 2010; Li et al., 2007). Twenty-first-century neuroscience makes it possible to measure the pharmacological effects of essential oils from plants. As an example, lavender scent has a sedative effect (Perry & Perry, 2006). As early as the 18th Century, however, Carl von Linné (1707–1778) was the first scientist to develop a classification system for plant scent (Zarzo & Stanton, 2006). Basically, he believed that the scent of plants could be divided into those that are either good or bad for vitality (Bresle, 1986).

Ole shows me a tree with a scent he especially likes. It is an Antarctic beech that releases an appetizing scent of cinnamon buns during springtime. It makes me think that a scent which gives a common association can be an obvious way to start a conversation. Perhaps scent can help the patients make contact with the therapist or other patients, and in that way facilitate the first steps towards improved health on a social level.

The Katsura tree is also part of the collection of trees in the Arboretum. In the autumn when the leaves drop, it smells of caramel. The mouthwatering scent reminds Ole of cotton candy and trips to the amusement park as a child. Scent can indeed trigger feelings, memories and daydreams in a very powerful way (Haug, Sand & Sjaastad, 2000). This potential may strengthen a patient's feeling of identity by provoking significant memories. But whether a scent triggers a pleasant or

FIGURE 9.4 The freshly cut wood strengthens the resinous scent which characterizes 'The Dark Forest Room'

Image: author's own, 2015.

an unpleasant feeling may vary greatly from person to person. As an example, the smell of soil caused a patient to have a flashback of a traumatic event. In this respect, scent can also be considered as a dangerous substance that can affect patients without warning. Fortunately, the therapist was able to use the reaction to help the patient cope with the feelings linked to the stressful memory. In this way, scent may support the therapist's practice.

We are now no longer in the forest, but out in the meadow again. In this part of the forest garden, it is possible to grow vegetables and berries. Ole hands me a leaf from a red currant. It smells like most leaves. Then he hands me another leaf. Its sweet, spicy scent reveals that it is black currant. Smell is a useful trick that makes it possible to tell the difference between red and black currant when there are no berries on the plants. Ole and I head back to the wooden terrace where we started our tour. We enter the building, and Ole finds me a jar of honey made by bees in the Arboretum. In this way, I can take a little piece of Nacadia home with me to see if I can taste its many scents.

Towards a More Targeted Use of Plant Scent

The perspectives on how plant scent may support treatment in Nacadia show a great potential for exploring how olfaction affects patients. Although the use of plant scent in the forest garden is quite elaborate, a first step would be to register the many sources of scent in Nacadia and its nearest surroundings. Then it might be relevant to accentuate certain scents, tone down others and integrate new ones in order to further structure the use of plant scent in the treatment. There are many uncertainties when using scent from living plants in a design, but still it is possible to regulate plant scent on at least three different levels. The first is by supporting the species-specific factors that promote the release of scent from the plant. For example, some plants release most scent in hot sunshine, while others need to be touched to release a smell. The second is by regulating how the scent is experienced, by adjusting its concentration in the air. This could be done by regulating the size and the number of plants and by adjusting accessibility to them by potential smell recipients. The third is by controlling the scent when it has been released from the plant. This could be done with the help of shelter from the wind, or by trapping the scent inside a greenhouse (Lacey, 1995; Rawlings, 1998).

A more targeted use of plant scents could form a good basis for exploring how they affect patients. However, it is important to keep in mind that plant scent is just one of numerous types of sensory stimuli in Nacadia that all influence the way the patients perceive their surroundings and are affected by them. Having a multisensory approach would greatly benefit this research. Results could contribute to the development of both healing garden design and nature-based therapy for patients with stress-related illnesses. In the end, the results may give us a better understanding of the importance of plant scent for our health and well-being.

Acknowledgments

Thanks to Arboretum Supervisor Ole Rasmus Byrgesen, Psychologist Dorthe Djernis and Professor Ulrika K. Stigsdotter, who shared their knowledge and experience.

References

Bresle, Å. (1986). *Näsa för lukter – handbok om lukter och luktsanering*. Malmö: Ingenjörs Förlaget.

Corazon, S. S., Stigsdotter, U. K., Moeller, M. S., & Rasmussen, S. M. (2012). Nature as therapist: Integrating permaculture with mindfulness- and acceptance-based therapy in the Danish healing forest garden Nacadia. *European Journal of Psychotherapy & Counselling*, *14*(4), 335–347.

Diehl, E. R. M. (Ed.). (2007). *Definitions and positions*. Seattle, WA: American Horticultural Therapy Association.

Engberg-Pedersen, A., & Meyhoff, K. W. (2004). *At se sig selv sanse: Samtaler med olafur eliasson*. Copenhagen: Informations Forlag.

Frosell, P. H. (1982). *Fra duftenes verden: Parfumer og deres historie*. Copenhagen: Hernovs Forlag.

Haug, E., Sand, O., & Sjaastad, Ø. V. (2000). *Menneskets fysiologi*. Copenhagen: Gads Forlag.

Jo, H., Fujii, E., & Cho, T. (2010). An experimental study on physiological and psychological effects of pine scent. *Journal of the Korean Institute of Landscape Architecture*, *38*(4), 1–10.

Kaiser, R. (1993). *The scent of orchids: Olfactory and chemical investigations*. Basle: Editiones Roche.

Kelman, L. (2007). The triggers or precipitants of the acute migraine attack. *Cephalalgia*, *27*(5), 394–402.

Lacey, S. (1995). *Scent in your garden*. London: Frances Lincoln.

Li, Q., Morimoto, K., Nakadai, A., Inagaki, H., Katsumata, M., Shimizu, T. . . . & Kagawa, T. (2007). Forest bathing enhances human natural killer activity and expression of anti-cancer proteins. *International Journal of Immunopathology and Pharmacology*, *20*(2 Suppl. 2), 3–8.

Marcus, C. C., & Barnes, M. (1995). *Gardens in healthcare facilities: Uses, therapeutic benefits, and design recommendations*. Martinez, CA: Center for Health Design.

Perry, N., & Perry, E. (2006). Aromatherapy in the management of psychiatric disorders. *CNS Drugs*, *20*(4), 257–280.

Pollatos, O., Albrecht, J., Kopietz, R., Linn, J., Schoepf, V., Kleemann, A. M. . . . & Wiesmann, M. (2007). Reduced olfactory sensitivity in subjects with depressive symptoms. *Journal of Affective Disorders*, *102*(1), 101–108.

Raven, P. H., Evert, R. F., & Eichhorn, S. E. (1995). *Biology of plants*. New York: Worth Publishers.

Rawlings, R. (1998). *Healing gardens – aromatherapy, feng shui, holistic gardening, herbalism, color therapy, meditation*. Minocqua, WI: Willow Creek Press.

Sherman, S. A., Varni, J. W., Ulrich, R. S., & Malcarne, V. L. (2005). Post-occupancy evaluation of healing gardens in a pediatric cancer center. *Landscape and Urban Planning*, *73*(2), 167–183.

Stigsdotter, U. (2014). Nacadia healing forest garden, Hoersholm Arboretum, Copenhagen, Denmark. In C. C. Marcus & N. A. Sachs (Eds.), *Therapeutic landscapes* (pp. 198–205). Hoboken, NJ: John Wiley & Sons.

Ulrich, R. (1984). View through a window may influence recovery. *Science*, *224*(4647), 224–225.

Verderber, S. (1986). Dimensions of person–window transactions in the hospital environment. *Environment and Behavior*, *18*(4), 450–466.

Zarzo, M., & Stanton, D. T. (2006). Identification of latent variables in a semantic odor profile database using principal component analysis. *Chemical Senses*, *31*(8), 713–724.

10

THE CREATION OF SEOUL SCENT

Invisible Beauty

EunSook Kwon

Introduction

The legacy of the Olympic Games has the power to deliver lasting benefits. Amongst these is the construction of new venues built in the host cities, which can be used for sports or comparable community activities after the Games have ended. Maintaining these lasting benefits for the community of the host city is difficult relative to their long-term use after the Games, and the venues for the Seoul 1988 Olympic Games are not an exception. In 2008, I was appointed Director General of the *Seoul Design Olympiad 2008* (SDO 2008), the first comprehensive landmark design event organized by the Seoul Metropolitan City Government celebrating the city of Seoul, which had been designated World Design Capital 2010 by the International Council of Societies of Industrial Design. The biggest challenge, as the Director General, was to transform the historic Seoul Olympic Stadium into an accessible living space for creative activities. The stadium facilities had deteriorated during the past two decades when they had been little used, and it had to be reopened and reclaimed as the venue for the new celebration. This chapter is a case study describing the creation of an antidote to the architectural odor that lingered throughout the site, and also describes the making of a new smellscape in an abandoned building. It charts the development of new olfactory experiences based on historical and cultural associations.

Transforming the Seoul Olympic Stadium

The Seoul Olympic Stadium, also known as Jamsil Olympic Stadium, was designed by SwooKeun Kim, South Korea's most prominent architect, and built in 1984 using a synthesis of new materials and form, which profiled the elegant curves of Korean ceramics from the Joseon dynasty. During the two decades after the Games, Jamsil Olympic Stadium had been used for elite athletics, major sporting events and concerts. Without being converted for alternative uses, the venue has served as a narrative symbol of rapid economic growth, a voice of esteemed national pride and connectivity to the global village, and a critical understanding of Seoul's urban modernity. The stadium's narrative symbols of fluidity and harmony have provoked continuous dialogue and ideas, which have been transported around the world. As Barthes (1977) observed, "narrative is international, transhistorical, transcultural; it is simply there like life itself" (p. 79).

Towards the end of the first decade of the 21st Century, however, Jamsil Olympic Stadium was going to open to the public for the first time in its history not for sports, but for the creative and cultural activities of the *Seoul Design Olympiad 2008* (see SDO, 2008). To reuse the Olympic Stadium as a cultural space was a major challenge, posing many questions, such as: how could we transform a former sports site into a venue of creative inspiration without losing its essential character so important to the community, and how could we reconstruct the monumental concrete structure for the enrichment of personal experiences and the generation of new narratives?

While visiting the stadium in January 2008, we were immediately awed by its beauty. The void created by curved steel and concrete was powerful in its physical poetry, and the open structure allowed for a dynamic mixture of airflow, lighting and shadow, adding to the creation of intense sensory and spatial experiences. The fragmented space which had been designed for stadium foot traffic and to facilitate human flow could be utilized to create a dynamic mixture of old and new, West and East, tangible products and intangible experiences. But even with the welcoming visuals embodied in the built environment, an uninviting, unexpected problem met us head-on as we entered the space. This problem was the odor of a stadium which had been unoccupied for 20 years. The odor embedded in the space was inevitably distressing, and we quickly recognized that the issues of smell had to be addressed in a timely manner for the preparation of SDO 2008, which was to be held in October of that year. As we walked down the steps into the stadium, we realized that the scent increased with humidity and temperature. Multiple site visits reinforced the importance of incorporating smell rectification into the redesign process. The odor of the stadium became terrible in the spring, and dreadful in the summer due to the increased humidity, especially within the awesome void of the stadium.

Transforming this historical space required us to remove unwanted layers of association with disrepair and disuse. Strategic plans for the cleaning and sanitization process were thoroughly applied to the worst odorous areas. We then monitored and managed air circulation throughout all rooms in the stadium so as to preserve indoor air quality. A team of 200 worked daily to create proper air maintenance and control the odor in the stadium during SDO 2008 itself. In addition to the sanitization process, a new strategy was developed focusing on olfactory design and its application to SDO 2008, as a final layer of odor removal and to enhance spiritual, creative and spatial qualities existing within the concrete structure. In combination with vision and tactility, this strategy was supported by Porteous (2006), who emphasized that "smell and the other non-spatial senses provide considerable enrichment of our sense of space and the character of place" (p. 91). By adding more desirable layers to people's sensory experiences, we sought to create individual, invisible olfactory encounters that could allow people to feel strong emotional responses based on personal memories of scents (Larsson & Willander, 2013). From these contextual needs, the idea of creating a new scent was developed over the next eight months.

Case Study: Creation of the New Scent

This case study presents the design thinking used to identify courses of action taken during scent creation, and also reports on the design development of the stadium smellscape. From olfactory research in psychology, Porteous (2006) summarized that "most smell preferences are learned, which shows the importance of cultural adaptation and insideness" (p. 90). To find general preferences for scents, we first investigated the literature relating to scent products and culture in Korea.

Scent in Korea

The history of incense in Korea originated from Buddhism, where it is believed that incense can reach sacred places as the aroma rises through the ritual of burning (Park, 1998). The materials and practice of burning incense for use in religious ceremonies, as medical cures and as a supplicatory

FIGURE 10.1 Celadon incense burner, National Treasure 95.

Downloadable image from the National Museum of Korea.

medium arrived from China during the Shilla dynasty (57 BCE–935 CE). The development of incense culture during the Shilla period was signified by the appearance of diverse forms of incense burners made of bronze, clay or celadon. From the lotus flower to the dragon and the phoenix, these incense burners created an art of aroma which depended on the management of airflow floating through the shape of the burners. One of the best examples of the excellence of incense culture in Korea is a celadon incense burner created in the Goryo dynasty (918–1392) that followed the Shilla dynasty. This 12th-Century incense burner, shown in Figure 10.1, is composed of three parts: a base held by three rabbits, a burner represented by lotus blossom leaves, and a top sphere with large openings to allow the incense smoke to flow.

Another function of the burning of incense signified cleanliness and purification of its surroundings. Koreans burned incense not only during ancestral rites, but also in places where the mind should be purified. Living Buddhist and Confucian life accompanied by incense throughout the day perpetuated the growth of the incense business in the Goryo and Joseon dynasties. Diverse methods of incense production have been developed to provide users with a choice of incense design, such as sticks, round pills and powdered forms. The unique character of incense culture in Korea is embedded in the mixture of incense types, shapes, volumes and the choice of incense burners.

New Scent Development for the Seoul Design Olympiad 2008

Adding a new layer of olfactory experience with a newly created scent in the historic space was explored as a design tool for the successful transformation of the Olympic Stadium. For this challenging task, I approached a cosmetic company where I had conducted color and design research for a decade: Amorepacific,[1] a leading cosmetic company in Korea. They were determined to support SDO 2008 with a team of experts in scent creation. From the literature review, we collected representative symbols for Seoul and Seoulians' lifestyles. Symbols of historic Korean significance were also collected via brainstorming and surveys involving Amorepacific customers to identify Seoulians' lifestyles and values over the last 600 years. Survey questions included historic and cultural associations of representative significances and symbols related to the quality of lives in the Joseon dynasty. Research showed that the Seonbi ('virtuous scholar' in Korean) could be one of the most representative symbols of significance for Koreans. The Seonbi had come to represent, for most Seoulians, an appreciation of nature and the virtue of living with nature, in the period since Seoul had been named as the capital of the Joseon dynasty (1392–1910). The Seonbi signified the spirituality and integrity of Seoulians' Confucian lifestyles. For example, to remain faithful to one's principles and to maintain one's integrity, a Seonbi practices four virtuous acts: lighting incense, drinking tea, making paintings and planting flowers such as the orchid. This rhetorical image of the Seonbi conveys a simple and harmonious lifestyle with its aesthetics of smell, taste and sound.

Seonbi passed over wealth and power for lives of study and integrity, and signified the conscience of society and humanity. From the narratives of the Seonbi and their modest lifestyle, we took a historical perspective to the aesthetics of smell to embody the Seonbi spirit for the creation of the new scent. In poetry and paintings during the Joseon dynasty, the spirit of Seonbi had been denoted as bamboo and pine trees. The pine tree is one of most abundant trees in Korea's mountains, and has been closely tied to the lives of Koreans with its versatile uses, such as for furniture and housing. It is also embedded within the South Korean National Anthem's lyrics as the strong pine upon Mount Namsan:

> As the pine atop Namsan Peak stands firm, unchanged through wind and frost,
> as if wrapped in armor, so shall our resilient spirit.
> The autumn skies are void and vast, high and cloudless;
> the bright moon is like our heart, undivided and true.
>
> *nationalanthems.info, n.d.*

As pine tree metaphors were frequently related to Seonbi, it was decided that the pine tree would be the base of a Seoul Scent associated with a long and healthy life. For the creation of new scent for Seoul, we took aromas from the pine tree and blended them to create a synthesis of aromatic forms to suit the contemporary lifestyles of Seoulians. A mixture of the Seoul Scent could deliver a clean, fresh and calming odor through the elemental attributes of woody, dry, fresh, light and soft. This new perfume was named *Incense of Seoul: Pine Tree of Namsan.*

Scent-embedded Materials Study

Unlike traditional incense burning or spraying scents, an alternative method to deliver the power of olfactory experience might be innovation with new incense materials in the contemporary built environment. Eastman,[2] a chemical company producing a broad range of materials, and Rotuba[3] had both led advanced materials research with embedded scents. Using Eastman's cellulosics, a plastic polymer primarily derived from wood pulp, Rotuba had developed a scent-encapsulating polymer material that would be easily extruded and molded. In 2005, they showcased *Breeze of Iguacu*, as part of Brazilian-scented plastics, featuring synthetic colors of yellow and green embedded with the fragrance of an Amazonian scent (Special Chem, 2005). It was a good example of synesthesia, the response of one sense to a stimulus applied to another sense (Hope & Walch, 1990). Both companies in the United States welcomed the idea of developing Seoul Scent plastics to create one of the first examples of designing materials for a city.

Based on the previously conducted research relating to Korean myths and symbols, Rotuba explored diverse odor mixtures for a Seoul Scent and produced a prototype of scented plastic sticks. This greenish transparent cellulosic material emitted the scent of a fresh pine tree for up to several years, and when spent, it could be recycled. Pine tree-scented plastics could be incorporated into diverse forms, such as waste bags, containers and furniture, to help replace unpleasant odors in the stadium. It was a radical idea to develop new materials and demonstrate the effectiveness of its olfactory features. The applied study of designing products of Seoul Scent plastics for SDO 2008, however, could not be further developed, due to constraints of the manufacturing schedule and logistics of facilities in Korea and the United States. Although the mission was not completed, it did show, through discourse and experiment, how the study of scented material development could complement the potentials for odor management in small- and large-scale environments while maintaining a commitment to sustainability.

New Narratives of the Stadium Smellscape

The *Seoul Design Olympiad 2008* was a comprehensive and challenging event where contemporary design works were exhibited and discussed across 86 different programs. Two million visitors participated in 21 days of design exhibitions and festivals. Before the opening of SDO 2008, the idea of spraying the Seoul Scent into the stadium was discussed, but was not executed due to the volume needed to cover the stadium's massive structure. Instead of filling the void of the stadium with scent, we decided to devise new sensory experiences with the Seoul Scent to complement the environments and conjure up scents that could be expressed through color and mood in forms such as sprays and incense. A pavilion was built inside the stadium where authentic and sensory images of trees were projected with the airy scent. The placement of perfume within the void of three-dimensional space surrounded by only four walls stimulated odor perception and generated a unique spatial and temporal experience. The perceived Seoul Scent would become the sign of spirituality and trigger a memory of SDO 2008. Amorepacific produced 5000 Seoul Scent perfumes with a signature package design signifying harmony and oneness. Perfume design was inspired by the symbol of SDO 2008, and represented the intrinsic value in the form of the scent which had been discovered in the ancient incense burner shown in Figure 10.1. A perfume, *Incense of Seoul: Pine Tree of Namsan* (see Figure 10.2), was distributed as a gift to the people who visited the pavilion.

Their olfactory experiences of the new scent and people's encounters with narratives associated with its propagation are visualized in Figure 10.3.

The transformation of the smellscape of the stadium triggered a new design strategy and tool for the creation of the event. 'Design is Air' was the theme conceived for SDO 2008. This theme was used to set the goals and principles for the planning and execution of all the events. The theme addressed contemporary design issues of sustainability, humanity, empathy and ubiquity. Accommodating the diverse design activities in a large-scale open structure required adding new layers of spatial components to the existing iconic structure. An air dome, which was 97 m long × 27 m wide × 14.7 m high, was built on the track of the empty main stadium as a symbol of SDO 2008, and represented this theme (see Figure 10.4).

Jamsil Olympic Stadium was also transformed into the 'Plastic Stadium,' where 1.7 million discarded plastic items covered the entire skin of the structure. With these added spatial elements, we accommodated a series of lectures and exhibitions within and outside the stadium, attended

SEOUL DESIGN
OLYMPIAD 2008

FIGURE 10.2 *Incense of Seoul: Pine Tree of Namsan.*

Image: author's own, 2015.

FIGURE 10.3 Seoul Scent experience.

Image: author's own, 2015.

FIGURE 10.4 SDO 2008 Air Dome.

Image: author's own, 2015.

throughout the fall by students, designers, artists, citizens and tourists. Driven by companies, universities and city governments from around the world, this mode of collaborative experimentation and the participatory programs continuously transformed the space, and added another layer of human scents and energy to the stadium. The ambitious 86 program schedules further activated the space and connected the stadium depicted in Figure 10.5 to people and the community.

FIGURE 10.5 Transformed Seoul Olympic Stadium for SDO 2008 Plastis Stadium.
Image: author's own, 2015.

Conclusion

A number of important lessons were learned during the SDO 2008 project.

Foremost are the challenges of designing new olfactory experiences and the ways we sought to represent the value and impact of smell. Adding a fresh and calming scent of Seoul to the stadium was invisible, but certainly enhanced the pleasure of the airflow crossing the rooms and diverse activity areas. For the entire duration of SDO 2008's creation, we observed how the newly created smell transformed the abandoned space and enhanced the intrinsic value of human experiences in the stadium smellscape. The quality of the presence of the Seoul Scent signified the completion of the preparation of the event and spoke to the dynamic energy of the design festival in ways that have begun to be used in other festival and event settings across the world (Roubin, 2006).

Notwithstanding the major achievement of redesigning the stadium smellscape within eight months, this case study is also important for methodological reasons. We did not develop our scent by following classical research methods of designing and validating a newly created aroma. Instead, it was based upon historical and cultural research in addition to analyzing individual and group differences of sensory responses to smell, and as such was strongly compatible with perfumery practice in the region (Jung, 2013). Data relating to the Seoul Scent was collected before and during the SDO 2008 event. More could have been done in relation to impacts of smell in the stadium. The new smellscape could also have been analyzed by the collection of data concerning personal interpretations of the smell, combined with people's experiences in spatial activities. This additional information would have helped to validate and explore the ways a sense of smell can create deeper personal understandings and interpretations of our presence in time and space.

Notes

1 www.group.amorepacific.com/ Retrieved March 1, 2017
2 www.eastman.com/Pages/Home.aspx Retrieved March 1, 2017.
3 www.rotuba.com Retrieved March 1, 2017.

References

Barthes, R. (1977). *Image music text.* New York: Hill and Wang.
Hope, A., & Walch, M. (1990). *The color compendium.* New York: Van Nostrand Reinhold.
Jung, D. (2013). The cultural biography of agarwood: Perfumery in Eastern Asia and the Asian neighborhood. *Journal of the Royal Asiatic Society, 23*(1), 103–125.
Larsson, M., & Willander, J. (2009). Autobiographical odor memory. *Annals of the New York Academy of Science, 1170*(1), 318–323.
nationalanthems.info. (n.d.). *South Korea.* Retrieved March 1, 2017 from: www.nationalanthems.info/kr.htm
Park, H.-J. (1998). Discovering Korea: Incense culture. *Koreana, 12*(4), pp. 62–65.
Porteous, J. D. (2006). Smellscape. In J. Drobnick (Ed.), *The smell culture reader* (pp. 89–106). Oxford: Berg.
Roubin, L. A. (2006). Fragrant signals and festive spaces in Eurasia. In J. Drobnick (Ed.), *The smell culture reader* (pp. 128–136). Oxford: Berg.
SDO. (2008). *Seoul Design Olympiad 2008.* White Paper. Unpublished.
Special Chem. (2005, April 14). Eastman creates first memorable national plastic for Brazil: "Breeze of Iguacu" latest scent in material difference sample program. Retrieved March 1, 2017 from http://omnexus.specialchem.com/news/industry-news/eastman-creates-first-memorable-national-plastic-for-brazil-breeze-of-iguacu-latest-scent-in-mat-7640

11

DESIGN WITH SMELLS

The Environmental Scientist's Perspective

Michael Bull

When environmental scientists become involved in a new design, their aim is to preserve or enhance the environment. In the case of odors, an environmental scientist's aim in the design process is to avoid creating exposure to any new odors, by containment or by attempting to remove them before they are released to the environment. This is because odor is considered to be a negative contributor to the environment, and no distinction is made between pleasant and unpleasant odors. This is exemplified in guidance produced in Europe, New Zealand and Australia (see, e.g., Environment Agency, 2007), which controls levels of new odor exposure. Of course, odors are an acknowledged feature of the environment; for instance, agricultural odors are, to an extent, an accepted quality of the 'countryside.' However, planning guidance in the UK (Department of Communities and Local Government, 2012) sets expectation of high-quality environments through the protection of 'amenity' for the whole population. Consequently, a science has been developed to assess the nature and quantity of odors, and then to determine whether these are acceptable.

This chapter examines our knowledge in this area, how it is applied, and the consequences of its application for our living conditions.

Odors as Nuisance

Odors have long been an environmental issue in cities (Van Harreveld, 2003), but the speed of development of waste disposal did not match the speed of expansion of urban areas. During the industrial revolution, sewage and waste disposal was rudimentary and rivers were used to dispose of wastes. As a result, the quality of the main rivers in cities such as London deteriorated during urban growth, to the status of open sewers. Contemporary descriptions of environmental conditions demonstrate the extent of the problem; John Evelyn (1661) wrote that the city was "so full of Stink and Darknesse." Charles Dickens (1857) noted that the Thames had become "a deadly sewer . . . in the place of a fine, fresh river" (p. 28), It was reported that 3 million Londoners emptied their sewage onto the streets in 1855, and this was recognized as an unhealthy situation by the Victorians, to the extent that Parliament was forced to take action (Van Harreveld, 2003), and so the city invested in a new drainage and sewerage system. One consequence of this new sewage system was an improved odor environment for the population of London.

The same situation still arises today: an activity causes an odor problem, and the operator of the activity then needs to invest to reduce or remove odors. For instance, commercial kitchens are now required to meet a minimum level of odor control even though many would consider that the odor produced is pleasant (Department for Environment, Food and Rural Affairs, 2005), and the European Union has produced guidance on best techniques for odor control for several industries, including waste (European Commission, 2006) and intensive animal rearing (European Commission, 2007), to reduce odorous emissions to an acceptable level.

UK law recognizes that odors can cause problems to the general population; for instance, the concept of statutory nuisance is provided in Section 79 1d of the Environmental Protection Act. This describes statutory nuisance as "any dust, steam, smell or other effluvia arising on industrial, trade or business premises and being prejudicial to health or a nuisance" (UK Government, 1990). Similar concepts of nuisance exist elsewhere in Europe, and it is reported that some of the earliest legislation addressed the nuisance caused by odors (Van Harreveld, 2003). Where a nuisance is proved to exist, the operator must take steps to remove or reduce odors to a level where the nuisance is removed. Although the law includes reference to smells being prejudicial to health, there have not been any cases to the author's knowledge where this has been the basis for any nuisance action. Local authorities have the responsibility to determine where a statutory nuisance exists. Whilst some guidance exists (Department for Environment, Food and Rural Affairs, 2012), it is largely the opinion of the enforcing officer that determines if a statutory nuisance exists. Local authorities have considerable powers available to require that odors are reduced where a statutory nuisance exists, and can serve an 'abatement order' (UK Government, 1990) to reduce odor emissions.

Odor is one of the more frequent causes of complaint received by local authorities. In a survey conducted on behalf of the UK Government (Temple Consultants, 2012), it was reported that local authorities in England had received nearly 7000 complaints relating to odor between 2009 and 2011. This was around three times more than complaints relating to dust, and more than twice as many as complaints relating to the generic 'Fumes and Gases' category, suggesting that odors are the leading cause of complaint in relation to atmospheric pollution. Whilst many of the complaints related to unpleasant by-products arising from activities, such as sewage treatment or sludge or slurry spreading, odors from commercial kitchens were responsible for 10% of the total, even though cooking odors would often be considered to be very pleasant.

Measuring Odor and Odor Standards

The issue for an environmental scientist is at what point odors reach a level where they are considered to be a problem that requires action, and then, how this problem can be removed. Regulators throughout the world have examined this issue and have taken various approaches to assess nuisance or to prevent it occurring. Odors can be considered in the design of land uses: separating odorous (or offensive) industries from sensitive land uses is one approach to avoid nuisance. This concept was suggested, for instance, by Milyutin (Bocharov & Khan-Magomedov, 2007) in his linear city proposal, where housing was located near to industry, but separated by a park strip. It is also embedded in guidance in countries such as Australia (Environmental Protection Agency Victoria, 2013).

The planning system can be used to avoid conflicting land uses being located together; in the case of odor, this almost always manifests itself as avoiding placing residential properties near to odorous processes (and most frequently, sewage works) (Department for Environment, Food and Rural Affairs, 2006). A variation of this approach is to place undeveloped zones, or so-called *cordons sanitaires*, between an odorous activity and sensitive land uses when considering the design of new development. This approach is used for planning in some countries, with defined separation distances dependent on the type of activity. For example, in the Australian state of Victoria, acceptable distances between development and odorous sites are defined for various odorous activities: some examples of Australian practice are provided in Table 11.1. This approach is based

TABLE 11.1 Acceptable distances for development in relation to odor sources

Source	*Separation distance (meters)*
Abattoir	500
Bakery	100
Coffee roasting	250
Pet food manufacturing	500
Rendering	1000

Source: Environmental Protection Agency Victoria (2013).

on the concept that odors become diluted until they are no longer a nuisance by the time they reach sensitive land uses.

Whilst this idea of acceptable separation is a very simple concept to apply in practice, there are two main problems with the use of buffer zones or *cordon sanitaires*. Firstly, they assume that a standard distance is applicable to every situation regardless of the nature and scale of the activity. The second is that it sterilizes areas of land for possible future development.

To decide whether a particular situation is acceptable, a range of assessment approaches have been developed by environmental scientists (Bull, 2014); however, they all are largely based on criteria known by the acronym FIDOL (Environment Agency, 2011):

- **Frequency** – The odors must be experienced in sufficient frequency to be considered a nuisance.
- **Intensity** – The odors must be experienced above a threshold intensity; that is, they must be sufficiently 'strong' odors to be considered a nuisance.
- **Duration** – The odors must be experienced for a sufficient length of time on each occasion they are detected.
- **Offensiveness** – The more 'offensive' odors are, the more likely they are to cause offense.
- **Location** – Where the odors are being experienced is important; the expectations for the area need to be considered when determining whether a nuisance exists.

The FIDOL approach is widely applied in guidance (Department for Environment, Food and Rural Affairs, 2012; Environment Agency, 2011; Ministry for the Environment, New Zealand, 2003) to determine acceptability of a proposed development, and it can be adapted to be used in a more quantitative manner. Most people find it surprising to learn that it is possible to measure odors; however, this has been carried out since at least 1848 (Valentin, 1848). European standards exist, and a number of specialist laboratories provide this service (e.g. Verein Deutscher Ingenieure, 1994).

There is a straightforward current approach in European standards that is used to measure odors (see, e.g., British Standards Institution, 2003). A sample of the air to be analyzed is collected in a non-reactive plastic bag and taken back to a laboratory. Here the sample is presented to a trained group of four people at various dilutions. When only half the panel can detect the presence of the odor, the concentration of odor is defined as 1 odor unit. If the sample had to be diluted by 200 times before half the panel could detect the odor, then the original concentration of odor would have been 200 odor units.

Having the ability to describe odors numerically allows them to be placed in context; for instance, a concentration of 5 odor units is described as a faint odor, and 10 odor units as a distinct odor (Department for Environment, Food and Rural Affairs, 2012). We do not live in an odor-free environment, and it has been reported that in a rural area, we are typically exposed to an odor concentration of 30 odor units (Ministry of Agriculture, Fisheries and Food, 1998).

TABLE 11.2 Hedonic odor scale

Hedonic tone score	*Description*
+4	Very pleasant
+3	Pleasant
+2	Moderately pleasant
+1	Mildly pleasant
0	Neutral/no odor
–1	Mildly unpleasant
–2	Moderately unpleasant
–3	Unpleasant
–4	Offensive

Source: Environment Agency (2011).

TABLE 11.3 Hedonic scale of different smells

Odor type	*Hedonic tone*
Cadaverous (dead animal)	–3.75
Fishy	–1.98
Garlic, onion	–0.17
Black pepper	+0.19
Tea leaves	+1.40
Lavender	+2.25
Bakery (fresh bread)	+3.53

Source: Environment Agency (2002b).

Another important consideration is the offensiveness of the odor; this can be assessed by the lab-based measurement of 'hedonic tone,' where an odor panel rate the perceived offensiveness on a nine-point scale shown in Table 11.2.

Hedonic tones have been reported for many odors, and examples of lab-based ratings are provided in Table 11.3.

Hedonic tone is subjective, and it has been noted that the same odor can be considered to be both pleasant and unpleasant (Henshaw, 2014) by different groups of people, or by the same people, but when experiencing the odor in different contexts. One point of interest is that, from a nuisance point of view, even odors that are considered to be very pleasant can cause a nuisance. There have been several cases where statutory nuisance action has been taken as a result of pleasant odors, such as food flavorings, being emitted by manufacturers. A nuisance arises because the odor is unwanted at the frequency and level of exposure, as can be seen from the basis of the FIDOL approach.

A further complication in assessing hedonic tone is that it depends on the concentration of the odor. For instance, at low concentrations, coffee is perceived as a pleasant smell, but at very high concentrations it is considered highly offensive, with a hedonic tone of -4 (i.e. worse than a rotting animal) (Environment Agency, 2002a). As a result, the standard method for assessment of hedonic tone requires that it be assessed at different odor concentrations (Verein Deutscher Ingenieure, 1994).

Using this approach, environmental regulators have proposed numerical standards to define where odors are unacceptable (Environment Agency, 2007). The use of the FIDOL approach is essentially embedded within these standards. In particular, they reflect that odors must be present at a particular frequency and intensity to be a nuisance. Therefore, the standards have not been set as absolute single values that cannot be exceeded. Instead, they are set as percentiles that acknowledge that odors can exist for a period without resulting in a nuisance. The most common approach used in Europe (Environment Agency, 2007) is to set the standard at a level that can be exceeded for a defined percentage of the year. In design terms, that means that the aim is for an environment where smells can be present intermittently, but that constant or persistent odors will cause annoyance even if they are considered to be pleasant.

These standards have been derived from research that examines the relationship between reported annoyance in a community and its variation with exposure to odors (Van Harreveld, 2004). Most of this research was carried out in Holland in the 1980s and 1990s (e.g. Bongers, Van Harreveld & Jones, 2001), and standards that have emerged since have been mainly based on experience around pig farms, although the research also considered other industries. The standards

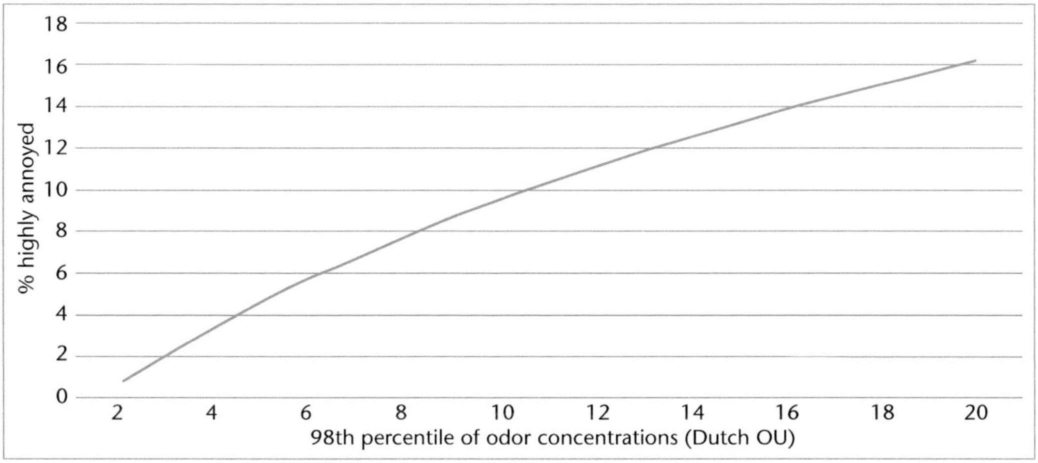

FIGURE 11.1 Population annoyance and odor concentration.

Image: author's own, 2016.

derived are very unusual compared with others used to protect the environment. Normally, an environmental standard is derived from observations of the effects of exposure to a pollutant, based on measured concentrations. However, it is not possible to measure odors continuously over a year, and therefore exposure has been derived from modeled, rather than measured, odor concentrations.

Annoyance was determined by community questionnaires undertaken in Holland (Environment Agency, 2002a). If a respondent noted that they were annoyed by particular odors, then they were asked to rank their annoyance on a scale: for instance, 'not annoyed,' 'annoyed' and 'very annoyed.' One of the problems with the research was that different rankings were used for different studies, making comparison difficult. However, responses can be standardized to an extent, and relationships derived between the exposure to odor and the percentage of the community that is 'annoyed.' Figure 11.1 has been derived from the outcome of this research, and shows how the percentage of the population reported to be highly annoyed increases with increasing exposure to odors.

The UK Environment Agency has taken the view that odor standards should be set at a level that permits 10% of the population to be annoyed (Environment Agency, 2002a). This level is where research has shown that a clear effect can be detected, and takes account of the facts that there are some individuals who are highly sensitive to odors, and that it would be unreasonable to set standards at a level where the whole population was protected.

Odor standards can also take into account how offensive odors are, and there is some evidence that more offensive odors result in a nuisance at a lower concentration (Environment Agency, 2002a). Although the evidence on the effect of offensiveness is not conclusive (Both, Sucker, Winneke & Koch, 2004), the UK Environment Agency advises on 'benchmark levels' as follows (all based on a value that can be exceeded for 2% of the year):

- 1.5 odor units for the most offensive odors;
- 3.0 odor units for moderately offensive odors;
- 6.0 odor units for less offensive odors.

Where predicted odor concentrations are above these benchmarks, this would indicate the likelihood of unacceptable odor pollution (Environment Agency, 2011).

Although these benchmarks appear fairly straightforward, there are issues with their application, and in particular with the definition of the offensiveness categories for a specific odor, and the accompanying guidance only provides a few examples, with no reference to hedonic tone. As a result, when a new development is proposed, there is often still disagreement between experts about whether it is an acceptable situation or not, particularly when there are several odor sources on site (e.g. at a sewage works, where there are several different types of processes used).

Current standards in the UK do not consider the expectations of the population exposed to odors. Research has shown that when there is an expectation that an odor will be present, people will be more tolerant (Bongers et al., 2001). Although the context of the area is important and there is some suggestion that odor standards could be modified to take account of expectations in an area (Environment Agency, 2002a), this is not described in any guidance to date. For some years, some environmental scientists operated differing standards for urban and rural areas when assessing odors from intensive animal rearing, and considered that standards in a rural area should be more tolerant. This is reflected in research that shows that to achieve the same level of 10% annoyance in the community, exposure had to be more than doubled in an area where pig farming was common, as against areas where it was not expected (Bongers et al., 2001). Furthermore, if people exposed have a clear economic association with the activity, they are prepared to tolerate even higher levels of odors: levels need to be 10 times higher than the normal population to give the same level of annoyance. The tolerance of communities to odors was also noted in the research on statutory nuisance complaints (Temple Consultants, 2012). It is not certain whether this higher level of tolerance is a result of adaption via habituation in the community to the odors, or whether the community itself is self-selecting, with those finding the odors intolerable moving away.

Modeling

As more information regarding the response of communities to odors has become available and measurement techniques have improved, so environmental scientists have increasingly turned to the use of modeling techniques to examine whether proposed designs would be acceptable in terms of their exposure to odor. The principles of this approach are to build up what is known as an 'emissions inventory' for the site emitting odors. This is a compilation of the details of all the odor sources on a site, including their location and emission rate.

This requires measurement of the emission rate of odor from each of the sources on the site. This is relatively easy if odors are being emitted from a well-defined and constrained source such as a vent or chimney. The odor concentration and the flow rate of the emitted gas can be measured, and hence a total amount of odor emitted can be calculated. However, many odorous activities do not allow this approach, so techniques have been developed to measure odors from the more common open sources such as tanks at sewage works, composting piles and even manure heaps (Bull, 2014). These approaches use an inverted open-top box, the open side placed onto the surface of the liquid in the tank or a solid surface. Clean air is passed through the box, simulating the wind blowing over the surface, and the odor emission rate can be calculated from measuring odor concentration in the air leaving the box and the flow rate of the air through the box.

The information in the emissions inventory can then be placed into an atmospheric dispersion model together with information about local weather conditions and the nature of the terrain. The results from such a modeling exercise can be seen in Figure 11.2, where the predicted odor levels are presented as contour plots, and these can be used to define the area of acceptable odor conditions for development. Some care needs to be taken with the use of this approach, so as to acknowledge the uncertainties in the process. There are several aspects that can give rise to uncertainty, particularly as most odors arise from biological processes that are cyclic or unstable. Seasonal effects,

FIGURE 11.2 Odor concentrations from an emission source.

Image: author's own, based on Microsoft product screenshot reprinted with permission from Microsoft Corporation.

particularly temperature, mean that odorous emissions can change throughout the year although there is no consensus on the scale of this effect (Bull, 2014).

Conclusion

The role of environmental scientists has been to minimize or eliminate odors from the environment, and they generally do not consider any beneficial effects that odors can bring to a user's experience of an area. Research does provide some backing for this approach, as it has found that people can find any odor offensive and that people's reactions to the same odor can be very different. There is a considerable body of research that has tried to establish the level of odor that is acceptable to the general population, and a range of techniques are available to measure and model odors to determine whether they would be acceptable.

This research shows that people begin to find exposure to odors unacceptable when they experience them sufficiently frequently, above a certain threshold, for a sufficiently long period. Odor assessment is based around these principles. However, there is no similar research available that examines how people respond to lower levels of odor exposure, where their experience is likely to be more positive. There now appears to be a shift in thinking in other environmental topics such as ecology, where enhancement of the environment is an additional aim during examination of new developments. It can be argued that odors within the environment are a feature that can also provide a positive experience and a sense of place for an area. So there is an opportunity for scientists and planners to work together to consider how odors can contribute towards a positive experience for people visiting and living in an area.

References

Bocharov, Y., & Khan-Magomedov, S. (2007). *Nikolay Milyutin* [In Russian]. Moscow: Fond Russky Avangard/Arkhiterkura-S.

Bongers, M. E., Van Harreveld, A. P., & Jones, N. (2001). Recent developments in research supporting pig odour policy reviews in the Netherlands and in Ireland. In J. Jiang (Ed.), *Proceedings of the 1st IWA International Conference on Odour and VOC's: Measurement, Regulation and Control Techniques*. London: International Water Association.

Both, R., Sucker, K., Winneke, G., & Koch, E. (2004). Odour intensity and hedonic tone – important parameters to describe odour annoyance to residents? *Water Science & Technology*, *50*(4), 83–92.

British Standards Institution. (2003). *Air quality: Determination of odour concentration by dynamic olfactometry (BSI EN 13725:2003)*. London: British Standards Institution.

Bull, M. A. (2014). *Guidance on the assessment of odour for planning*. London: Institute of Air Quality Management. Retrieved March 1, 2017 from www.iaqm.co.uk/text/guidance/odour-guidance-2014.pdf

Department of Communities and Local Government. (2012). *National planning policy framework*. London: Department of Communities and Local Government.

Department for Environment, Food and Rural Affairs. (2005). *Guidance on the control of odour and noise from commercial kitchen exhaust systems*. London: HMSO.

Department for Environment, Food and Rural Affairs. (2006). *Code of practice on odour nuisance from sewage treatment works*. London: HMSO.

Department for Environment, Food and Rural Affairs. (2012). *Odour guidance for local authorities*. London: Defra.

Dickens, C. (1857). *Little Dorrit*. London: Bradbury and Evans.

Environment Agency. (2002a). *Assessment of community response to odorous emissions*. R&D Technical Report P4-095/TR. Bristol: Environment Agency.

Environment Agency. (2002b). *Draft horizontal guidance for odour part 1 – regulation and permitting*. London: HMSO.

Environment Agency. (2007). *Review of odour character and thresholds*. Science Report: SC030170/SR2. Bristol: Environment Agency.

Environment Agency. (2011). *H4 odour management: How to comply with your environmental permit*. London: HMSO.

Environmental Protection Agency Victoria. (2013). *Recommended separation distances for industrial residual air emissions*. Carlton, Victoria: EPA Victoria.

European Commission. (2006). *Integrated pollution prevention and control reference document on best available techniques for the waste treatments industries*. Brussels: The Commission.

European Commission. (2007). *Integrated pollution prevention and control (IPPC) reference document on best available techniques for intensive rearing of poultry and pigs*. Brussels: The Commission.

Evelyn, J. (1661). *Fumifugium: Of the inconvenience of the aer and smoak of London dissipated*. London.

Henshaw, V. (2014). *Urban smellscapes: Understanding and designing city smell environments*. New York: Routledge.

Ministry for the Environment, New Zealand. (2003). *Good practice guide for assessing and managing odour in New Zealand: Air quality report 36*. Retrieved 29 March, 2016 from www.sinia.cl/1292/articles-55480_ONSL_0305.pdf

Ministry of Agriculture, Fisheries and Food. (1998). *Code of good agricultural practice for the protection of air*. London: Ministry of Agriculture, Fisheries and Food.

Temple Consultants. (2012). *Local authority statutory nuisance survey*. London: Defra.

UK Government. (1990). *Environmental Protection Act*. London: HMSO.

Valentin, G. (1848). *Lehrbuch der Physiologie des Menschen*. Braunschweig: Friedrich Bieweg und Sohn.

Van Harreveld, A. P. (2003). Odour regulation and the history of odor measurement in Europe. In N. a. V. Office of Odor (Ed.), *State of the art of odour measurement*. Tokyo: Environmental Management Bureau, Ministry of the Environment, Government of Japan, pp. 54–61.

Van Harreveld, A. P. (2004, March 23). How much odour is annoying? In *Current best practice techniques for odour control*. London: Royal College of Physicians.

Verein Deutscher Ingenieure. (1994). *VDI 3882: Part 1, olfactometry: Determination of odour intensity. Part 2, determination of hedonic odour tone*. Berlin: Verein Deutscher Ingenieure and Beuth Verlag.

PART IV
Retail, Scent and Service Design
INTRODUCTION

Charles Spence

There is a growing realization of the potential power of scent in many commercial/public settings. Engineering or designing the right smell or fragrance for a given space, be it the interior of an aircraft, a shopping center, or an individual store premises, can, in a way, be seen as a natural extension of Kotler's (1974) early ideas around store atmospherics. This view was subsequently taken up and extended with Pine and Gilmore's (1998, 1999) promotion of 'the experience economy.' The profound influence of this, more sensory, approach to marketing and design is highlighted, I think, by the fact that wherever you look nowadays, people are increasingly trying to sell you 'the experience' (e.g. Hilton, 2015; Spinney, 2013). Aradhna Krishna recently talked of this as 'a sensory explosion' (JWT, 2012).

The experience, of course, is not just about 'the tangible product,' in Kotler's words, but rather 'the total product offering.' The latter incorporating the atmospherics of the store or, for that matter, any other public space, and that, increasingly, is starting to include everything from hospitals and other healthcare facilities through museums and shopping centers. It is important to note that *all* of the senses can feed into creating an atmosphere that might encourage approach or avoidance behaviors, that might stress us out (or arouse us) or else help us to relax, that might encourage us to linger longer and perhaps spend a little more than we otherwise might. Practitioners and researchers have long thought about the visual attributes of design, and also about the sound of space. However, far less attention has been given over to the smell of the spaces we inhabit.

Smell rarely takes center stage in our everyday awareness. We are, after all, primarily visual creatures (Gallace, Ngo, Sulaitis & Spence, 2012). Crucially, therefore, some of the most dramatic influences of scent occur when we are not actually aware of smelling anything at all (e.g. see Li, Moallem, Paller & Gottfried, 2007). The increasing availability and affordability of technologies that can help to scent spaces both large and small is making olfactory marketing a much more practicable possibility for many (see Chapter 12 by Dominic Medway and Gary Warnaby). No surprises, then, that the question 'What does your brand smell like?' is heard increasingly often around the trendy marketing agencies in East London these days. And workshops that help brands to think about how they would present themselves through smell are occurring with a regularity and popularity that would have been unimaginable even a few years ago.

At the simplest level, scents can be associated with specific brands (see Chapter 15 by Samantha Warren and Kathleen Riach). These are what are known as 'signature scents.' Here, one can think of everyone from Singapore Airlines, whose cabin staff used to all wear the same fragrance, through

Westin Hotels with their *White Tea* scent that one experiences in their hotels/spas and so on and can then be purchased in candle form to recreate 'the experience' when you get home (Lindstrom, 2005; Spence, Puccinelli, Grewal & Roggeveen, 2014; see also Chapter 12 by Dominic Medway and Gary Warnaby).

Scents can also be functional, though. That is, there are some who use scent and fragrance to help relax people, such as the relaxing scents released into the lobbies of airport hotels that receive a lot of long-haul passengers whose journeys tend to be stressful (Spence, 2002). Here, one can also think of the growing interest in the use of scent to mask other more stressful aromas that we associate with dental surgeries, hospitals and so on (for a review, again see Spence, 2002). And as highlighted in Chapter 14 by Claire Sokell Thompson, scents can be used to tell stories around a brand, as well as to prolong the experience. Just think of the way that the emotionally laden smell of the sweet shop is used to prolong the diner's experience around a meal at the three-Michelin-starred Fat Duck restaurant (see Spence & Piqueras-Fiszman, 2014).

However, by far the biggest use of scent currently has to be in the food and beverage category. Some stores release different food aromas in different parts of their stores. Other chains selling fragrant and calorific treats often pick locations in shopping malls near the bottom of staircases, and apparently choose the weakest extractor fans possible so more of the delicious scent of the product will hit the nostrils of more of the shoppers. We have all, no doubt, had the experience of bookshops and railway stations being increasingly imbued with the scent of the coffee shop.

In fact, according to a recent *Wall Street Journal* article:

> Cinnabon, the bakery chain, places ovens near the front of its stores so the enticing smell of warm cinnamon rolls escapes when oven doors open, says Kat Cole, president of Cinnabon, a unit of Focus Brands Inc. The bakeries are intentionally located in malls or airports, not outside, so smells can linger. Over time, the company has recognized that aroma is a huge part of its formula, Ms. Cole says. Putting ovens in the back of stores at a test location "significantly" lowered sales, she says. Cinnamon rolls are baked at least every 30 min. Some store operators heat additional sheets of brown sugar and cinnamon to keep the aroma in the air, she says.
>
> *Nassauer, 2014*[1]

I do, though, have to wonder about whether the recent rise in olfactory marketing, especially our increasing exposure to food scents, especially in Out Of Home advertising, just the kind of thing that Prask Sutton discusses in Chapter 13. Could our increased exposure to all those food-related aromas be having a negative effect on our health and well-being (see Spence, 2015), leading us by the nose, as it were, into eating more foods that aren't so healthy than we otherwise might? This is not to say that I don't like the scent that greets one on arrival at Hilton Doubletree hotels when you are given an aromatic cookie on check-in. The lobby is filled with the sweet (and 'natural') smell of cookies (see Demetros, 1997). In this case, a desirable food aroma is combined with an unexpected (at least on your first visit) gift. This is a good idea from the sensory marketing perspective, that's for sure. I, for one, am certainly a regular visitor. That said, I can't help but worry that my exposure to the high-energy sweet food scent encourages me to consume a cookie that I might otherwise not.

I have also heard of hotel chains who have considered releasing the scent of breakfast from behind the reception desk in order to try to encourage guests to add a breakfast to their bill before they eventually check out. Ultimately, though, one always has to be careful with olfactory marketing. After all, we inhale the aromas, and that internalization of the signal holds the potential for consumer backlash, as when the London Underground tried releasing a pleasant scent without doing anything to improve the cleanliness of the surroundings (see Spence, 2002). Another example comes from a study we conducted for a global burger chain a little over a decade ago which was trying to release a fragrant fruity/floral scent into its stores to mask all those unhealthy smells. The press eventually picked up the story and turned it into a potentially damaging exposé. Periodically, too, consumers may claim to be allergic to scent (see also Sutton's Chapter 13 for one oft-discussed example).

Ultimately, and as brought out by several of the chapters in this part of the volume, one should not think of the olfactory component of our experience of various commercial and public spaces in isolation, but rather contextualize it as but one element, or part, of the total multisensory experience or atmosphere. As soon as one takes that multisensory perspective, then the question of congruency crops up, making sure that one's scent conveys the same message, or effect, as the signals that are being sent – a challenge, to be sure, but one that holds the promise of both delivering more memorable and stimulating multisensory experiences on the one hand, while at the same time potentially leading us into less beneficial behaviors on the other.

Note

1 Interestingly, however, airport terminals, at least in the UK, seem to be much more olfactorily neutral spaces – this despite the fact that one often finds many of the same stores that are simultaneously filling other public spaces with their scent.

References

Demetros, V. A. M. (1997, April). The sweet smell of success. *The Crafts Report.* Retrieved May 23, 2017 from *Internet Archive* website https://web.archive.org/web/20061020170908/www.craftsreport.com/april97/aroma.html

Gallace, A., Ngo, M. K., Sulaitis, J., & Spence, C. (2012). Multisensory presence in virtual reality: Possibilities and limitations. In G. Ghinea, F. Andres, & S. Gulliver (Eds.), *Multiple sensorial media advances and applications: New developments in MulSeMedia* (pp. 1–40). Hershey, PA: IGI Global.

Hilton, K. (2015). Psychology: The science of sensory marketing. *Harvard Business Review, 93*(3), 28–31.

JWT. (2012). JWT: 10 trends for 2013 – Executive summary. Retrieved July 19, 2017 from https://www.slideshare.net/jwtintelligence/jwt-10-trends-for-2013-executive-summary

Kotler, P. (1974). Atmospherics as a marketing tool. *Journal of Retailing, 49*, 48–64.

Li, W., Moallem, I., Paller, K. A., & Gottfried, J. A. (2007). Subliminal smells can guide social preferences. *Psychological Science, 18*, 1044–1049.

Lindstrom, M. (2005). *Brand sense: How to build brands through touch, taste, smell, sight and sound.* London: Kogan Page.

Nassauer, S. (2014, May 20). Using scent as a marketing tool, stores hope it – and shoppers – will linger. *The Wall Street Journal.* Retrieved April 10, 2016 from www.wsj.com/articles/SB10001424052702303468704579573953132979382

Pine, II, B. J., & Gilmore, J. H. (1998). Welcome to the experience economy. *Harvard Business Review, 76*(4), 97–105.

Pine, II, B. J., & Gilmore, J. H. (1999). *The experience economy: Work is theatre and every business is a stage.* Boston, MA: Harvard Business Review Press.

Spence, C. (2002). *The ICI report on the secret of the senses.* London: The Communication Group.

Spence, C. (2015). Leading the consumer by the nose: On the commercialization of olfactory design for the food and beverage sector. *Flavour, 4*, 31.

Spence, C., & Piqueras-Fiszman, B. (2014). *The perfect meal: The multisensory science of food and dining.* Oxford: Wiley-Blackwell.

Spence, C., Puccinelli, N. Grewal, D., & Roggeveen, A. L. (2014). Store atmospherics: A multisensory perspective. *Psychology & Marketing, 31*, 472–488.

Spinney, L. (2013). Selling sensation: The new marketing territory. *New Scientist, 2934.*

12

DESIGNING SMELL INTO THE CONSUMER EXPERIENCE

Dominic Medway and Gary Warnaby

Introduction

> Even in the buffet the ladies began to smell the perfume department; the penetrating scent of sachets pervaded the gallery. There was quite a struggle over a particular soap, the Paradise soap, a speciality of the shop . . . . The salesmen had used their ingenuity to decorate the display with all their white china pots and all their phials. The customers were delighted by a silver fountain in the centre, a shepherdess standing in a harvest of flowers, from which a continuous trickle of violet water was flowing, tinkling musically in a metal basin.
>
> *Zola, 2012, p. 418*

This excerpt is from Émile Zola's 1883 novel *Au Bonheur des Dames* (The Ladies' Paradise), which depicts a mid-19th-Century department store of the same name, modeled on Le Bon Marché in Paris. At that time, department stores represented an emergent retail form in Western Europe. They marked a transition from a type of retailing where merchandise was on limited display (sometimes hidden away in drawers behind a counter) to one where the customer was able to interact with goods – walk amongst them, touch them, sample, test or try them, and indeed smell them. In this sense, department stores marked the beginning of shopping as 'being with things,' and a shift towards retailing as a multisensory consumer experience. We see in Zola's narrative above that at least three senses are at play in stimulating the shopper as consumer. The smell of perfume piques the olfactory senses and draws shoppers in from the buffet like flies into a spider's web. Visual interest comes through shape and form (the arrangement of pots and phials) and color (the silver and violet in the fountain), and aural stimulation in the musical sound of water.

A series of technological advances from the mid-19th Century onwards undoubtedly assisted the growing incorporation of such multisensory stimulation in retail merchandising (Nava, 1996). Not least was the emergence of iron-framed building techniques. These allowed for the easy construction of large interior sales spaces that could provide suitable environments for the transmission of sound and smell. In addition, the introduction of both electric lighting and plate glass for display windows held enormous potential for visual engagement. Arguably, therefore, the development of the 19th-Century department store represented a turning point in thinking about how the senses of the retail consumer might be stimulated. Of particular interest in this chapter, however, is just one of the five senses – namely, olfaction. In the following discussion, we first chart how smell has been developed as a tool for enhancing and maximizing the consumer experience: predominantly in retailing, but also in other consumer service environments, or 'servicescapes' (Bitner, 1992), such as hotels and even airlines. We then summarize some key benefits that designing smell into service environments might provide for both the consumer and service provider. Finally, we conclude by detailing three critical factors that practitioners should consider when attempting to enhance the consumer experience through olfaction.

Retail Service Environments and Beyond – Olfaction and the Consumer Experience

The depiction of smell in the above excerpt from Zola's novel suggests a serendipitous olfactory occurrence. Here, products like perfume and soap, which smell pleasant to many individuals, naturally help to attract customers and draw them in. Thus, whilst the owner of *Au Bonheur des Dames*, Octave Mouret, is depicted in Zola's novel as a great manipulator of consumers, especially women, olfaction does not seem foremost in his armory of tools. When smell does enhance the consumer experience, and perhaps even encourage them to spend more money, as appears to be the case in Zola's description of the perfume department, then it would seem to be down to the pure happenstance of the product itself smelling nice. In this sense, smell is not really being used or designed into the environment; it is simply there – or not there – as an essentially unplanned and ambient environmental element. In many ways, it could be argued that this represents the way in which smell in retail environments was regarded, at least until the late 20th Century.

Then rumors began to emerge amongst the general public and press that supermarkets were manipulating the smell of bread within their stores in at least two ways: first, by placing bakeries at the back of the shop so their smell drew customer footfall through all aisles where potential purchases could be made, and second, by intentionally pumping baked bread smells into stores supposedly to whet shoppers' appetites and make them buy more groceries as a result. Thus, as a schoolboy in the 1980s, one author can remember this being taught as fact in the subject of geography when learning about the spatial layout of supermarkets. The French grocery chain Carrefour, with its short foray into the UK market in the 1970s and 1980s, was cited as an example of such 'manipulative' practices. But looking back, there seems no evidence to support these claims, and one wonders, with hindsight, whether it was an assertion rooted more in a fear of all things French and new, rather than having any genuine connection to reality. Equally, more recent suggestions that supermarket operators introduce artificial, rather than real, bread smells into their stores have been identified as a 'persistent online conspiracy theory' by some commentators (Hayward, 2010).

Similar rumors about smell in retailing also emerged in relation to US fast-food chain Subway from the mid-1980s onwards, as it started to expand its international operations. Indeed, Roston (2011) notes that there is still plenty of online chatter regarding Subway's supposed intentional venting of its baking bread smell onto the pavement to pull in passers-by, as well as suggestions that the bread smell is itself fabricated or modified in some way, presumably to heighten its consumer appeal. However, senior Subway representatives are keen to debunk such ideas as myths (Roston, 2011).

The hyperbole and rumor surrounding the topic of bread smells in both supermarkets and Subway suggest that the mere idea of odors emanating from service providers feeds consumers' perceptions that they are being subjected to potentially underhand, manipulative and unethical corporate practices, even if this may not be the case. Interestingly, the kind of conspiracy theories mentioned above appear to crop up less in relation to the other four senses, even though there are many accounts of these all being variously used to affect consumer behavior in service environments such as retailing (see Spence, Puccinelli, Grewal & Roggeveen, 2014). It suggests that the mere idea of using smell as a tool to try to enhance the consumer experience is an especially sensitive topic from the perspective of potential recipients, perhaps because of this particular sense's strong links to emotion (Aggleton & Waskett, 1999; Holloway, 1999).

Thus far, our discussion of smell and consumer experience has focused principally on retailing. In academic work on consumers and olfaction, this is also the service environment attracting the main focus of attention. There is, therefore, a substantive canon of literature that examines the use of olfaction, both actual and potential, in retail settings. This typically draws on theoretical perspectives from areas such as psychology and consumer behavior, using, where possible or relevant, real-life examples to illustrate discussion (Ballantine, Jack & Parsons, 2010; Soars, 2009; Davies, Kooijman & Ward, 2003; Ward, Davies & Kooijman, 2003; Spence et al., 2014). Within this broad stream of work, some studies have taken a single-case approach, looking at the role of smell for a given retailer from the consumer's perspective (Wade Clarke, Perry & Denson, 2012). Other authors have adopted experimental designs to explore consumers' evaluations of retail environments and merchandise under various ambient olfactory conditions that are not intrinsic to the products being sold. These investigations include ascertaining perceptions of contrasting olfactory situations in terms of, for example: fragranced and un-fragranced retail stores (Spangenberg, Crowley & Henderson, 1996; Ward, Davies & Kooijman, 2007); using scents with supposed congruency or incongruency with the merchandise being sold in stores (Schifferstein & Blok, 2002); and conditions in which the in-store smell is moderated by different background music (Spangenberg, Grohmann & Sprott, 2005).

A relatively new phenomenon, but pertinent to our discussion, is the emergence of scent marketing (or 'smarketing') companies such as brandaroma[1] and ScentAir.[2] These organizations are able to chemically design and manufacture scents, and then, where necessary, provide the devices for releasing them in service environments for the apparent experiential benefit of consumers (see Figure 12.1). Consequently, this moves the scenting of service environments into a new era of automation. It means that artificially – and deliberately – introducing ambient odors into service environments in a controlled and sustained manner becomes an easier task. Certainly, it is less necessary for employees to undertake deliberate scenting practices such as sporadic manual spritzing – a practice reported in the retail stores of Abercrombie & Fitch and Hollister (Bhasin, 2014).

The automation of artificial scenting has undoubtedly broadened industry perspectives on what types of service environments can be enhanced through olfaction for the apparent benefit of the consumer experience. Within retail outlets, the last few years have seen the automated release

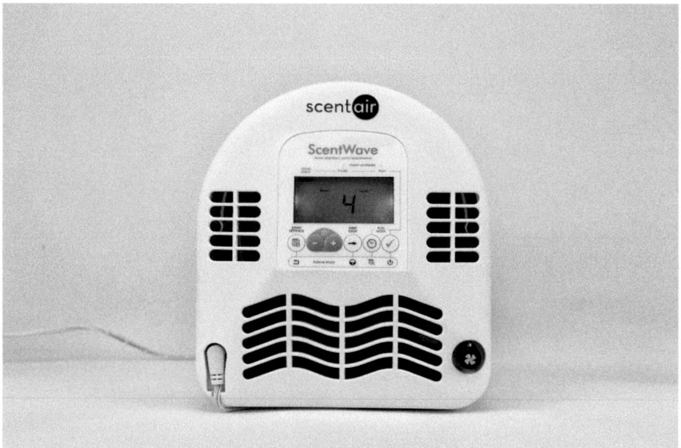

FIGURE 12.1 ScentAir's ScentWave device for releasing odor.

Image courtesy of ScentAir UK.

of designed and manufactured scents by operations as diverse as Sony Style stores, with their vanilla and mandarin orange aroma (Fetterman & O'Donnell, 2006), and Bloomingdales, which delivers the ScentAir-manufactured fragrances of baby powder, lilac and coconut in its infant, intimate apparel and swimsuit departments respectively. Beyond the confines of retailing, there has been a growth in the practice of automated scenting in the service environments of hotels (especially their lobby areas), gyms and even hospitals, with so-called 'signature' fragrances.[3] Indeed, the Westin Hotels chain is so happy with the signature *White Tea* scent that ScentAir produces for its 120 properties that it offers its customers the chance to buy online candles and oils that produce the smell. The really committed can even buy a domestic version of the ScentWave odor delivery machine for €185 (at the time of writing), to help automate Westin's olfactory consumer experience in their own home.[4]

Beyond the consumer service environments discussed above, companies such as Singapore Airlines have begun to deploy the use of chemically manufactured scents with the aim of enhancing the in-flight experience. Specifically, the airline has introduced a specially designed scent called *Stefan Floridian Waters*, reportedly impregnated in the hot flannels offered to passengers (see Figure 12.2) and present in the perfume of flight attendants (Lindstrom, 2005). Detailed academic examination and analysis of these manufactured scents being introduced within a variety of service environments beyond retailing is still nascent, emphasizing the relative novelty of such practices. There are exceptions however; thus, as far back as 1995, Hirsch (1995) explored how the introduction of manufactured odors into Las Vegas casinos increased the size of bets placed. The potential of smell to change behavior in this manner has led some authors to examine artificial

FIGURE 12.2 The Singapore Airlines scented hot towel.

Image: authors' own, 2015.

scenting through an ethical lens. Such work recognizes that if manufactured scents are deliberately but covertly introduced into service environments, without a consumer's knowledge and below his/her threshold of consciousness, then the organizations undertaking this are potentially open to accusations of manipulating consumers and violating their freedom of choice (Bradford & Desrochers, 2009).

Some Key Benefits of Designing Smell into Consumer Service Environments

What are the advantages of designing smell into consumer service environments, either through the manipulation of existing smells emanating from areas such as the service production process (e.g. the tactically venting the appealing aromas of, say, coffee in a coffee shop), or through the introduction of designed and manufactured ambient smells? The answer, at a surface level, is straightforward: there is a belief that such practices improve the bottom line of those organizations who own or manage the service environments in which olfaction takes place. Bradford and Desrochers (2009) encapsulate such thinking succinctly in the title of their article "The use of scents to influence consumers: The sense of using scents to make cents."

However, beyond this simple overview lie a more complex series of explanations relating to how scenting service environments can influence both affective state and cognition amongst consumers, and in turn, how this feeds through to impact on consumer approach and avoidance behaviors (Mehrabian & Russell, 1974). This further highlights the link between smell and emotion in various transactional and/or relational marketing contexts, for both consumer and service provider.

Starting with affective state, Spence et al. (2014) provide a useful summary of how consumer responses to smell in service settings such as retailing are likely to be hedonically charged, either positively or negatively. In respect of the former, this suggests a smell introduced into a servicescape can be liked by the consumer, making him/her *feel* happier or more optimistic or more relaxed about life at that moment – that is, a positive hedonic charge to affective state. In turn, the resultant affective state may impact on consumers' 'approach' behavior to deliver a transactional and/or relational marketing benefit. From a transactional perspective, if consumers like a smell, then research suggests that this may translate into more positive product evaluations and/or increased buying intentions in service environments such as retail stores (Spangenberg et al., 1996; Miller, 1991; Morrin & Ratneshwar, 2000). In turn, this could lead to higher consumer spending than under unscented conditions. The 'liked' smell may also encourage a consumer to stay longer in a service environment such as a retail store (Knasko, 1989), thereby increasing the likelihood of a product purchase being made. Such examples can amount to a transactional marketing benefit to both consumers (as they hopefully receive some level of satisfaction and/or utility from the product purchased) and service providers (as they receive the consumers' business).

Turning to relational marketing benefits, if the scent deployed within a service environment induces a positive affective state in consumers, then this may feed through to positively impact on the image of the service provider (Spangenberg et al., 1996; Mattila & Wirtz, 2001), its brand positioning in the consumer's mind, and as a result, the potential likelihood of future visits. The scents deployed by Singapore Airlines and Westin Hotels, highlighted above, are arguably an example of brand-building and encouraging the repeat business of a given consumer through the means of olfaction. These longer-term interactions between service provider and service consumer provide one of the fundamental elements of a relationship marketing approach (Baron, Conway & Warnaby, 2010), in which value in the marketing exchange may ultimately be co-created by both parties (Vargo & Lusch, 2004).

Beyond the initial moment of consumers' embodiment of an odor in a service environment, affective responses to smell are intimately connected to cognition, and specifically the processes of association. As Spence et al. (2014) note:

> Olfactory cues may have their greatest impact through associative means. Indeed, scent memory is thought to be the strongest of all senses showing significant memory for odors decades after the first smelling of the scent . . . . Many of the managerial approaches in this domain use scent to fuel specific consumer thoughts related to their products.
>
> *p. 477*

In this respect, memory associations relating to a given odor may have a beneficial effect on transactional and relational marketing exchange activity. For example, travel agents Fred Olson and Thomson have both used variants of a coconut smell in their retail outlets to try to boost sales of holidays to those Caribbean destinations a consumer might typically associate with such odors (Gordon, 2009). A Thomson marketing executive explains the thinking behind this approach in transactional marketing terms: "When customers come into a scented shop they say it's like walking into a holiday. It helps them relax and encourages them to stay longer, so it helps sales" (Gordon, 2009). With the movement of services such as travel agents online, the opportunity for leveraging consumers' odor associations in this manner is likely to become more reliant on the emerging and innovative technologies of digital olfaction, which have been discussed elsewhere (Henshaw, Medway, Warnaby & Perkins, 2016).

However, caution should arise from the fact that the associations consumers make with smell can be negative as well as positive. Such variations may be influenced by consumers' previous experiences with a smell and their individual characteristics as smell recipients – for example, gender and age. Thus, older people may have different smell associations to those who are younger. In illustration of this point, Zola's novel *Au Bonheur des Dames* describes an instance in which scented white violets are given to customers at their point of purchase in the store (see Zola, 1886, p. 354) – a practice echoed in modern-day luxury stores with fragrant flower bouquets on counters (Parsons, 2009). However, there is evidence to suggest that any positive impact such actions might have on affective state might be offset amongst older consumers by a negative association made between the smell of flowers and funerals (Bone & Jantrania, 1992). Is this sense, a smell which has negatively perceived associations might stimulate avoidance rather than approach behavior (Mehrabian & Russell, 1974) in a given service environment – an outcome also identified by Bone and Ellen (1999).

The cognitive associations consumers make with smell also raise the issue of olfactory congruency with other sensory stimuli. For example, Spangenberg et al. (2005) demonstrate through an experimental design that the use of a manufactured 'Christmassy' smell in a retail store (specifically a room spray called *Enchanted Christmas*) produced more positive evaluations of the store and its merchandise, and a greater likelihood to visit the store amongst consumers when combined with Christmas music. However, those evaluations dropped when the same scent was combined with non-Christmas music. Where a Christmas scent is concerned, the smell could also have more associative congruency with distinct temporal contexts; so if this particular smell is experienced in store in September, then it might be as annoying as Christmas decorations that are installed too early. The relative congruency of a smell with a product being marketed may also be important (for a detailed discussion of this issue, see Schifferstein & Blok, 2002). Thus, the smell of fish in a fish market might seem perfectly appropriate because of the associative congruency between smell and product, but the same smell in a bank would not appear to be an obvious fit with the service product provided, unless perhaps the bank's target segment was trawlermen! Such examples might seem fanciful, but they help to illustrate the potentially complex and subtle interplay between affective state and associative cognition in relation to olfaction where consumers are concerned.

Conclusion

On the basis of the above discussion, we conclude by providing three critical factors that practitioners need to consider when attempting to enhance the consumer experience through olfaction, all of which have different implications for olfactory management: namely, intent, ethics and congruency.

Beginning with intent, at the broadest level, it is evident that a smell within a consumer service environment can be unintentional or intentional in nature. Unintentional smells are those that are naturally present. Typically, these occur as a result of some aspect of the service production process. A key management decision is whether those smells are hidden from the consumer, or openly revealed. A smell which the majority of consumers deem unpleasant (something which could be easily ascertained through market research) might be hidden or suppressed so as to not negatively impact on their experiences, typically through olfactory management tactics such as vented dispersal, or masking/deodorization using smells deemed more pleasant (Henshaw, 2014). Examples of this practice are the active deodorizing used in the toilets of service providers such as hotels and retail stores, or the venting of stifling cooking smells from an open-kitchen restaurant so as not to negatively impact on the experience of diners. Alternatively, smells which consumers deem pleasant can remain unmanaged, or even actively manipulated through venting practices to ensure they are more likely to reach the consumer, and in doing so, create positive associations. There are potential paradoxes in the management of unintentional smells. Thus, a smell deliberately vented from inside a restaurant, because it is deemed to negatively affect the diner experience, might be positively received when it is diffused in open air outside the premises and encountered by pedestrians passing by.

By contrast, intentional smells are purposely introduced into service environments. At the most basic level, this can involve active manual odorizing or spritzing of consumer service environments, such as the Hollister retail store example outlined above, or the offer of the scented hot towel to passengers on Singapore Airlines flights. Alternatively, the active spritzing process can be automated using the kinds of chemically manufactured smells and electronic devices produced by companies such as ScentAir and brandaroma. Sometimes, however, the intentional introduction of scent into consumer service environments is arguably more passive than active in nature: picture the bowl of potpourri left to linger and gather dust on the hotel reception counter or retail display unit from one week to the next.

The intentional introduction of smells into any consumer service environment brings into focus the second critical factor of ethical practice. Those managing such spaces need to decide whether consumers should always be informed about any smells being used, and of the rationale for doing so, or indeed, whether this is an unnecessary degree of information. On the one hand, it could be argued that the purposeful introduction of any sensory stimulant with intent to alter the behavior of consumers demands their informed consent. Conversely, it could be argued that by entering a retail store or booking a plane flight with a company that intentionally uses olfaction in this manner, then the consumer is providing a tacit agreement to be subject to the relevant consumer experience, whatever legal actions or sensory stimulants that might entail. Indeed, those charged with providing that experience might coherently argue that any required declaration relating to the use of smell as a tool to influence consumers' behavior necessarily demands similar admissions regarding the use and purpose of other sensory stimulants, such as music, or even color. Yet it is easy to envisage that consumers might also find a constant bombardment with sensory stimulant declarations from service providers rather tiresome.

One caveat to the above is where an olfactory stimulant has the known potential to cause an allergic or physiologically unwelcome reaction in certain people. At this point, there is an ethical responsibility for those managing consumer service environments to give prior warning of potentially harmful olfactory interventions, or indeed to avoid them altogether. Analogous precedents are found in respect of the visual sense, with, for example, strobe lighting and epilepsy sufferers. Thus, appropriate warnings are usually provided in those consumer settings (e.g. a nightclub, the screening of a certain film) in which strobe lighting might be used, typically with an explanation of the problems this might cause for those affected by it.

The final critical factor for practitioner attention relates to congruency. This chapter has emphasized the importance from a psychological perspective of an olfactorily congruent consumer experience. However, a key point to understand is that this congruency has at least three dimensions. Specifically, when smell is used to try and enhance the consumer experience, then it needs to be congruent with both the product and also the brand with which that smell is supposedly associated. In addition, there should be congruency between a smell used to promote or communicate the product or brand and any other sensory stimulants employed for the same purpose.

In conclusion, this chapter has demonstrated that designing smell into the consumer experience is a complex business. The ultimate aim of any such action is to increase profits by stimulating greater consumer spending in transactional situations within shopping/service environments, and/or to improve repeat patronage by making the consumer experience a pleasant one (the latter aim is also relevant in non-profit contexts such as museums or art galleries). To optimally achieve these benefits, the use of olfaction in consumer settings needs to be fully planned, in a manner that is cognizant of the range of issues outlined above. Moreover, this needs to be characterized by an iterative approach, involving an ongoing revision of the smells that stimulate consumers' approach and avoidance behavior.

Notes

1 www.brandaroma.com Retrieved March 1, 2017.
2 www.scentair.com/ Retrieved March 1, 2017.
3 www.scentair.com/why-scentair/testimonials/ Retrieved March 1, 2017.
4 http://europe.westinstore.com/en/white-tea/home-fragrance/white-tea-scent-diffuser.html Retrieved March 1, 2017.

References

Aggleton, J. P., & Waskett, L. (1999). The ability of odours to serve as state-dependent cues for real-world memories: Can Viking smells aid the recall of Viking experiences? *British Journal of Psychology*, *90*(1), 1–7.

Ballantine, P. W., Jack, R., & Parsons, A. G. (2010). Atmospheric cues and their effect on the hedonic retail experience. *International Journal of Retail & Distribution Management*, *38*(8), 641–653.

Baron, S., Conway, T., & Warnaby, G. (2010). *Relationship marketing: A consumer experience approach*. London: SAGE Publications.

Bhasin, K. (2014, May 30). Why Abercrombie's smell makes you anxious. Retrieved March 1, 2017 from www.huffingtonpost.com/2014/05/30/abercrombie-cologne_n_5412020.html#

Bitner, M. J. (1992). Servicescapes: The impact of physical surroundings on customers and employees. *Journal of Marketing*, *56*, 57–71.

Bone, P. F., & Ellen, P. S. (1999). Scents in the marketplace: Explaining a fraction of olfaction. *Journal of Retailing*, *75*(2), 243–262.

Bone, P. F., & Jantrania, S. (1992). Olfaction as a cue for product quality. *Marketing Letters*, *3*(3), 289–296.

Bradford, K. D., & Desrochers, D. M. (2009). The use of scents to influence consumers: The sense of using scents to make cents. *Journal of Business Ethics*, *90*, 141–153.

Davies, B. J., Kooijman, D., & Ward, P. (2003). The sweet smell of success: Olfaction in retailing. *Journal of Marketing Management, 19*(5–6), 611–627.

Fetterman, M., & O'Donnell, J. (2006, September 1). Just browsing at the mall? That's what you think. *USA Today*. Retrieved March 1, 2017 from http://usatoday30.usatoday.com/money/industries/retail/2006-09-01-retail-cover-usat_x.htm#

Gordon, S. (2009, June 29). Smell of success: Travel agents use holiday scents to sell trips. *Mail Online*. Retrieved March 1, 2017 from www.dailymail.co.uk/travel/article-1196232/Holiday-smells-pumped-travel-agents-sell-trips.html#

Hayward, T. (2010, October 22). The smell of excess [Blog post]. Retrieved March 1, 2017 from www.theguardian.com/lifeandstyle/wordofmouth/2010/oct/22/smell-excess-shopping-food

Henshaw, V. (2014). *Urban smellscapes: Understanding and designing city smell environments*. New York: Routledge.

Henshaw, V., Medway, D., Warnaby, G., & Perkins, C. (2016). Marketing the "city of smells." *Marketing Theory, 16*(2), 153–170.

Hirsch, A. R. (1995). Effects of ambient odors on slot-machine usage in a Las Vegas casino. *Psychology & Marketing, 12*(7), 585–594.

Holloway, M. (1999). The ascent of scent. *Scientific American, 281*, 42–44.

Knasko, S. C. (1989). Ambient odor and shopping behavior. *Chemical Senses, 14*(5), 719.

Lindstrom, M. (2005). *Brand sense: How to build powerful brands through touch, taste, smell, sight and sound*. London: Kogan Page.

Mattila, A. S., & Wirtz, J. (2001). Congruency of scent and music as a driver of in-store evaluations and behavior. *Journal of Retailing*, 77(2), 273–289.

Mehrabian, A., & Russell, J. A. (1974). *An approach to environmental psychology*. Cambridge, MA: MIT Press.

Miller, C. (1991). Research reveals how marketers can win by a nose. *Marketing News*, 25, 1–2.

Morrin, M., & Ratneshwar, S. (2000). The impact of ambient scent on evaluation, attention, and memory for familiar and unfamiliar brands. *Journal of Business Research, 49*(2), 157–165.

Nava, M. (1996). Modernity's disavowal: Women, the city and the department store. In M. Nava & A. O'Shea (Eds.). *Modern times: Reflections on a century of English modernity* (pp. 38–76). New York: Routledge.

Parsons, A. G. (2009). Use of scent in a naturally odourless store. *International Journal of Retail & Distribution Management, 37*(5), 440–452.

Roston, T. (2011, December 7). What's behind the Subway bread smell? The fast-food chain's bread smell, dissected. Retrieved March 1, 2017 from www.foodrepublic.com/2011/12/07/whats-behind-the-subway-bread-smell/

Schifferstein, H. N., & Blok, S. T. (2002). The signal function of thematically (in)congruent ambient scents in a retail environment. *Chemical Senses, 27*(6), 539–549.

Soars, B. (2009). Driving sales through shoppers' sense of sound, sight, smell and touch. *International Journal of Retail & Distribution Management, 37*(3), 286–298.

Spangenberg, E. R., Crowley, A. E., & Henderson, P. W. (1996). Improving the store environment: Do olfactory cues affect evaluations and behaviors? *Journal of Marketing, 60*, 67–80.

Spangenberg, E. R., Grohmann, B., & Sprott, D. E. (2005). It's beginning to smell (and sound) a lot like Christmas: The interactive effects of ambient scent and music in a retail setting. *Journal of Business Research, 58*(11), 1583–1589.

Spence, C., Puccinelli, N. M., Grewal, D., & Roggeveen, A. L. (2014). Store atmospherics: A multisensory perspective. *Psychology & Marketing, 31*(7), 472–488.

Vargo, S. L., & Lusch, R. F. (2004). Evolving to a new dominant logic for marketing. *Journal of Marketing, 68*(1), 1–17.

Wade Clarke, D., Perry, P., & Denson, H. (2012). The sensory retail environment of small fashion boutiques. *Journal of Fashion Marketing and Management: An International Journal, 16*(4), 492–510.

Ward, P., Davies, B. J., & Kooijman, D. (2003). Ambient smell and the retail environment: Relating olfaction research to consumer behavior. *Journal of Business and Management, 9*(3), 289–302.

Ward, P., Davies, B. J., & Kooijman, D. (2007). Olfaction and the retail environment: Examining the influence of ambient scent. *Service Business, 1*(4), 295–316.

Zola, É. (1886). *The ladies' paradise: A realistic novel*. London: Vizetelly.

Zola, É. (2012). *The ladies' paradise* (B. Nelson, Trans.). Oxford: Oxford University Press.

13

THE USE OF SCENT IN OUT OF HOME (OOH) ADVERTISING

Prask Sutton

Introduction

In an industry which has been traditionally dominated by visual stimuli, organizations using Out Of Home (OOH) advertising are increasingly utilizing olfactory elements to entice, engage and entertain target audiences. OOH advertising has been defined as referring to the various forms of advertising which occur within the wider urban environment (Koeck & Warnaby, 2014), which can include advertising on billboards, street furniture such as bus shelters, free-standing panels in pedestrian/shopping zones and so on, as well as 'transit' advertising, such as that on buses and taxis (Cronin, 2010; Fill, 2009). However, in its efforts to disrupt, and create memorable experiences, is the advertising industry fully considering the potential impact of olfactory elements on consumers? Given the absence of guidelines and regulations governing the use of scent marketing more generally, the need for self-regulation is a pressing issue. What is best practice, and how might the industry avoid, or at the very least negotiate, the inevitable backlash arising from the olfactory overkill that could be a consequence of uncontrolled use of olfaction in this context?

This chapter will explore the rationale behind what I term Scented Out Of Home (SOOH) advertising, outline the pitfalls that practitioners and clients should be aware of in its use, and propose a quantitative and qualitative approach with a view to establishing best practice, including a basic code of conduct, as a point of reference for the industry. Industry perspective will be provided through insights from a number of the United Kingdom's OOH advertising thought leaders, with analogies drawn and lessons learned from proven policies already in place.

The Use of Scent in Advertising

Scent has affected people's perception of products, services, environments and other people for a very long time, leading to both conscious and unconscious judgments being made, based on the aromas accompanying both initial introduction of – and subsequent contact with – specific scents. However, the explicit and proactive use of scent to raise consumers' awareness and influence their opinions – the very essence of marketing and advertising – is a relatively new phenomenon.

Many things inherently give rise to scents. The smell of freshly baked bread wafting out from a bakery has not been artificially added in for marketing or promotional effect. It is simply a by-product of the production process that just so happens to be considered to be rather agreeable by the majority of people. The fact that the smell is generally well liked means that rather than try to remove it, it is usually allowed to persist, as it alerts people to the fact that freshly baked bread is available, and thus promotes a desire for it. Other aromas, however, are categorized as unpleasant, and as a result, efforts are made to remove them, often with a view to replacing them with something preferable. This idea of odor control (for more detail, see Henshaw, 2014) – the notion of removing 'bad' smells and/or substituting them with something perceived to be more pleasant – has led to the practice of artificially adding scents where they would not naturally occur, or at least not in such high concentrations.

Over time, this has evolved to a point where scent is now introduced to all sorts of products and also retail/service environments with a view to positively influencing consumer perceptions (for more detail, see Lindstrom, 2005, 2010). A food store selling bread may have it baked elsewhere and then delivered to the site, yet still seek to propagate the scent of freshly baked bread to entice consumers, alerting them to the presence of the product and creating a host of impressions, including that the product is fresh out of the oven. This sort of product-congruent scenting, whereby a smell inherently associated with something is added to further accentuate the positive connotations arising, has more recently been accompanied by a rise in what might be termed 'incongruent scenting,' where the goal of positioning something in the minds of consumers is achieved through the introduction of scents that do not actually have anything to do with the thing *per se*, but rather are added in the hope of creating an association with other things that will hopefully frame that something in a better light. A prime example of this is the scenting of hotels, or advertisements for hotels, which often make use of incongruent scenting, such as Westin Hotels' signature white tea-based scent, as actual white tea has little or nothing to do with the Westin brand. And so an industry has been born and built up around the idea of, often artificially, creating scents and using them to market and advertise pretty much everything you have ever exchanged money for, and you may not have even noticed.

The remainder of this chapter will briefly outline the digitization of OOH advertising (i.e. Digital Out Of Home (DOOH) advertising), explore the motivation behind the addition of scent to OOH advertising (i.e. Scented Out Of Home advertising), highlight some of the common mistakes made in SOOH practice, and consider those guidelines relating to the subject which are already in place before finally postulating what the future may hold for this aspect of advertising activity.

OOH Advertising and the Progression to DOOH Advertising

The category of OOH advertising is just that – advertising that takes place outside of people's homes. It includes, as noted above, billboards, advertising on street furniture, and transit-related advertising, on and around public transport. Most of us living in towns or cities are exposed to this type of advertising from the moment we leave home. It really is everywhere. As a result, we have, to a certain degree, become desensitized to it, as if a kind of 'ad-blindness' has set in. This is in part because traveling past the same advertising day-in, day-out has allowed us to get used to the placement of these ads; which is not great for advertisers, given that human beings' attention is naturally drawn to *changes* in our environment.

Recently, more and more OOH advertising has become digitized, with the old 'paper and paste' posters gradually being replaced with digital screens (for more detail, see Kinetic, 2011; Koeck & Warnaby, 2014). This digital signage, known as Digital Out Of Home, is able to garner considerable attention because of its ability to display *moving* images, providing the change (in this case, visual) that grabs our attention as we pass. So through DOOH, advertisers are able to successfully pique our interest and get us to look in the right direction. Although this is a considerable improvement from the advertisers' perspective, the fact remains that in order for their message to be viewed, these advertisements need to be seen. Given the busy 'always on' lives many of us now lead, the majority of time spent journeying from point A to point B in urban space is consumed by a variety of things constantly vying for our attention. If we are not in company and talking amongst ourselves, we are usually talking to someone on the phone, or with our heads buried in a book or magazine, or staring at a screen of our own, be it a smartphone, tablet, e-reader or laptop. Whatever we are doing, nowadays it is rare for us to be consciously scanning around, looking at the advertisements we pass by.

So if our eyes are otherwise occupied, how can advertisers get our attention as we go about our daily routines?

The Rationale behind the Nascent Industry of Scented Out Of Home Advertising

It is often said that we are multisensory beings, experiencing the world through all our five senses (Porteous, 1990). The more of these senses that can be engaged simultaneously, the more immersive, engaging and memorable an experience will be. From a brand's perspective, this is incredibly important, as in a world absolutely brimming with competing brands – all vying for our attention – the ability to stand out, be noticed and remembered is key.

Traditionally, OOH advertising has primarily been a visual experience. The artwork and copy – the pictures and words – that make up advertising's 'creative' content have been solely responsible for turning our heads. Every time a person passes such an advert, the Opportunity To See (OTS) of the ad is either capitalized upon or lost, often based solely on this creative content. However, this assumes that passers-by are looking in the right direction at the right moment in time, and as we have already noted, this may not always be the case. Enter, therefore, the idea of Multisensory Marketing and Advertising (MsMA), which seeks to build upon traditional creative content with additional non-visual communication. Given the subject matter of this book, we will not be exploring the use of sound or touch, but rather focusing on smell and, to a lesser extent, taste, in as much as it is a component of flavor.

Regardless of how busy or otherwise engaged one might be, and as easy as it is to walk past a billboard without it catching one's eye, it is far more difficult to walk past one that is giving off a particular smell without it being registered by the brain. So, as easy as it is to miss a visual cue by looking the other way, a smell emanating from the same direction will make it through to our perception nonetheless. The all-pervading nature of scent makes it a great tool in the advertiser's arsenal with which to cut through the visual noise of the oft-missed and ignored posters and digital screens, thereby grabbing the attention of passers-by who might otherwise have missed their messages. This has led to the birth of Scented Out Of Home advertising.

But which ad formats are best suited to SOOH? Those aimed at vehicular traffic are usually too far from the noses of those passing, and moreover, said noses are often passing at far too great a velocity for any meaningful olfactory experience to be provided. Formats aimed at pedestrian traffic, however, or those placed in areas of high dwell-time where people linger longer, are prime real estate from an SOOH perspective.

One of the most popular SOOH advertising formats is the 'Scented 6-Sheet'. The term '6-Sheet' is a reference to its size, approximately 1200 mm wide × 1800 mm tall, and is a remnant from the days when this size of poster was made up of six individual sheets of paper. A Scented 6-Sheet is simply one that has been enabled to deliver scent. If you are having trouble picturing a 6-Sheet, these are probably most recognizable as the advertising formats attached to many bus shelters.

There are several ways in which scent can be delivered from these formats in such environments, but the trend has been away from relatively low-tech options, such as scratch-and-sniff posters, to more hi-tech solutions involving computer-controlled pumps and compressors, which can also be incorporated into DOOH campaigns and used to deliver scent synchronized to audio-visuals comprising the ad's creative content with to-the-second accuracy.

For the most part, scent-enabled advertisements can be categorized as delivering either ambient scent or Point-of-Scent (PoSc). Ambient scent pervades the entirety of an environment, albeit subtly, with the scent usually delivered non-stop, or at least regularly enough to scent the area immediately surrounding the ad. PoSc, however, tends to be far more directional, with scent being tactically delivered to individuals or groups of individuals passing by. An advert with an accompanying ambient scent has the benefit of being able to attract the attention of passers-by – even those whose field of vision it might initially escape – but it is worth noting that the efficacy of such an ad requires the accompanying scent to be experienced in the presence of other sensory cues, such as visuals, in order for people to associate the scent with the product, service or brand being advertised. Otherwise, unless the scent is so familiar that people automatically know what it relates to, or the advertising strategy plans to subsequently re-introduce the scent in the presence of the overt reference required, people will not know what they are smelling, and the ad serves little purpose. For example, a well-known brand that is a household name supplies its frozen French fries to a large fast-food chain. The former has dabbled in ambient scenting, but quickly moved over to PoSc, because anyone smelling the scent of their fries without a visual clue was as (if not more) likely to associate the aroma with the fast-food chain instead. Indeed, the fact that the person experiencing an advert shrouded in an ambient scent has to be standing near to it in order to get a whiff of the scent being delivered is its major drawback. If this is not the case, and the scent is encountered without the accompanying visuals, then the connection of the olfactory experience back to the product/brand in question needs to be explicit.

Advertisements featuring PoSc, however, tend to deliver scent only to the space directly in front of the ad. As a result, by the very nature of the set-up, people have to be in the presence of the visual, or audio-visual, component of the creative content in order to be able to smell the scent being delivered. The olfactory experience is therefore connected back to the ad. There remains the scenario where somebody is standing in the right place, but their attention is focused elsewhere, preventing that all-important (audio)visual connection from being made. However, there is a solution to this potential problem, which is proactive engagement.

This brings us to the various methods used to precipitate the delivery of PoSc. An in-depth account of all the technologies currently available is beyond the scope of this chapter (although more detail is provided by Henshaw, Medway, Warnaby & Perkins, 2016), but we regularly use a whole host of techniques to trigger scent. The simplest triggering mechanism is a traditional push-button, similar to those found on myriad electronic devices, which is worked by being depressed by a finger. With digital infrastructure, such as digital 6-Sheets, the use of now-commonplace touchscreen technology can be considered an evolution of the push-button approach, with bespoke touchscreen games being a natural extension of this. Passive infrared detectors (PIRs) and pressure pads are sometimes employed to automatically trigger scent delivery, with the former sensing the motion of people, and the latter detecting the physical footfall of pedestrians.

More cutting-edge are 'machine vision' solutions, such as facial, gesture and thermal recognition, which involve computers analyzing digital camera feeds to determine whether people in view of the cameras match set criteria. With such setups, scent can, for example, be delivered to those smiling at an advert, passing it while wearing blue jeans, or with particularly warm armpits, or even combinations thereof. The prevalence of smartphones has also heralded the use of social media platforms such as Twitter as a bona fide engagement mechanism. Finally, at the leading edge of mobile marketing are new digital engagement technologies like Wi-5, which have been designed to allow people to fully engage with immersive digital systems, such as those in DOOH, from any wi-fi-enabled device.

However, despite the plethora of scent-triggering solutions available to the modern OOH advertising practitioner, the prime differentiator is that those methods requiring proactive interaction by the public are the ones that most often succeed in cementing in people's minds the relationship between what they are smelling and what they are seeing, making for the most effective ads.

Pitfalls to Be Aware of

Regardless of the method, and whether what is being delivered is a flavor or fragrance, a single aroma or a selection of scents, the smell of a new dog food or a highly anticipated fine fragrance, something subtle designed to speak to the subconscious mind or highly concentrated to tingle the nostrils, there are rules of thumb that, if followed, can avoid potential pitfalls that can negatively impact upon consumers.

Whatever fragrance or flavor is being delivered, it is usually prudent to deliver scented air, as opposed to vaporized oil. There are several reasons for this. Vaporized oil can stain, be flammable and possibly act as an irritant, thereby potentially creating all sorts of health and safety issues. There is also the issue of cost, as many more grams of fragrance are generally required to deliver vaporized oil-based scent compared to scented air over the same given time period. However, the delivery of vaporized oil can often be desirable, given the more intense user experience this method can readily create, by virtue of the sheer quantity of fragrance being delivered. This is particularly relevant in OOH advertising, given the degree of competing air movement that is often present in outdoor environments. In such environments, it may be desirable to deliver as high a concentration of scent as possible, as quickly as possible, before the wind or other atmospheric conditions disperse it. This does not negate the issues previously raised, but is presented here, given that the practice of using vaporized oil is still relatively commonplace.

Outside of environmental considerations, the question of the concentration at which to deliver scent is a complicated and oft-debated one. There is much talk of the 'power of scent,' equating it to some sort of magic that can be invoked to convert browsers to buyers against their will. Whether one would regard this as a good or bad thing, the fact of the matter is that this simply is beyond the ability of humble scent marketing. Nevertheless, many consumers have come across such bold claims, and as a result, can sometimes have a tendency to react negatively to overt, highly concentrated scent delivery when it forms part of an advertisement. Rather than the gentle persuasion in a specific direction, people can sometimes feel a sense of manipulation and that they are being advertised *at*. Lessons can be learned here from several high street retailers who have been heavily criticized after trialing highly concentrated scent delivery.

Conversely, scent marketing that takes a subtler approach is often less likely to raise the suspicions of those at which it is aimed. I would also argue that the subtle approach, whereby one may not consciously notice the presence of a particular scent, but would certainly notice something missing were it to be removed, is more likely to succeed in amplifying the other sensory elements present, without unduly dominating and distracting from the whole. As stated by Hultén, Broweus & Van Dijk (2009), "*Intensity* as a sense expression, which refers to whether a scent is distinct or subtle, can affect the opportunity to accomplish a positive smell experience. A subtle scent that is weakly perceived can have a subliminal impact on customers" (p. 49). Indeed, this sort of subtlety is regularly employed when ambient scenting is used in OOH advertising. In 2014, we worked on Tennent Caledonian's launch of Lemon T, a light lager mixed with lemon soda, ambiently scenting parts of the Glasgow Subway with a product-congruent aroma of sweet lemon, subtle enough to not distract commuters, but perfectly poised to amplify and pique interest in the backlit posters advertising the new Lemon T. However, this is certainly not a case for 'secret scenting,' as if – or when – found out, the consumer's sense of being duped is rightly aggravated.

In my experience, it is usually clients – and here I specifically refer to those directly paying for the activity – that are most likely to ask for the concentration to be 'turned up,' if not taking matters into their own hands and whacking the dial 'up to 11' when nobody is looking. This desire most frequently stems from a need to get more 'bang for their buck,' coupled with a lack of understanding as to the underlying methodologies at play. Any fears a client may have of not getting value for money or involving the efficacy of subtle scenting can usually be assuaged by clearly explaining the principles involved.

Lastly, from a technical standpoint, the subject of heat is relevant in this context. When heated, most fragrance – and often, when present, its carrier medium – has a tendency to break down on a molecular level. As such, scent delivery mechanisms that require heating to occur at some point in the delivery process can suffer from a loss of quality in terms of the end olfactory experience, with anything from subtle nuances to entire facets of a fragrance potentially being altered or destroyed. As can be imagined, this is almost always going to be to the overall detriment of the original scent. It is thus preferable to store scent as advised by its manufacturer and opt for scent delivery solutions that do not raise the temperature of the fragrance they are delivering. Exceptions to this rule are situations where the ambient temperature of an environment is significantly low, such that scent is required to be brought *up* to room temperature in order to prevent freezing, but this is very much an extreme scenario enforced by local environmental factors (such as when scenting a hotel made of ice in subzero temperatures!), and not one that many engaging in SOOH will regularly face.

Existing Guidelines and Regulations Governing the Use of Scent Marketing

The ability for scents to be delivered in an OOH advertising context in such a way or at such concentrations as to not be consciously detectable, understandable or physically avoidable by consumers certainly presents an opportunity for either intentionally or accidentally deceptive trade practice. Although there are laws relating to the use of subliminal marketing, under the remit of which such deceptive trade practice would fall, to date no laws have been passed specifically relating to scent marketing. Scent-free and so-called 'air aware' policies have been petitioned for and in some cases attempted to be enforced, albeit with great difficulty and with arguably little success, with the most common reaction being a raft of complaints and claims that said policies are an infringement of personal freedoms. A well-documented example took place in Sheet Harbour, Nova Scotia, where a teacher attempted to charge a student with assault for wearing scented toiletries to school. The Royal Canadian Mounted Police were brought in, but unsurprisingly, no charges were laid. More can be read about what was referred to as 'Halifax Hysteria' as well as the failure of similar policies implemented elsewhere in Canada, including Ontario, on the website of the Scented Products Education and Information Association of Canada,[1] which is interestingly hosted by the Canadian Cosmetic Toiletry and Fragrance Association.

One could assume that were olfactory-based marketing to become sufficiently prevalent, specific laws relating to it would eventually be drawn up and enacted. But, in lieu of such legally enforceable regulations, the need for self-regulation should be viewed as paramount, as any and all practices that engender negative reactions from the world at large harm the industry and hasten the arrival of scent-specific laws, which, if anything like those imposed on the perfume industry, will severely restrict what is permissible, regardless of common sense or best practice.

With self-regulation, it could be argued that a little common sense goes a long way, but there are also comprehensive and regularly updated guidelines on safety standards that are readily available. For example, if practitioners ensure that all scents used are approved by the International Fragrance Association (IFRA), they can assume that the ingredients and compounds contained therein feature only within established 'Safe Use Levels,' having been rigorously assessed according to 186 criteria, from the perspective of both people and the environment, for toxicity and allergens. The IFRA's global compliance program is also in place to ensure producers and their products comply with its Code of Practice, which contains the full set of IFRA Standards as well as outlining the behavior expected of its members, including fragrance supplier companies which are direct members or whose membership is by association through other national bodies.[2]

The Bright Future of Scented Out Of Home Advertising

It was only a few years ago that I was having to plead with colleagues, clients and friends in the industry to give scent marketing a chance in an OOH context. I spent an inordinate amount of time extolling the virtues of multisensory marketing and advertising, even founding an MsMA agency to create a one-stop shop, just to encourage uptake. It was recent enough to still be very clear in my mind, and I recall what an uphill struggle it was. However, I lived in hope that persistence would pay, and was strengthened in my resolve by the belief in how efficacious a multisensory approach is. Slowly but surely, silence turned into 'no's,' which became 'maybes' and gradually formed a trickle of 'yes's.' Nowadays, the majority of OOH campaigns I am involved in have some multisensory dimension, with a significant number featuring an olfactory element. Where a Scented 6-Sheet was once newsworthy in its own right, this sort of activity is now a regular occurrence. Although still classed as a 'special build,' a term used to cover anything other than the most basic of poster and digital campaigns, Scented 6-Sheets now feature on many media owners' rate cards, which detail prices and descriptions for the most commonplace ad placement options available. As such, they are almost as easy for brands to run as a traditional poster or digital campaign.

I recently spoke with Rosh Singh, Director of Digital Innovation and Head of Kinetic Active, the technology division of Kinetic Worldwide, an OOH media planning and buying company, about the use of multisensory in OOH. Rosh is a fellow futurist, and can always be relied upon to see potential where it exists. As such, he has been an early adopter of the use of scent in OOH and a strong advocate of multisensory marketing communications in general:

> OOH remains the only media channel that we can use to target and trigger all five of the public's senses and until the Internet starts to get really weird, it will remain so. We have an inherently visual medium in posters and to a greater extent digital screens, but our thinking remains somewhat shuttered when it comes to engaging the other senses and while we do dabble here and there, through experiential and one-off campaigns, it is not yet something that represents a core part of our arsenal.
>
> That being said, OOH has seen a lot of great uses of scent throughout the years. Whether it is the scent of a freshly baked and buttered jacket potato wafting alluringly from a bus shelter, or the smell of a washing powder drifting through a shopping mall, scent-emitting posters are a mainstay in the bag of tricks that any special projects department has. Stacking the senses and using them to complement each other is where the real opportunity lies within OOH, allowing us to leverage our multisensory trump card and use it to create an advertising experience that makes people *feel* the brand and not just see it.
>
> While scent and OOH have a history and are natural companions, the real excitement comes from what the future holds.
>
> *extract from discussion between the author and Rosh Singh, September 2015*

I am under no illusions. All parties are not yet convinced, but I am confident that we will get there! The use of scent in OOH is already a familiar enough concept to most for it to be raised in a meeting without a concurrent raising of eyebrows. Being at ease with the concept means that stakeholders are becoming more sophisticated in their tastes, which is being reflected in the sort of scented campaigns they sign off and allocate budget to. Having won the confidence of clients and having an ever-greater understanding of what works has allowed for the subtler and increasingly intelligent use of scent in campaigns. But this is just the beginning, and I believe that the best is still to come. Olfactory marketing will continue to be utilized in such ways as to benefit brands and the people they are seeking to appeal to. As long as we work wisely, maintaining suitable awareness and perspective, these multisensory elements will serve to contribute to, rather than detract from, the environments our ads exist within. If this considered approach can be maintained, scent, in its many forms, will no doubt become a regular feature in the bright future of OOH.

Notes

1 www.cctfa.ca/scented/index.html Retrieved March 1, 2017.
2 For more information, see www.ifraorg.org/ Retrieved March 1, 2017.

References

Cronin, A. M. (2010). *Advertising, commercial spaces and the urban*. Basingstoke: Palgrave Macmillan.

Fill, C. (2009). *Marketing communications: Interactivity, communities and content* (5th ed.). Harlow: Prentice Hall Financial Times.

Henshaw, V. (2014). *Urban smellscapes: Understanding and designing city smell environments*. New York: Routledge.

Henshaw, V., Medway, D., Warnaby, G., & Perkins, C. (2016) Marketing the city of smells. *Marketing Theory*, *16*(2), 153–170.

Hultén, B., Broweus, N., & Van Dijk, M. (2009). *Sensory marketing*. Basingstoke: Palgrave Macmillan.

Kinetic. (2011). *On the threshold of change: The future of out of home media in the UK – the industry, consumers and technology to 2020*. London: Kinetic.

Koeck, R., & Warnaby, G. (2014) Outdoor advertising in urban context: Spatiality, temporality and individuality. *Journal of Marketing Management*, *30*(13–14), 1402–1422.

Lindstrom, M. (2005). Broad sensory branding. *Journal of Product and Brand Management*, *14*(2), 84–87.

Lindstrom, M. (2010). *Brand sense: Sensory secrets behind the stuff we buy* (2nd ed.). London: Kogan Page.

Porteous, J. D. (1990) *Landscapes of the mind: Worlds of sense and metaphor*. Toronto: University of Toronto Press.

14

SCENT AND BRAND STORYTELLING

Claire Sokell Thompson with contribution from Jo Barratt

Introduction

I used two bottles from the collection of travel minis in the bathroom. "You smell of The Savoy!" exclaimed my children (aged five and seven). As a family, we had stayed at the hotel for one night the previous year, and the smell of Miller Harris toiletries from our hotel room elicited excited chatter and reminiscences about The Savoy. The next half an hour was spent sharing stories: the memories were detailed and personal, and the conversation concluded with adamant insistences that we return immediately. The olfactory power of recall and emotional engagement is undeniable:

> Scent reaches us in ways that elude sight and sound but conjure imagination in all its sensuality, unsealing hidden worlds. A whiff of a once familiar odour, and memories surge into consciousness on a sea of emotion, transporting us – first to a camping trip, steeped in the smell of pine and burning wood; to the steamy windows and vanilla laced air of a winter kitchen where cookies are baking.
>
> *Aftel, 2001, p. 11*

It is one of the reasons I started my business.

The research and statistics support the power of olfactory branding: examples include a scent campaign for Mitsubishi selling out the car in two weeks and increasing company sales by 16%; retail sales rising by 33% in direct response to the addition of smell (Lindstrom, 2010); anxiety levels in emergency wards decreasing by 63% in response to vanillin (Fox, 1997); and gambling increasing by 45% in Las Vegas casinos when particular scents were introduced into the environment (Hirsch, 1995). Smell is undoubtedly powerful, but our relationship with smell has changed dramatically: from the human reliance on our sense of smell for survival, to our desire to eradicate or mask smell as a result of increasing density of population, urbanization and industrialization (Henshaw, 2014). Our sense of smell has simultaneously been assaulted and dulled, and its status has been diminished (Drobnick, 2006). The majority of communications rely almost entirely on visual and linguistic communications: "Western history was shaped for some three thousand years by the introduction of the phonetic alphabet, a medium that depends solely on the eye" (McLuhan, 1967, p. 44). Another consideration is the impact of language. The vocabularies of most languages fail to offer us a shared vocabulary around smell (Wnuk & Majid, 2013), and if we do not have a shared lexicon, then we are unable to share or understand it, and so its cultural significance decreases (Howes, 2005).

It is interesting that despite these limitations, mankind has desired and created scent since early civilization for use in homes, religions, rituals, relationships and commerce. From the earliest commerce, scent has been a valuable commodity, and has evolved into a multi-billion-dollar industry. As an instantaneous and evocative medium with a direct connection to the limbic system, smell demands attention and has the power to conjure stories and instantly transcend time and space, but it can also repel, evoke disgust and alienate.

Reactions to scent pertain to two subjects: design and context. Using smell as a brand dimension offers unrivaled opportunity to enchant and engage customers, strengthen the relationship between brand and customer, and frequently opens up increased revenue opportunities, but there are many challenges to executing it successfully. The first half of this chapter explores the role of scent within the wider context of branding, and the second half outlines the challenges and pitfalls of designing and using scent in this context, and offers guiding principles for olfactory design.

Brands and Storytelling

So what is a brand? Rolling back to the Wild West, when cattle were herded across the plains of America, their hides were seared with branding irons to mark ownership. A practice initially established to protect and distinguish the goods of a supplier also turned out to be useful for buyers at the cattle markets. The mark enabled buyers to identify the product and connect it with its provenance and quality. Since that time, competition and commercial evolution have necessitated that brands also need to develop aspirational and emotional resonance to stand out and succeed.

A powerful brand has a strong sense of self at its heart: a clear, unique central motivator which drives and permeates everything it does – the basic brand proposition, communications, channel strategy, customer service, identity, corporate and social responsibility policies, human resources, internal culture, tone of voice, services, pricing, technology infrastructure and business processes (Earls, 2002). Although the process of branding has become more complex, the basic functions and benefits remain the same: to protect and distinguish a company's goods, services and attitude, and to allow a customer to identify and choose those. However, it is not enough just being unique and consistent. In a world where we are constantly exposed to hundreds of advertisements and messages through multiple channels, attention is increasingly hard to both attain and retain. Companies spend millions on campaigns about products and services, but in order to engage people, you need to capture their imagination and make them *feel*. Storytelling is a device employed as a way to communicate what brands believe in, who they are, who they are for and to create involvement and desire.

Why is this? According to Booker (2004), storytelling is an intrinsic part of culture. Since before the written word, tales have permeated our lives and provided the basis of communication. From cave paintings to installation art, and from campfires to movie theaters, stories have been a critical element of how we make sense of the world and relate to each other. Whether factual or fictional, stories provide a vehicle for articulating fears and aspirations, and are the way we continue collective knowledge. Everything we do is recounted in the form of stories: news, history, science, exploration, art, consumerism and media: "They are far and away one of the most important features of our everyday existence, the most natural way we know to describe almost everything which happens in our lives" (p. 2). Stories depend on the imagination and skill of the storyteller, but they also require active participation and interpretation to be understood: a writer may lead the way, but we bring the wealth of our personal experience and our imagination to them

(Gottschall, 2012), and that involvement results in a story emerging as part of your own personal universe and experience. Powerful stuff.

Storytelling in Scent

Clever brands realize that seduction is emotional and sensual, and that you need to appeal to all the senses to bring stories to life (Lindstrom, 2010). For centuries, perfumers have told their own stories through scent, recreating magical experiences, evoking memories and emotions, and transporting us to other worlds. In the same way that writers create "structured sequences of imagery" (Booker, 2004, p. 3), perfumers assemble an arrangement of moments, playing with ingredients, combinations and structure. As Ellena (2011) states: "Perfume is a story in odors, sometimes a poetry of memory" (p. 48).

Our sense of smell is unique. Unlike our response to our other senses, olfactory neuro-anatomy is intertwined with primary emotional areas. Incoming odors first trigger an emotional response, followed by cognitive recognition (Krusemark, Novak, Gitelman & Li, 2013). When we 'log' a smell in our brain, we attach our emotional response to the smell, so smell not only triggers memories, but also how we were feeling when we first smelled it. As such, smell has the potential to become a powerful force in the narrative of the brand and in the emotional and commercial relationship between customer and brand.

The luxury fragrance industry offers great insight into the potential of storytelling through scent, full of suggestion and enchanting stories: scents such as *Neroli Portofino*, *Bubblegum Chic*, *French Lover* and *Un Jardin sur le Nil* conjure an image, a lifestyle and a story. For example, perfumer James Heeley describes his fragrance *L'AMANDIÈRE* as:

> A portrait of Spring. Green almond trees in blossom, bluebells, hyacinths, spring flowers, rose, green grass, jasmine and tilia (linden blossom) . . . . The result is like an olfactive interpretation of an imaginary Spring orchard in which everything is simultaneously in flower. Kissing in a French country orchard.[1]

Sadly, I have never been kissed in a French almond orchard, but I am instantly able to imagine and romanticize the sensory decadence of standing with cool grass between my toes, sunlight dappling my face, as I inhale late spring flowers and blossoming almonds. I haven't experienced that moment in reality, but the story and scent allow me to relate, and consequently when I next wear it, that 'memory' almost becomes my own: becoming a part of my identity. In this manner, fine fragrances demonstrate how beautifully scents can be constructed to tell a story, a process that isn't always extended into designing and using scent as part of a brand experience.

Olfactory Design

If you have noticed the smell of baked bread blasting out in front of a supermarket only to find the bread section at the back, you will realize that artificial means are being employed to encourage you to follow your nose. There is a growing olfactory industry offering retailers the opportunity to scent their brands in some way.

As far back as 1965, one of the first industries to employ smell as a deliberate and effective intervention was the car industry. As materials and production processes changed, the Rolls-Royce Silver Cloud lost its unique aroma. In response to complaints, Rolls-Royce 'recreated' the smell (Lindstrom, 2010), a practice subsequently employed by the car industry at large and constantly cited by customers as a positive experience. Other industries have a lot to learn.

Brands spend millions of dollars on creative designers, photographers, directors and writers to develop identities, tone of voice, interiors, campaigns, movies, editorial content, events and ads – all designed specifically as an expression and extension of their brand. The way to achieve an olfactory brand is to approach it in the same way, by commissioning the creation of a bespoke smell or a palette of smells to suit the brand attributes and personality.

Having spent my career in the creative and commercial sectors, it seems bewildering that brands buy 'off-the-shelf' or customized scents with the aim of making a place smell 'nicer.' They have recognized the power of smell and are investing in technology to distribute it, but have not considered that scent should be *designed* as a unique identifier. The reason for this appears to be twofold, attributable to both clients and suppliers.

Clients do not consider themselves to be 'smell-literate'; they do not possess the knowledge or language to brief or judge scent well; and they do not have the correct resources (staff, budgets, support) to manage the process. On the supply side, the ambient scent industry is relatively young. The primary commercial suppliers entered the market with effective distribution machines, maintenance contracts and a library of fragrances. However competent their perfumers, the commercial motive is to sell and distribute stock fragrances alongside the machines and maintenance contracts. Whilst this satiates the desire for a smell, it is not a unique olfactory footprint, and it is not an expression of the brand.

Let us compare that model to established creative service industries such as design and advertising. With foundations built on creativity and professional craft, the industries offer established processes, excellent creative personnel (or 'creatives'), tools, and 'translators' (strategy and account teams who understand both the clients' and the creatives' languages). The necessity and value of the creative work is understood and appreciated at board level, with budgets and staff allocated to manage the process. This creative content is distributed through 'channels.' Currently, in the commercial ambient scent model, the suppliers are currently more akin to 'channels' without creative agencies creating the content.

Arguably, this state of affairs is attributable to the traditional fragrance industry, which has historically been dominated by impenetrable global conglomerates cloaked in mystery (Burr, 2007). Consequently, it is hard to find the right perfumer for a project, and particularly those who will work on individual projects, because the industry standard model is to pay price per kilo (rather than for the creative service) and high minimum order quantities. There are now a few independent creative agencies, working with a handful of perfumers, chosen to suit each client/project, who understand brands, fragrances and implementation, but they are still in the minority.

Designing in Context: Principles of Olfactory Design

There is now more public awareness of the use of scent in commercial environments. The fact that smell can be used to create mood or alter behavior in some way can be concerning to customers, and they are quick to notice 'gimmicky' smells. We do not want to feel unwittingly manipulated (Henshaw, 2014), so when approaching scent for a brand and a physical environment, it is critical to ensure the scent is appropriate for both. Building on research and commercial experience, the design principles my company adheres to are: authenticity, intensity, quality, suitability and multisensory.

Authenticity

Authenticity, as discussed here, is the feeling that something is appropriate, and not contrived. This response can be achieved in various ways: firstly, by allowing the environment to smell of what it actually does, or to 'smell what you see.' One of my favorite shops is Kilwin's, St. Armand's Key, Florida. You are enticed inside by the heady, velvety aroma of melted chocolate: mouth-watering, sensuous and utterly seductive. In the window, chocolatiers stir huge vats of molten fudge and chocolate, pour it onto white marble slabs, cut it in front of you, leaving 'offcuts' in a tasting bowl. This could be perceived as manipulative, but most customers thoroughly enjoy the theater of sensory retail and want to return.

In stark contrast, enter a Thorntons chocolate shop on the UK high street and there is nothing to smell or seduce. Undoubtedly, there is a commercial and quality requirement for Thorntons as a high street chain to package at the point of production, but there are ways of re-introducing smells which could feel authentic and provide a welcome addition to the retail experience. A vat of hot chocolate or a chocolate river would smell divine whilst also adding theater, taste, interaction (and potential new revenue streams).

When the product itself doesn't smell, designing scent appropriate to the brand, product and customers is critical to ensure a sense of authenticity. Jasmine and lemongrass would suit Thai Air, but could be perceived as nauseating, artificial and disingenuous for Virgin Atlantic. Another example is Apple. The 'smell' of a new Apple product is familiar to many customers, even if on a subconscious level. New products have a distinctive scent, and given such a large portfolio of products, manufactured in different factories, sold in countries across the globe, it is hard to believe this is incidental. Community forums describe the smell like plastic, metal, silicon, printed cardboard, paper, plastic sleeve, ozone, factory, rubber and glue. Even if it is artificial, much like the smell of new cars, the scent is consistent with electronics and packaging, and feels authentic. For me, it is synonymous with the brand, triggers positive emotional response, and has become a part of the excitement of a new Apple 'toy.'

Size, function, context, location and season are also all essential considerations, and not only from a technical perspective. Take a trampoline park in a lofty out-of-town warehouse retail context. Firstly, the function of the space needs to be addressed: given the sweaty nature of bouncing, the primary challenges designing for a trampoline park might necessarily include the first three processes of Henshaw's (2014) processes of odor management: separation, deodorization and masking (i.e. air circulation and extraction, heat and humidity control, cleaning, etc.). Having neutralized the environment, consider the scent: florals or fresh grass in an out-of-town warehouse in the middle of winter would seem completely incongruous. More appropriate olfactory cues would be found in smells related to cleaning products, the surroundings and the occupants (rubber, ozone, metal, paint, chewing gum, pheromones). In addition, consider the opportunity to influence mood and performance: adding oxygen and stimulating scents can influence energy levels and mood (Warrenburg, 2005).

In some countries, there is now an anti-olfactory trend: public bans of scent; stories of hospitals and schools being shut in response to smell; fragrance-free organizations, offices, buses and zoning within restaurants. This trend not only nullifies potential offense, but also the potential joy offered by power and mastery of this sense (Fumento, 2000): "The intensity of smells of perfumes and perfumed products was highlighted as a main reason for disliking specific types or brands . . . sometimes provoked reactions such as disgust" (Henshaw, 2014, p. 158).

Intensity

Intensity is effectively controlled by dilution and delivery. Ambient scent can be delivered in various ways, each providing varying levels of control and intensity to suit the space and application (e.g. units in air conditioning, electronic diffusers, candles, reeds, room sprays, heated wax). Smell can sometimes be too powerful and intrusive, and in our experience, implementing smell is most effective when designed as a journey, with silences and crescendos in both time and space. While one hotel may choose to put olfactory units in the hotel-wide air conditioning system to create consistency, a hotel desiring a less intrusive experience might choose candles in specific spaces. Unique olfactory imprints are created when scent is delivered in layers, through experiences, spaces, staff, cleaning products, and the design should address the context as a whole.

Quality

Quality of scent is also a key factor to the success of smell in a commercial environment, and manifests in two forms: ingredients and construction. The majority of products comprising our smellscape are not only manufactured, but are also mass-produced fast-moving consumer goods, such as toiletries, cleaning products and skincare. Their aroma is frequently artificial, pungent and ill-constructed (Aftel, 2001). However, it is interesting to look at the niche fine-fragrance industry to understand the significance of quality. Wide-scale mistrust of globalization has led to the evolution of niche brands with a strong cultural and commercial trend towards quality, hand-crafting, ethics, provenance and authenticity, such as Frédéric Malle, Kilian and Diptyque. More recently, the art of the perfumer, the creator, the alchemist has re-emerged as the subject of wonder, and the quality of ingredients and construction is a trend across the industry. The media coverage, olfactory blogs, online communities and the rapid growth of this niche industry all attest to the altering of the olfactory landscape and consumer olfactory palate, sophistication and expectations.

Suitability

Suitability is another principle which can be challenging for international brands. Beauty is in the nose of the beholder, and scent should always be designed with sensitivity to the audience, but smell can be a very individual and sensory perception, specific to demographics, age, culture and regions (Herz, 2001). For example, generations born in the 1970s and after have a very different nostalgic olfactory landscape (e.g. Play-Doh, crayons, suncream, plastics) to those born in the 1960s or earlier (fresh bread, wood, *Yardley Lavender*). Audiences and cultural references, therefore, need to be carefully considered in the use of olfaction for marketing and branding purposes.

A creative interpretation of this principle is demonstrated by a notorious and controversial example of scent branding. Outside an Abercrombie & Fitch store, the male staff sport topless six packs and low-slung jeans, and a miasma of sickly-sweet scent clouds the entrance. Inside, the lighting is so low it feels like a nightclub, and the interiors are so heavily scented the smell clings to the clothes of customers well after they get home. The company has received much criticism. However, arguably, Abercrombie knew their audience and their approach to smell was in line with their overall brand approach: repel parents and create a counter-culture, like the 'KEEP OUT' sign on a teenager's bedroom. They designed and implemented smell with a clear strategy.

Multisensory

Multisensory design is essential. One of the biggest mistakes when scenting spaces is designing smell without considering the demands on the other senses. If all the senses are stimulated at once, the result is sensory overload. Design should consider the multifarious dimensions of experience: time, space, senses, emotions, function and location.

Conducting a sensory audit before the design process begins ensures that all aspects of the experience have been considered and that odor is being constructed in context. A space is never experienced in a moment, it is a journey, and the senses should be employed at different stages of the journey to support the required function. If you think about sensory design like having a graphic equalizer of the senses, these need to be adjusted in 'volume' relative to each other.

Hotel Costes are masters at this. In the heart of Paris, white marble steps lined with decadent buckets of deep-red roses climb to the unmarked glass-paneled doors, revealing a dark, sumptuous interior. Enter the hotel through the 'sound gallery,' the dark wood reflected in the carefully lit mirrors, and the scented passage brings you in. The plush furnishings in deep vibrant colors and low-level lighting are intimate and sensual. The soundscape is atmospheric and textural, with slow, gentle variance in volume and tone. Curated music tracks fade in and out, converging from all sides. An irresistibly warm, seductive scent dances rose and spices with a sultry wood base. Scent, sound, feel, theater, taste and impeccable design are the foundations of the brand, and have become part of its story. What is more, the hotel's unique olfactory identity was created specifically for the brand, leading into the next, and arguably most important, design principle: *signature*.

Designing Signature Scents

At the heart of a brand is its narrative – its heritage, purpose, promises, people, personality, identity, communications, ambitions and style – and when developing an olfactory brief, the focus should be on the heart of a brand.

I return to The Savoy hotel as a case study. It is an iconic hotel with a remarkable heritage laden with legendary stories, yet it hasn't yet developed its own signature olfactory identity. It continues to buy in 'branded' toiletries such as Miller Harris and LeLabo. Whilst there are benefits in choosing this route (such as brand association and group purchasing), this approach misses critical opportunities: firstly, extending your brand into a more memorable and evocative sensory experience; secondly, the ability to increase revenues and extend brand reach outside the confines of the hotel by selling candles toiletries and so on; and finally, it forgoes the opportunity to ensure that you own and retain the associated brand equity. Thus, my olfactory experience with The Savoy was related to Miller Harris: buy in another brand, and the brand equity and mnemonic has been lost.

What a shame. The Savoy is a brand with storytelling at its very soul. Offering unrivaled luxury, it attracted kings, queens, presidents, prime ministers and maharajas, and became Hollywood's home in London, hosting the likes of Marilyn Monroe, Laurence Olivier, Elizabeth Taylor, Charlie Chaplin, Liberace, Ava Gardner and Lauren Bacall. It possesses a myriad of tales, from The Other Club (a private club for British prime ministers) to Frank Sinatra tinkling tunes in the foyer at 3 a.m., from Fred Astaire dancing on the roof to Walter Hagan driving golf balls across the Thames from the roof. What a rich seam to mine, a veritable cornucopia of evocative olfactory references: Churchill's ashtrays; leather-lined corridors; theatrical greasepaint; Marlene Dietrich's pink roses; afternoon tea; Escoffier's Peach Melba; slings, juleps and cobblers; and Art Deco glamour. Designing a signature scent would bring these stories to life: extending the brand narrative; becoming part of the brand experience; welcoming guests and appealing directly to their emotions, evoking joy, familiarity, affection, nostalgia and the desire to return.

Conclusion

Successful brands of the future will need to be multisensory. Smell can be controversial and elicit polarized opinions. It can inspire disgust and mistrust, but it can also be a delight. It offers unique opportunities for brands to become more emotional, more engaging and more successful, creating much more symbiotic relationships between customers and brands. But the success or failure relies on design and application.

The signature smell and olfactory footprint of every brand should be considered, idiosyncratic and bespoke. Good design is the result of careful thinking and a good brief. As in any creative field, olfactory design should reflect the underlying narrative, carefully consider context, audience, and above all be a thing of beauty.

Working with perfumers and other smell professionals has opened an entirely new dimension to my life. There is little I do now which does not consciously include smell. It has changed the way I read, see and experience the world around me and the way I live. Used elegantly and playfully, scent seduces us willingly with stories and pleasures, adding new dimensions to our sensory landscape and lives.

Note

1 www.jamesheeley.com/en/extrait-de-parfum/28-l-amandiere.html Retrieved March 1, 2017.

References

Aftel, M. (2001). *Essence and alchemy*. London: Bloomsbury.
Booker, C (2004). *The seven basic plots*. London: Bloomsbury.
Burr, C. (2007). *The perfect scent*. New York: Picador.
Drobnick, J. (2006). *The smell culture reader*. Oxford: Berg.
Earls, M. (2002). *The creative age: Bananas, business and the death of marketing*. London: John Wiley.
Ellena, J. (2011). *Perfume: The alchemy of scent*. New York: Arcade.
Fox, K. (1997). *The smell report*. Oxford: Social Issues Research Centre. Retrieved March 1, 2017 from www.sirc.org/publik/smell.pdf
Fumento, M. (2000). Scents and senselessness. *The American Spectator*, April. Retrieved March 1, 2017 from http://fumento.com/mcs/scents.html
Gottschall, J. (2012). *The storytelling animal: How stories make us human*. New York: Houghton Mifflin Harcourt.
Henshaw, V. (2014). *Urban smellscapes: Understanding and designing city smell environments*. London: Routledge.
Herz, R. (2001). *Ah sweet skunk! Why we like or dislike what we smell*. New York: The Dana Foundation. Retrieved March 1, 2017 from www.dana.org/Cerebrum/Default.aspx?id=39206
Hirsch, A. (1995). Effects of ambient odors on slot-machine usage in a Las Vegas casino. *Psychology and Marketing*, *12*(7), 855–594.
Howes, D. (2005). *Empire of the senses*. London: Bloomsbury.
Krusemark, E. A., Novak, L., Gitelman, D. R., & Li, W. (2013). When the sense of smell meets emotion: Anxiety-state-dependent olfactory processing and neural circuitry adaptation, *Journal of Neuroscience*, *33*(39), 15324–15332.
Lindstrom, M. (2010). *Brand sense: Sensory secrets behind the stuff we buy*. London: Kogan Page.
McLuhan, M. (1967). *The medium is the message*. London: Penguin.
Warrenburg, S. (2005). Effects of fragrance on emotions: Moods and physiology. *Chemical Senses*, *30*(Suppl. 1), i248–i249.
Wnuk, E., & Majid, A. (2013). Revisiting the limits of language: The odor lexicon of Maniq. *Cognition*, *131*(1), 125–138.

15

OLFACTORY CONTROL, AROMA POWER AND ORGANIZATIONAL SMELLSCAPES

Samantha Warren and Kathleen Riach

Introduction

In this chapter, we reflect on the practice of aroma management in spaces where people interact, such as workplaces and other institutional settings, considering how issues of control and power may arise – inadvertently or deliberately – within smell practices. The 'smellscape' is an increasingly important part of organizations' physical and cultural design, underpinned by debates in architectural design (Pallasmaa, 2005), marketing and sensory branding (Brumfield & Gouldney, 2008) and the natural sciences (Barker et al., 2003). Through reflecting on a range of recent smell practices predominantly apparent in retail spaces and service sector environments, we argue that the biologically deterministic assumptions upon which these developments are premised – namely, that smell subliminally and predictably 'short-circuits' the brain to influence behavior – silences considerations of power and ethics surrounding hierarchy, control and individual sovereignty. We highlight the role these issues play in the interpretation of organizational olfactory experiences. Our intention is therefore to sensitize designers to the lived experience of smell through an alternative reading of a socio-cultural phenomenon. As such, smell cannot be taken as independent of either professional or occupational norms or the employee bodies that experience, negotiate and themselves contribute to organizational smellscapes.

The Power of Aroma in Organizational Environments

The commercialization of smell (e.g. the realization of added value using olfactory means) has a long history, and continues to be big business at the level of the body through deodorants and perfumes. However, increasingly smell is promoted more broadly as a means of evoking atmospheres conducive to productivity, and used to augment all kinds of products and market exchanges (Classen, Howes & Synnott, 1994). The fields of 'sensory marketing' and 'sensory branding' are gathering momentum, developed from a more general interest in 'experiential marketing' of the late 1990s (Schmitt & Simonson 1997; Schmitt, 2000; Pine & Gilmore 1997; La Salle & Britton, 2003). Subsequently, popular marketing texts, including Martin Lindstrom's *Brand Sense* (2010), Renvoise and Morin's (2007) introduction to neuro-marketing, Brumfield and Gouldney's (2008) treatise on what they refer to as the 'scent revolution' and Derval's *The Right Sensory Mix* (2010) all offer advice about, and accounts of, smell and the consumer. Examples of these practices within the retail and service sector in particular are numerous, and include Harrods, who have trialed a variety of aromas in different areas of their store (Aromaco, n.d.) and the formal clothing chain Pink, which adds the smell of freshly laundered cotton to their atmosphere (Lindstrom, 2010, p. 104). Lindstrom cites the automotive industry, shopping malls and retail outlets as examples where scents have been introduced to manipulate consumer responses. Bespoke scents are also becoming an increasingly prominent element of the 'branding mix,' with scent engineering companies such as The Aroma Company[1] and brandaroma[2] providing organizations the opportunity to create bespoke sensory environments for consumers and other stakeholders. Samsung and Sony have similarly commissioned smells as signature scents for their stores (Tischler, 2005), and it appears that the branding of smells is not limited to physical spaces either; French company Exhalia create and sell software and hardware smell solutions for marketing multimedia. According to their website, these are not just for companies where one would expect smell to be part of the product. For example, the municipal website of the city of Grasse, in France, and *Stations Nouvelle Vague de Bretagne* (a portal for rail travel in the French Brittany region) both have 'scented internet sites' (Exhalia, n.d.).[3] Even the Catalan Socialist Party have introduced a branded scent (Hamilos, 2004). These texts have been accompanied by academic commentary too, such as Hultén, Broweus and Van Dijk's textbook *Sensory Marketing* (2009) and Krishna's (2010) edited collection.

These writers believe that smell has a role to play in the design of consumer lives in ways that are just as important as the design organization and physical arrangement of space, lighting, and acoustics. This in itself is not a new idea: as Corbett (2006) notes, the management of smells via airflows around a building was a pressing concern for architects as far back as the Enlightenment, where *de*odorization was key and "the degree of stench [in a building] became the measure of the architect's efficiency" (p. 335). More recently, this has involved conquering odor by attempting to achieve a blank canvas of 'no' odor, through air conditioning, for example. However, Panos (2001, cited in Malnar & Vodvarka, 2004, p. 281) suggests that the smell of a building is now one of six dimensions which good sensory design should use to enrich users' journey through space beyond the physical mechanics of navigation. "[T]he act of building inevitably involves a metaphysical message – a 'metaphysical experience'" (Pallasmaa, 1980, p. 41), and may include a consideration of 'nostalgia' in the smell around the approach to a building or the enhancement/elimination of smells and differences between the smells of different rooms (Panos, 2001, cited in Malnar & Vodvarka, 2004).

As suggested above, the coordination and control of smells that appear as a by-product or 'naturally' appear in a space is only one component of smell management in terms of design. Smell management entails the deliberate introduction, choreography and manipulation of smell as a resource which will achieve ends beyond the smell itself. For example, Diaconu's (2007) tripartite framework for olfactory design in public space suggests that smell can be 'designed in' cosmetically, as an artificial addition to the structure and/or space: naturally – as would be the case when planting scented gardens, for example – or technically, which involves selecting building materials specifically on account of their smell – such as cedar wood. Examples of aroma management technologies intended to *re*odorize buildings have also begun to emerge. In 1990, a news article reported on Japanese-based Shiseido Co. Ltd's 'aromatic equipment,' which their brochure suggested could deliver a "lemon scent in the morning to wake workers up; a light floral scent to aid concentration at mid-morning; an odorfree [*sic*] lunch, and wood, lemon and floral scents in the afternoon" (Marx, 1990, p. 1). Likewise, Japan's Shimuzu Corporation patented a fragrance system as part of a building's overall design. This delivers doses of different aromas into air conditioning ducts, which one assumes are controlled centrally by facilities managers according to pre-defined quantities and fragrances in a similar way to the Shiseido example above (Gaygen & Hedge, 2009; Classen et al., 1994; Brumfield & Gouldney, 2008). In UK organizations too, a newspaper article published in the year 2000 made reference to the travel company Thomas Cook's provision of a 'sensorama' corridor in its workplace, intended to impart a 'holiday-type' feeling to its call center workers through – among other stimuli – the smell of coconut oil (Williams, 2000). The article focused on how call centers were trying to shake off their negative image as 'dark satanic mills' and one assumes sensorially 'duping' staff into feeling in a holiday mood is an acceptable way to do this. More recently, a primary school in the UK experimented with pumping the scent of peppermint into class to aid children's concentration (Shepherd, 2010).

Such examples suggest that smell is an important resource in workplaces that see their employee behavior as the key to for better results, greater outputs or higher productivity. Brumfield and Gouldney (2008) mention this in relation to workplaces, where they claim:

> There are many objectives that can be met with a scent-based solution . . . our lives are becoming more hectic (not easier, as we once conceived), so anything we can do to create positive attitudes in the workplace should be done. The implementation of aromas in the office can help to accomplish this goal.
>
> *p. 254*

In some senses, this workplace impact is already underway: while practitioner and academic focus has been on smell in marketing, retail and branding, it is important to remember that retail, hospitality and spaces of service provision *are* workplaces for those who are employed there. Others are experimenting with organizational dynamics in smell events themselves. For example, in a recent conference in the United States, *HEADSPACE: A Symposium on Scent and Design*, a scented card was distributed to delegates that was infused with aromas purported to encapsulate different times of the day (for a review, see Tischler, 2010).

And yet there is an irony, in that while those designing our environments are increasingly using aroma as a resource, stakeholders immediately experiencing those environments have for the most part overlooked smell, with other senses such as vision being seen as culturally more important (Corbett, 2006). Public sanitation and hygiene measures have improved over the years, to the extent that we have learned to forget smell and aroma management techniques can often pass unnoticed by those they are intended to shape. This may be part of the reason why, despite olfactory interventions designing the ways we live, they are rarely interrogated in terms of the power or controlling aspects they present.

I Sniff Therefore I Am: Subliminal Olfactory Control

In thinking through the control power behind olfactory interventions, it is important to remember that much of this power is attributed to its subliminal influence. Sensory management, and smell in particular, assumes the ability for aromas to 'short-circuit' the conscious cognitive apparatus of the brain and evoke instantaneous and subliminal emotional responses. Marketers, architects, and increasingly organizations, hope these will form associations or otherwise shape stakeholder behavior. The techniques reviewed above draw their legitimacy from scientific studies aimed at establishing the effect different fragrances and odors have on dimensions of human performance and behavior. Primarily psychological in nature, these studies typically involve experiments to isolate changes in participants' ability to carry out cognitive and behavioral tasks in response to the introduction of aromas in a controlled setting. For example, Zoladz and Raudenbush (2005) found that exposure to the scent of cinnamon and peppermint respectively raised attention, enhancing the operation of working memory and visual-motor response speed. This echoed earlier findings by Barker et al. (2003) establishing that the smell of peppermint improved clerical tasks, including typing, memorization and alphabetization. Ho and Spence (2005) have shown that peppermint scent is a factor in improving the performance of difficult tasks. However, they also found that easier tasks were unaffected by the introduction of the smell, suggesting that certain odors can improve performance only where sustained concentration or complex mental processing are necessary. All of these processes position smell as cognitively determined, with little or no discussion about individual preference or context being a potentially important component of the smell experience. For this reason, they are particularly attractive to organizations, since their (claims) of universalism transcend culture, group or individual and can therefore be applied to entire workforces.

Organizations are also likely to be persuaded by studies exploring the relationship between emotional behaviors and smell. Motomura, Sakurai and Yotsuya (2001) found that lavender reduced stress levels, and Lehrner, Eckersberger, Walla, Pötsch and Deecke's (2000) earlier research showed how anxiety levels of female dental patients were reduced on exposure to the scent of orange oil. In a slightly different study, McBurney, Shoup and Streeter (2006) identify the importance of partners' body scent left on clothing and smelled during periods of separation, while Liljenquist, Zhong and Galinsky (2010) experimented with the 'scent of cleanliness' and found participants more willing to engage in virtuous behaviors – trust, reciprocity and charity – in a room smelling of a freshly sprayed citrus cleaning product. Relationships with other people are also mediated through smell. Zemke and Shoemaker (2006) showed how a pleasant ambient smell promotes social interaction among strangers, while Moallem, Paller and Gottfried's (2007) research determined a positive effect of smell on social preferences when viewing photographs of other people, although they found that whilst *subliminal* odors shaped perceptions of others favorably, conscious and/or stronger smells had disruptive effects – making impressions hard to form. Yet such studies neglect larger questions that face organizations surrounding the ethics of manipulating people when broader circuits of power, politics and hierarchies exist. For example, there is no discussion of what may be seen as the sovereignty of emotions – in other words, who has the 'right' to provoke or evoke these feelings in individuals, and for what ends.

These connections between smell and behavior are far from infallible. Phillips and Cupchik (2004) noted a link between inhaling scents while reading and the reader's engagement with the text and subsequent recall, but only when the smell was congruent with the reader's liking for, or identification with, the events of the story. Gaygen and Hedge (2009) did not find any effect of smell on participants' lexical decision performance. In fact, in the second experiment they conducted, a fragranced room was found to *detract* from decision-making ability. They assumed smell was a distractor, as their research participants tried to work out what it was and where it was coming from – consciously or otherwise. Yet we do not see much evidence of this imperfect relationship considering different types, intensity or contexts of smell as an inherent part of smell experiences for consumer, employee, patient, student or other groups. If different scents have different effects on different tasks, then how does one manage the diffusion of aromas to the myriad of tasks undertaken in complex organizations, for example? Furthermore, the studies above are set up to investigate cause and effect, rather than how smell is perceived by the smeller. How do they make meaning from the scent, based on their manifold personal, biographical, psycho-social 'smell history'? In summary, the basis for a majority of aroma management techniques may fail to acknowledge the broader questions surrounding power and control inherent in subliminal influence, and may also not take into account the historical, biographical and cultural contexts that interplay with more cognitive processes surrounding smell.

Between You and Me: (Re-)socializing Smell as a Political Intervention

Given that aroma management is only likely to increase in popularity as new technologies enable the facilitation and control of smell over time and space, it is likely organizations will increasingly explore smell practices within the workplace. It is therefore important to begin considering what the consequences may be for those who are living in these practices every day. This is especially so given that smells are not bounded by physical structure, floating on whatever current of air catches them and therefore permeating architectural and symbolic boundaries despite people's best efforts to contain them (see Riach & Warren, 2015). By way of starting this debate, we outline three areas where olfactory practices are pertinent to larger debates surrounding organizational control.

First, there are very real health and safety considerations. Concern has already been raised about the 'chemical cocktails' that we associate with common aromas such as 'new car smell' (Travers, 2016), and these dangers may be exacerbated when any environment is deliberately altered. Gaygen and Hedge (2009) note that building fragrancing systems utilize complex mixtures of volatile organic compounds (VOCs) to synthetically produce aroma. These can cause the irritation, allergic reaction and other adverse health effects that have been linked to the so-called 'sick building syndrome.' As Wolkoff, Wilkins, Clausen and Nielsen (2006) state, based on research into the effects of VOCs:

> Certain odors, in addition to odor annoyance, may result in psychological effects and distraction from work. Some building materials continually cause perceivable odors, because the odor thresholds of the emitted compounds are low. Some oxidation products of alkenes (e.g. terpenes) may contribute to eye and airway symptoms under certain conditions and low relative humidity.
>
> *p.* 7

This indicates that the introduction of smell to workplaces may be hazardous to employee health, apart from the 'distraction from work' and 'annoyance' that is also recognized here. In support of this contention, albeit anecdotally, during the release of an ambient scent at the *HEADSPACE* conference discussed above, one delegate tweeted that she had to leave the event early as the smells were making her feel nauseous (Twilley, 2010). Similarly, as Worwood (1995) somewhat wryly observes, despite science starting to recognize what aromatherapists have been known for years (that smell has real effects on the body and mind), scientists are advocating artificial solutions, "where even more man-made, synthetic, non biodegradable chemicals will be pumped into the atmosphere" (p. 25). This point is even more significant as we move towards buildings where people work with centralized air control systems through, for example, air conditioning or windows that cannot be easily opened, and subsequently do not allow individuals control of the immediate organizational smellscape.

Relatedly, such biological concerns also raise the question of whether spaces where people are required to be for a non-smell-related purpose – such as the workplace – are an appropriate arena for aroma management in a social sense – natural or otherwise. Even without direct management olfactory intervention, workplaces are a political hotbed for smell politics. In a recent study we undertook to explore what might be at stake *socially* in the manipulation of office smellscapes (Riach & Warren, 2015), we found that smell may be one aspect of the communal experience of work, binding workers together. However, it could also differentiate individuals on account of ethnicity (e.g. what constitutes a 'smelly' lunch) or identity (e.g. what perfumes a woman of a particular age should wear to work). As such, smells were experienced through organizationally dominant normative or exclusionary ideas of what work 'should' be and what was deemed a 'professionally' acceptable smell. While Tan (2013), in her study of smokers in urban spaces, advocates for such antagonistic olfactory politics to be retained rather than making way for "antiseptic urban environments" (p. 68), in workplace environments there are always unequal power relations that have the potential to reproduce difference and inequality in and through smell practices. If you are the person deemed to have 'the smelly lunch,' then this not only carries symbolic, but material effects on your everyday lived experience of work and perceptions of 'professionalism.' It is likely, then, that explicit managerial intervention may only serve to further reproduce narrow idealizations of the 'model worker' in terms of what organizational smell practices are acceptable.

Finally, our study also leads to us to question the extent to which individuals *can* be fully duped by aroma management techniques. Some scientific research has already critiqued an a-contextual approach to investigating smell. For example, Kirk-Smith and Booth (1987) argue that perception of fragrance is largely a product of the situation in which it is encountered, and the odor is only one ingredient in the perceiver's response. Elsewhere, Dalton (1996) explains how much of the perception of smell is not a biologically determined 'top-down' process, but 'top-down' in that we make sense of smell according to conceptual schema and accumulated experience. It is this last point that Gaygen and Hedge (2009) reinforce in their observation in regards to the workplace that is of specific concern to us here:

> [Our] result does not support our original hypothesis that the exposure to a pleasant but unidentifiable fragrance will facilitate performance on a lexical decision task and challenges simple assumptions that the introduction of fragrant air into a building will necessarily improve worker performance.
>
> *p. 89*

Such accounts are valuable in reframing the politics of smell in organizational environments, not only through questioning their efficacy, but also allowing for a discussion of workers as individuals, who themselves are infused with various biographies, ideas, emotions and beliefs that allow them to creatively navigate the environments they find themselves in. This was just what Law (2001) found to be happening in her study of the 'sensory landscapes' created by Filipino women workers' weekly gatherings in a Hong Kong square. Law (2001, p. 279) eloquently shows how "a complex [sensory] articulation of nostalgia and desire" created situated resistances against dominant personal and social identities ascribed to the women on the basis of their economic and social status as domestic workers. In relation to workplaces, smell may be part of the ways employees actively connect with other workers, help transition between different tasks and duties, or simply relieve boredom during a long working day. It may be, then, that aroma management is a means of creating an environment conducive for work, but only when employees are considered as more than neural receptors, and consulted about their own ideas and beliefs surrounding the organizational smellscape.

Conclusion

The prospects for designing smellscapes related to organizational workspaces are undoubtedly exciting and appeal to a broad range of architects, marketers and organizations. All of these professions and agencies are keen to realize added value from their stakeholders' capacity for sensory engagement. As we have suggested in this chapter, the use of smell in organizational contexts takes varied forms and is likely to increase given technological developments and more sophisticated neuro-biological understandings of how aroma 'works on' the brain and psychological processes. However, it is perhaps the intersection between biological and socio-cultural experiences of smell that provides greater opportunities for not only advancing the effective use of smell as a resource, but also allowing a discussion of the consequences and implications surrounding power and control as well as the possibilities presented through aroma management practices. Smell both enters the body as physical particles and has strong personal and cultural resonances. Reflecting on these dimensions and how they are entangled with each other marks an important step for smell designers. Rather than conceptualize smell as acting *on* bodies in space, we would urge designers to see bodies as actively doing things *with* smell.

Notes

1 www.aromaco.co.uk/ Retrieved March 1, 2017.
2 www.brandaroma.com Retrieved March 1, 2017.
3 These technologies work using software that 'tells' a scent diffuser connected to the website visitor's computer what configuration of scent should be released. Thus, presently, extra equipment is needed for the process to work: both a diffuser and the concentrated aroma refills. However, given the speed at which technology develops, we can probably expect scent diffusers to be standard features of computers and other technologies in the future. Sony Ericsson (now Sony Mobile Communications Inc.) have already experimented with an aroma-releasing phone to reduce stress (Bielinis, 2007), although, at the time of writing, there was no sign of such a product advertised on their website, suggesting it was not a popular addition to their portfolio.

References

Aromaco. (n.d.). Harrods. Retrieved March 1, 2017 from www.aromaco.co.uk/portfolio/harrods/

Barker, S., Grayhem, P., Koon, J., Perkins, J., Whalen, A., & Raudenbush, B. (2003). Improved performance on clerical tasks associated with administration of peppermint odor. *Perceptual and Motor Skills*, *97*, 1007–1010.

Bielinis, S. (2007, January 23). NTT DoCoMo/Sony Ericsson SO703i Aroma Phone. Retrieved March 1, 2017 from www.unwiredview.com/2007/01/23/ntt-docomosony-ericsson-so703i-aroma-phone/

Brumfield, C. R., & Gouldney, J. (2008). *Whiff! The revolution of scent communication in the information age*. New York: Quimby.

Classen, C., Howes, D., & Synnott, A. (1994). *Aroma: A cultural history of smell*. London: Routledge.

Corbett, J. M. (2006). Scents of identity: Organisation studies and the cultural conundrum of the nose. *Culture and Organization*, *12*(3), 221–232.

Dalton, P. (1996). Odor perception and beliefs about risk. *Chemical Senses*, *21*, 447–458.

Derval, D. (2010). *The right sensory mix*. New York: Springer.

Diaconu, M. (2007). Olfactory design: Strategies and applications. Paper presented at *Aesthetics Bridging Cultures: XVIIth International Congress of Aesthetics*, organized by SANART in collaboration with the International Association for Aesthetics, Ankara, July 9–13

Exhalia. (n.d.). Genesis of the project: The "fragrant web" of France Telecom. Retrieved March 1, 2017 from www.exhalia.com/us/societe/

Gaygen, D., & Hedge, A. (2009). Effect of acute exposure to a complex fragrance on lexical decision performance. *Chemical Senses*, *34*(1), 85–91.

Hamilos, P. (2004, January 14). Aaah! The sweet smell of socialism. *The Guardian*. Retrieved March 1, 2017 from www.theguardian.com/world/2008/jan/14/spain.paulhamilos

Ho, C., & Spence, C. (2005). Olfactory facilitation of dual-task performance. *Neuroscience Letters*, *389*(1), 35–40.

Hultén, B., Broweus, N., & Van Dijk, M. (2009). *Sensory marketing*. London: Palgrave.

Kirk-Smith, M. D., & Booth, D. A. (1987). Chemoreception in human behaviour: Experimental analysis of the social effects of fragrances. *Chemical Senses*, *12*(1), 159–166.

Krishna, A. (2010). *Sensory marketing: Research on the sensuality of products*. London: Routledge.

La Salle, D., & Britton, T. (2003). *Priceless: How to turn ordinary products into extraordinary experiences*. Cambridge, MA: Harvard Business School Press.

Law, L. (2001). Home cooking: Filipino women and geographies of the senses in Hong Kong. *Cultural Geographies*, *8*(3), 264–283.

Lehrner, J., Eckersberger, C., Walla, P., Pötsch, G., & Deecke, L. (2000). Ambient odor of orange in a dental office reduces anxiety and improves mood in female patients. *Physiology & Behavior, 71*(1), 83–86.

Liljenquist, K., Zhong, C., & Galinsky S. (2010). The smell of virtue: Clean scents promote reciprocity and charity. *Psychological Science, 21*(3), 381–383.

Lindstrom, M. (2010). *Brand sense: Sensory secrets behind the stuff we buy*. London: Kogan Page.

Malnar, J. M., & Vodvarka, F. (2004). *Sensory design*. Minneapolis, MN: University of Minnesota Press.

Marx, G. T. (1990, July 6). Bosses should nix job in-scent-ives. *Newsday*. Retrieved March 1, 2017 from http://web.mit.edu/gtmarx/www/smell.html

McBurney, D. H., Shoup, M. L., & Streeter, S. A. (2006). Olfactory comfort: Smelling a partner's clothing during periods of separation. *Journal of Applied Social Psychology, 36*(9), 2325–2335.

Moallem, I., Paller, K., & Gottfried, J. (2007). Subliminal smells can guide social preferences. *Psychological Science, 18*(12), 1044–1049.

Motomura, N., Sakurai, A., & Yotsuya, Y. (2001). Reduction of mental stress with lavender odorant. *Perceptual and Motor Skills, 93*(3), 713–718.

Pallasmaa, J. (1980). The two languages of architecture: Elements of a bio-cultural approach. In A. Salokorpi, J. Pallasmaa, & M. Kärkkäinen (Eds.), *Yearbook 2 – Abacus* (pp. 57–90). Helsinki: Museum of Finnish Architecture.

Pallasmaa, J. (2005). *The eyes of the skin: Architecture and the senses*. Chichester: John Wiley.

Phillips, K., & Cupchik, G. C. (2004). Scented memories of literature. *Memory, 12*(3), 366–375.

Pine, J., & Gilmore, J. (1997). *The experience economy*. Boston, MA: HBS Press.

Renvoise, P., & Morin, N. (2007). *Neuromarketing: Understanding the buy buttons in your customer's brain*. Nashville, TN: Thomas Nelson.

Riach, K., & Warren, S. (2015). Smell organisation: Bodies and corporeal porosity in office work. *Human Relations, 68*(5), 789–809.

Schmitt, B. (2000). *Experiential marketing: How to get customers to sense, feel, think, act, relate to your company and brands*. New York: Free Press.

Schmitt, B., & Simonson, A. (1997). *Marketing aesthetics: The strategic management of brands, identity and image*. New York: Free Press.

Shepherd, J. (2010, April 24). School tests peppermint aroma to help students concentrate. *The Guardian*. Retrieved May 23, 2017 from https://www.theguardian.com/education/2010/apr/23/peppermint-boosts-students-concentration

Tan, Q. (2013). Smell in the city: Smoking and olfactory politics. *Urban Studies, 50*(1), 55–71.

Tischler, L. (2005, August 1). Smells like brand spirit. *Fast Company*. Retrieved March 1, 2017 from https://www.fastcompany.com/53313/smells-brand-spirit

Tischler, L. (2010, March 29). Sweet stank of innovation wafts across Parsons' scent symposium. *Fast Company*. Retrieved May 23, 2017 from https://www.fastcompany.com/1600543/the-scent-of-sexy-time-and-other-smelly-insights-from-parsons-scent-symposium

Travers, T. (2016, March 15). Is new-car smell bad for your health? Retrieved March 1, 2017 from www.bbc.com/autos/story/20160315-is-new-car-smell-bad-for-your-health?

Twilley, N. (2010, March 30). Ginger biscuits and deodorant guns. *Edible Geography*. Retrieved March 1, 2017 from www.ediblegeography.com/ginger-biscuits-and-deodorant-guns/

Williams, E. (2000, September 3). "Satanic mills" see the light. *The Independent*.

Wolkoff, P., Wilkins, K., Clausen, P. A., & Nielsen, G. D. (2006). Organic compounds in office environments – sensory irritation, odor, measurements and the role of reactive chemistry. *Indoor Air, 16*(1), 7–19.

Worwood, V. (1995). *The fragrant mind*. Reading: Bantam.

Zemke, D., & Shoemaker, S. (2006). A sociable atmosphere: Ambient scent's effect on social interaction. *Cornell Hospitality Quarterly, 49*(3), 317–329.

Zoladz, P., & Raudenbush, B. (2005). Cognitive enhancement through stimulation of the chemical senses. *North American Journal of Psychology*, 7(1), 125–140.

PART V
Smell Learning Environments
INTRODUCTION

Kevin Thwaites

The standout quotation from these four excellent chapters for me is this: "One simply cannot turn up one's nose these days about the role of scent in design" (Osman, Noppeney and Endrissat, Chapter 17, p. 173). In this respect, this part paints a vivid picture not only of the importance of a resurgence of scent awareness to a fuller appreciation of the human capacity for creative expression and environmental understanding, but also of the potentially pivotal role that 'smell study' has in bridging our currently obstructive disciplinary barriers.

Writing as an educator and researcher in the built environment disciplinary sphere, it is hard to overstate the importance of this. An increasingly urbanizing world brings with it an urgent societal need to develop better understandings of the nature of human–environment relationships within complex, dynamic urban settings. However, the quest for a more humanistic city, as Henshaw points out, is currently hindered by a prevailing failure to recognize the intrinsic connectivity of bodily experience and thinking processes in human–environment relations. Learning in general, and architectural and urbanist learning in particular, remains limited by a separation of rational and emotional experience: essentially the outcome of a persistent Enlightenment philosophical tradition increasingly seen as inadequate for dealing with contemporary human–environment challenges.

The need for a shift in mindset capable of challenging the dominance of the visual and overcoming a professional loss of connection with the human body in design decision-making is a theme set by Victoria Henshaw in Chapter 16 and recurrent in all the chapters in this part. A significant dimension in such a mind-shift is said to require changes to the traditional patterns of learning and the environments in which learning takes place that can re-introduce, rather than exclude, emotional and corporeal forms of knowing. Henshaw illustrates, through examples of her own educational practice, how this might begin to be facilitated by means of introducing 'micro-smell sources' into the lecture theater and through guided 'smellwalk' workshops which afford participating students 'a more experientially informed learning opportunity.' In this sense, making explicit the sense of smell in learning environments is associated with a release from the constraining impact of traditional rational forms of learning, enabling students instead to draw from more intuitive forms of knowledge. In so doing, they are then able to recover a full-bodied approach to learning that Henshaw believes is important to the delivery of more humanistic city design.

Osman, Noppeney and Endrissat (Chapter 17) then echo this point from the perspective of art and design education specifically. Like Henshaw, they also highlight the privileged position that the sense of vision (and sound) maintains in the art and design world, whilst pointing out that this is increasingly becoming out of step with recent trends in the culturalization of scent through its portrayals in literature and the establishment of sensory research centers. They argue that if scent is becoming an increasingly significant part of culture, then the art and design disciplines and their educational foundations must embrace scent more explicitly in order to retain compatibility with contemporary culture. They advocate progress in the educational establishment in the form of new programs that can elevate scent education from its current technical and fragmented roots in the perfumery industries to that of a broader intellectual field of sensory studies.

What is interesting here, in relation to Henshaw's concerns about the obstructive impact of a prevailing duality in rational and emotional forms of learning and the extent to which this reflects our currently self-referential and discrete disciplinary infrastructure, is that scent has become more successfully embedded in academia via trans-disciplinary programs. This opens up the intriguing possibility that scent study, although hard to accommodate into traditional divisions of human knowledge, may possess unique qualities that allow it to transcend disciplines.

The trans-disciplinary potential of scent is also evident in Chapter 18 by Norman and Pozzani, who demonstrate through three case studies ways in which scent-based educational provision can become operationalized in various aspects of art and design education. They show that when students are encouraged to experience fragrance as an art form, unique forms of creative expression can emerge. Their work also highlights that when programs of olfactory study are offered, they prove to be highly popular, attracting broad-ranging interest in studying and exploring scent as part of a palette of tools in the inspiration and generation of creative work. Again, this takes a distinctly cross-disciplinary form, demonstrating the capacity of scent to act as a catalyst in drawing together different art and design fields into more unified forms of investigation and creative expression.

Melanie McBride and Jason Nolan (Chapter 19) conclude the part by returning the focus of attention back to Henshaw's main concerns about the role of smell in the education and training of built environment professionals, contributing a pedagogically oriented conceptual framework that can support them. It is pointed out that smell is particularly significant for built environment professionals concerned with the design and delivery of public spaces. In this context, it is argued, smell awareness becomes a matter of socio-cultural significance. This seems a highly significant point, which stems from the capacity of smells in the public realm to influence how individuals perceive and use it in relation to individual predisposition, social and cultural traditions and preferences. As other chapter authors have highlighted, however, design of public space is almost exclusively driven by visual factors, operationalized through use of two- or three-dimensional imaging techniques and tools depending on scale and context. Olfactory considerations rarely form part of the planning and design process of public realm, other than in situations where environmental odors are captured by specialized equipment. As McBride and Nolan point out, however, this cannot account for the subjective dimensions of smell, which are inextricably linked with authentic lived experience and environmental context.

If, then, built environment professionals need a more explicit understanding of the olfactory dimensions of the environment in the interests of delivering socially inclusive public realm, how might that be understood in light of its inherent subjectivity? McBride and Nolan propose that this lies with taking a more embodied and humanistic approach to engage with principles and practices for designing with smell. It is suggested that this involves a five-point pedagogically oriented framework of considerations for intersensory design. Overall, this framework invites participants into a critical reflection of what influences attitudes and assumptions about smell in ourselves and others, followed by smell-focused tasks that raise awareness of the possibility of smell as a mode of inquiry, the embodiment of sensory knowledge and its communication throughout the professional community. Taken as a whole, such a pedagogical framework of intersensory design, with its embodied, holistic and experiential emphasis, makes a highly significant point about the potency of smell awareness that takes it into wider philosophical territory, beyond its importance in the art and design world.

What emerges here, and indeed is evident to a greater or lesser extent in all the chapters in this part, is that those who study and explore smell in relation to spatial experience and creative expression seem to do so from an innate and commonsense understanding that smell cannot be detached from the broad and holistic spectrum of human–environment relations. This stands in stark and significant opposition to the way in which much of the built human–environment realm is conceived, designed and made. As Henshaw points out, this is in large part a consequence of an Enlightenment philosophical position of rationalistic duality and separation. If, as would seem to be the strong implication throughout this part, new mindsets are required, then the whole-body implications that are intrinsic to intersensory inquiry may very well be signaling the direction.

Perhaps, therefore, those of us looking for more holistic human-oriented forms of place-making better able to integrate the spatial, social and ecological dimensions of human–environment relations, an increased awareness of the potency and cross-disciplinary nature of smellscape study, may well offer fertile ground for progress.

16

EXPERIENTIAL LEARNING AND OLFACTORY ARCHITECTURES

Accommodating Smell in Teaching Practices in the Built Environment

Victoria Henshaw

Introduction

Experiential learning, or put simply, the act of learning or creating meaning by doing, has existed in human societies for many thousands of years, used prior to the written word as a way of passing on skills, knowledge and understanding from generation to generation. However, throughout the 19th and 20th Centuries, learning in the higher education (HE) environment became more commonly associated with a behaviorist approach involving a teacher who passed on knowledge to a learner in a classroom environment, with much of the learning thought to occur within the mind of the learner and illustrated in resulting behaviors or outcomes (see, e.g., the work of behavioral theorists such as Pavlov, 1903; Watson, 1930; Skinner, 1935).

Experiential learning was re-introduced to the HE environment in the 20th Century as a theory of learning, or 'philosophy of education' as it is termed by Kolb and Kolb (2005, p. 193), and can be drawn upon by teachers from across the disciplines as one way of enhancing student learning. However, experiential learning theory as it is currently understood is criticized by Michelson (1998) as separating thinking processes from human experiences of real life:

> Like many Western knowledge-practices, experiential learning is seen to require what is, in effect, an act of self-dismemberment, a splitting of the head from the body that denies the connectedness of knowledge to the body and thus loses sight of knowledge as a product of corporeally and emotionally grounded life.
>
> *p. 217*

In this chapter, I draw from the work of key theorists to explain the theoretical underpinnings of experiential learning and to evaluate Michelson's arguments with respect to the use of experiential learning within my own disciplinary field of architecture and urbanism. Moreover, I will focus upon a specific aspect of Michelson's criticisms of experiential learning theory in relation to my own research-led teaching area of sensory environmental design, and will outline examples of learning opportunities where I have encouraged learners to draw from their experiences of smell in the built environment. In doing so, I seek to reveal current limitations in experiential learning theory and highlight opportunities presented within architectural teaching practices to better recognize the multisensory nature of experiential learning.

Experiential Learning

Theories of experiential learning as an educational concept are founded in the work of theorists such as pragmatic philosopher John Dewey and psychologist Kurt Lewin, both of whom acknowledged the role of the experiences of the individual learner in creating new ideas and knowledge. Dewey (1963) highlights a relationship between motivation (or impulse), observations, knowledge and judgments in informing future practice. Lewin (1947) similarly progresses his theories of learning as being informed by observations and reflections upon concrete experience in developing and testing new ideas, with the resulting insights being used to inform new experiences. Perhaps the theorist most closely associated with experiential learning is Kolb (1976, 1981, 1984; Kolb & Kolb, 2005), who draws from both Lewin and Dewey, as well as Piaget's (1970) model of cognitive and learning development. Piaget's model is based upon the division of human learning into four different stages, ranging from that of infancy, where the child learns directly from a phenomenon such as an experience of a particular sight, sound or taste, through to an internal reflexive way of knowing that is experiences in adulthood. In drawing from these models, Kolb (1984, p. 30) argues that experiential learning is best thought of as a process rather than an outcome, and suggests that effective learners require four different kinds of abilities which he uses in proposing his own experiential learning cycle. He identifies these four abilities as: concrete experience, reflective observation, abstract conceptualization and active experimentation. In contrast to Piaget, who proposes a developmental process with one stage leading into the other (Piaget was, after all, highlighting learning styles according to childhood development as opposed to focusing upon adult learners), Kolb considers each of these four learning abilities as being necessary for effective learning.

Illeris (2002, 2003) similarly references Piaget as an influence in developing his Theory on the Processes and Dimensions of Learning, including three learning dimensions: cognition, emotion and society. Illeris (2003) argues that in order for learning to occur, two processes must be in place: *interaction* processes between the individual and the environment, identified as primarily social and cultural in nature, and inner mental *acquisition* processes, argued as psychological, enabling individuals to mediate cognitive and emotional dimensions. Illeris describes the cognitive side of the acquisition process as the rational dimension of learning, where knowledge, skills and understandings are formed, with the emotional dimension including the development of motivations and attitudes. Illeris suggests the relationship between these two dimensions changes as the individual develops, and outlines: "During the pre-school years the two sides of the acquisition processes gradually split away from each other, but they are never totally separated" (p. 170).

Honey and Mumford (1982) add an additional perspective, suggesting that learners have different learning styles, and identify these as the activist, theorist, pragmatist and reflector. They identify attributes associated with each of these styles, and suggest different activities that might stimulate learning accordingly. Like Kolb, Honey and Mumford argue that effective learners require abilities in all four areas, but suggest there to be innate preferences for particular learning styles deep within us all.

However, Michelson (1998) argues there to be a strange dualism in these theories of experiential learning, in that they are tied both to "the empiricist valorisation of the evidence of the senses" and to the "pragmatist regrounding of learning in 'real life,' and thus to the belief that bodily experience is indeed material for knowledge" (p. 218). To explain Michelson's point in further detail, it is useful to look back in scientific thought to understand the epistemological context within which theories of experiential learning were developed. The Enlightenment and post-Enlightenment periods witnessed a move towards a more civilized and humane society, placing an emphasis on the value of rational and objective thought. In effect, society sought to differentiate humankind from the wider animal kingdom, and as part of this process, prioritized the 'noble' senses of sight and hearing above the 'lesser' senses of smell, touch and taste. These internal bodily chemical senses were viewed animalistic by figures such as Hegel and Freud (Mavor, 2006), and as a result, their role as a valid means of knowing, understanding and thus experiencing the world was rejected by science in favor of the detached external sensorium (Classen, Howes & Synnott, 1994; Drobnick, 2006). The Enlightenment period also brought a view of the natural role of women as being determined through physical and biological factors, with the chemical senses considered feminine and 'witchlike' (Classen, 2005) and sight and sound as more reasoned masculine senses.

As part of Enlightenment thinking, the place of science became the sterile, unpolluted and objective environment of the laboratory, with a focus upon what could be observed. This emphasis upon objective observation contributed towards the rise of behaviorist theories of learning in the early 1900s. Although experiential learning signifies a move away from a positivist/empiricist approach towards a more human-centered pedagogy, the early theories of Dewey and Lewin maintain a focus upon observation; it is, however, the observations of learners upon their own experiences and their resulting modifications to their own knowledge, and thus resulting learning that occurs, which is central. Michelson (1998) further argues that experiential learning theories separate thinking processes from bodily experiences as though the two occur independently of one another in what the feminists term 'an abstract masculinity.' She argues that instead, experience should be understood as "the place where personal and social meet" (p. 227), and rather than being excluded from the classroom, emotional and corporeal knowing should be incorporated into learning.

Experiential Learning in Architecture and Urbanism

In my own disciplinary field of architecture and urbanism, experiential learning by necessity plays a central role in developing the essential skills and knowledge of future architects and city designers. Architecture is defined as "the art or practice of designing or constructing buildings" (Oxford Living Dictionaries, 2017), and therefore involves a combination of both the real and the imaginary. Over six years, students combine taught lectures, where they learn architectural construction, traditions and regulation, with studio practice, where they hone design and creative skills. This organized learning, some of it teacher-led, but much of it led by the learners themselves, is supplemented by a one-year placement in architectural practice, required by the Royal Institute of British Architects before an individual can be fully qualified as an architect in the UK.

Learners at the Manchester School of Architecture – where I was employed before joining the Department of Town and Regional Planning at the University of Sheffield – are also required to undertake research studies and analysis out in the city itself, with the built environment replacing the classroom for such learning. This is based upon the principle that to truly understand buildings or the urban environment, one cannot afford a distance between the focus of the enquiry and the environment within which that enquiry is carried out. In learning architecture and urban design, the built environment, and frequently the city itself, undertake a dual role, becoming both subject and the site for learning (for a fascinating examination into the use of the city as site for knowledge by the Chicago School of Urban Studies, see Gieryn, 2006). As an important part of their learning experience, students are encouraged to visit, study and evaluate buildings and public spaces in towns and cities around the world. However, just as experiential learning theories have been accused of displaying an ambivalent relationship with bodily experience as a valid site for knowledge, everyday architectural practices have been accused of failing to look beyond visual aspects of architectural design. As Sennett (1994) described in his examination of the place of the body in Western civilization and cities:

> When I first began to explore sensory deprivation in space, the problem seemed a professional failure – modern architects and urbanists having somehow lost an active connection to the human body in their designs. In time I came to see that the problem of sensory deprivation in space has larger causes and deeper historical origins.
>
> *p. 16*

It is this failing that I have attempted to respond to in my own research-led teaching practice of recent years, and I do so specifically with respect to the sense of smell. In the following section, I outline some of those methods, techniques and tools I have used in introducing odor into the learning experience. I will later go on to explore the position of these in relation with experiential learning theory.

Methods, Techniques and Tools for Introducing Smell into Learning

Smell in the Lecture Theatre

In addition to learning sessions undertaken in the studio, I deliver more traditional lectures on sensory urbanism and sensory environmental research methods as a contributor to a range of modules, such as architectural research methods, and to student year-groups of undergraduate and postgraduate levels. I do so in large lecture theatres, and occasionally to groups of 100-plus students. In these instances, I deliver slide presentations highlighting the dominance of vision in current practices and advocate consideration of the wider senses, and odor specifically, in architectural analysis, conception and practice. Yet it has always struck me as ironic that in using words and images, this vocally orientated style of teaching, to use Kolb's (2015) term, is symptomatic of the very abstracted masculinity of traditional learning and architectural practices which I seek to address. In response, I have explored different methods of introducing odor into the learning experience, which, given the constraints of most indoor, enclosed lecture theater or classroom environments, is not without its complications.

Within these scenarios, I use what might be termed here 'micro-smell sources': small-scale localized and predominantly containable learning aids, including: *odorous substances* in pots, such as manufactured 'smellpots' produced by sensory suppliers and targeted at children with sensory disabilities, or pots that I fill with different items such as vegetation, foodstuffs or products; *perfumes and essences*, some being place-related odors that I have created myself, described in further detail below; and *scratch-and-sniff* books. Scratch-and-sniff is a process involving the micro-encapsulation of odor molecules onto paper (Charbonneau & Relyea, 1997), and is most commonly used in both the commercial marketing industry and the creation of books for children. It is rarely used in the production of books for adults or, by implication, adult learners, with the exception of a small number produced to accompany rare exhibitions of odors, such as the *If There Ever Was* exhibition at the University of Sunderland (Blackson, 2008).

When teaching smaller groups in the lecture theater or studio, I also use these tools in the stimulation of two-way dialogue. I ask learners to discuss their own experiences of place- or building-related odors, to identify their most and least favorite odor preferences and to discuss what they think about their own sense of smell.

In using these odor tools in the teacher-led environment of the lecture theater or classroom, I aim to facilitate opportunities for learners to make a connection between the materials I am presenting visually and aurally with their own daily experiences and intuitive knowledge gained through their sense of smell. In doing so, I seek to provide a powerful means of encouraging learners to understand the relevance of this sense to their work, and to begin to engage with it in their thinking.

Smell in the City and Studio

In an effort to provide the students with a more experientially informed learning opportunity, I have also organized and delivered workshop sessions, including 'smellwalks' with students in the city. In these sessions, I lead students on a pre-planned walk in the city center, where we explore the odors detected by learners and investigate whether these meet their expectations of the urban olfactory environment. To support this activity, I usually provide participants with maps of the smellwalk route and ask them to note down any smells detected. Where feasible, I follow up the walks with an organized presentation and design session back in the studio, where learners are asked to reflect upon their experience and produce their own work in response.

Occasionally, these learning sessions involve learners in my own research-driven activity, focusing upon and attempting to resolve in some way the complex challenges faced when dealing with smell, including those encountered when attempting to classify, define and represent odors. In May 2012, I organized an intensive one-day workshop session in Manchester entitled *Representing Smell in the City*, as part of a wider project exploring *Smell and the City*.[1] The workshop was advertised across universities in the North of England and to students of a range of disciplines, including built environment, the social sciences, arts and natural sciences. As such, it included 20 learners and six academic staff from eight different universities, with learners ranging from first-year undergraduates through to final-year doctoral candidates. All learners were tasked with the explicit goal of investigating and developing ideas for capturing, recreating and representing smells in, and of, the city.

The session commenced with a smellwalk, with learners exploring and making note of the odors they detected in city streets and spaces. Following the smellwalk, the teaching staff delivered short presentations back in the studio on relevant factors to be considered, including the psychology of smell, environmental smell perception and artistic representations of smell. The students were divided into four inter-disciplinary groups, provided with a maximum budget of £25 per group and allowed three hours to develop their ideas before presenting back to the wider group and teaching staff from across the Manchester School of Architecture. Each group was allocated a tutor, with whom they could consult as they wished, and set about resolving their task as they saw fit.

The four groups developed and presented very different experimental results. The first focused upon synesthetic elements of sensory experience by conducting a number of experiments. One of these included the collection of a water sample from the Manchester Ship Canal, with this being placed it in a covered container and workshop participants asked to identify what color they associated with the liquid using their sense of smell alone. The results varied, but all fell within the earthy category, for example green, black and brown.

The second group recorded interviews with people out in the streets of Manchester, asking them about their favorite smells; their associations with particular smells, such as cinnamon; and the smells of their local environments and homes. The group developed a short film[2] and artwork displaying the results. The group also collected local materials and purchased products and branded goods, and placed these in covered containers for workshop participants to smell and attempt to identify. All participants correctly identified some very familiar odors such as coffee, but few correctly identified some odors such as that of fresh limes, although they were able to classify the smell as citrus.

The third group examined smells associated with specific genders and visited the local chemists to purchase different perfumes and products. After infusing fabrics with odors, the students placed the fabrics in a lift in one of the university buildings and observed how people responded to them. One observation the group made was that the selected odors were predominantly associated with people, such as perfumes and toothpaste. They believed this influenced the findings, in that people did not want to comment on the smells for fear of offending others in the lift.

The fourth group explored the impact of sensory deprivation upon smell perception by blindfolding one another and leading each other around the city center. Each blindfolded participant described his/her smell experiences as he/she was led, with this being filmed and presented back to all workshop participants at the end of the session. Group members described enhanced smell observations as a result of being deprived of their sight, and were able to detect environmental odors often far away from the source, such as when approaching the canal. They also reflected upon their limited vocabularies when describing smells, falling back to using the same words that described the intensity of the smell, such as weak, strong and so on, or relating these to other sensory modes, for example sweet, sour, spicy.

Overall, the workshop was structured to provide a learning experience that enabled students to draw from their own experiences in identifying, working through and responding to some of the challenges faced when researching and working with smell. Ultimately, it also stimulated new ideas and opportunities for learners to draw from this knowledge in applying it to their own work and delivering more full-bodied design and creative practices in the future.

Discussion

Smell has traditionally been excluded from the learning environment of built environment disciplines, and architecture and urbanism specifically; it has been confined to the emotional domain, and thus dismissed as having little to contribute towards rational thought processes, and overlooked in resulting practice as a result. However, as the case studies I have examined have illustrated, in introducing smell back into the teaching of these disciplines, new opportunities are presented to the teacher to devise, structure and facilitate effective learning opportunities.

These approaches, such as the incorporation of micro-smell sources like pots of odors and scratch-and-sniff books, can easily be incorporated into teaching in the indoor environment of the classroom or studio. However, given the limitations of such off-the-shelf tools, for example scratch-and-sniff books being predominantly targeted at children and 'smell pots' being primarily targeted at people with learning disabilities, the teacher is required to engage his/her imagination and to either devise tools or experiments him/herself or to proactively seek out potential odor tools beyond his/her usual materials suppliers for use with students.

Smellwalks out in city streets, or alternatively contained within a specific building, provide another option for the teacher in providing "a transformation of experience" (Kolb, 1984, p. 41), encouraging learners to draw from their own bodily experiences of the environment and to reflect upon these in their architectural design practice and appreciation. However, an important issue faced when encouraging this more full-bodied approach to experiential learning with respect to smell, specifically, is that odor training is not something that we, as a society, teach our children in schools. As a consequence, adult learners are generally poorly equipped with the knowledge, ability and language to begin to identify and think about the chemical senses to the maximum potential in their learning and practice. Odor tools and smellwalks provide the opportunity for learners to begin to appreciate the range of odors that exist within the built environment; to understand the role of the various built environment professionals in influencing these; and to make steps towards accommodating a more proactive approach to odor in their designs. They do not, however, overcome the basic foundational issue of a limited ability to identify and describe odors. To create a more satisfying built environment aesthetic in the future, the learner can draw from their own experiences in combining what Illeris (2002, 2003) describes as the two sides of the knowledge acquisition process: the cognitive and emotional, or what Michelson (1998) would describe as revisiting and making sense through the act of re-membering.

My argument is not just a theoretical one; it is also pragmatic. I am keen to witness the integration of new experiments and projects with smell and smell training into the learning experience. I suggest that elements of scent training be incorporated into wider architectural and urban design education. To do so is not as straightforward a task as introducing odor tools into the lecture theater or undertaking smellwalks. It requires the will by teachers and the support of their departments in the form of training that can be used to inform the students' learning, as well as access to materials with which the learners might experiment. As such, this requires a shift-change in built environment teaching over time, influenced by the "sensorial turn" (Howes, 2003, p. xxi), such as is underway in social science practice, if not in the teaching methods themselves.

It is this issue that I would like to address further with our architecture and urban design students in the future. I hope to develop a more full-bodied approach to experiential learning in teaching practice: one where bodily sensation and knowing are one and the same thing; where, as Michelson (1998) describes, "The re-membering of mind and body enables a very different knowing subject, a being who remembers: a being of passion, thought, activity, engagement, conflicts and alliances" (p. 228).

Conclusion

Academic practice in the disciplines of architecture, urbanism and across the wider built environment professions currently draws from and applies experiential learning theories. Without them, it would be impossible for students to acquire the skills, knowledge and experience required to design and construct buildings and urban environments of the future. However, current teaching practices can be criticized in the same way that Michelson (1998) and others critique experiential learning theory, developed and applied with influence from a positivist/empiricist paradigm that can be traced back to the Enlightenment period. In recognition of this limitation, I suggest that rather than experiential learning in architecture being limited to what Robinson (1980) described as being about what architects "see and what they do with what they see" (p. 27), a more full-bodied corporeal approach to experiential learning and teaching be developed, one which places value upon information received through non-visual means.

In this chapter, I have outlined examples where odor has been introduced into indoor and outdoor learning experiences as part of my work advocating the consideration of smell in built environmental design practices. I have illustrated that theories of experiential learning which separate bodily experience or emotion from rational thought are misleading, undervaluing knowledge gained through the sense of smell and reducing the potential of learners to draw from their intuitive knowledge in delivering more humanistic designs of the future.

Acknowledgments

I would like to thank Bernard Lisewski for his encouragement in developing this chapter and his comments on an early draft, and Chris Perkins, Dominic Medway, Gary Warnaby, Kate McLean, Mark Sergeant, Olivia Glasser and Raymond Lucas for their input into the *Smell and the City* project work and workshops. Finally, I would like to thank my dear friend and colleague Deljana Iossifova for her work on assisting in the editing of this chapter.

Notes

1 See the project blog at http://smellandthecity.wordpress.com/ Retrieved March 1, 2017.
2 The film can be viewed at www.youtube.com/watch?v=jqxJxJV82CA Retrieved March 1, 2017.

References

Blackson, R. (2008). *If there ever was: A book of extinct and impossible smells*. Sunderland: Art Editions North.

Charbonneau, J., & Relyea, K. (1997). The technology behind on-page fragrance sampling. *Drug and Cosmetic Industry*, *160*(2), 48–52.

Classen, C. (2005). The witch's senses: Sensory ideologies and transgressive femininities from the Renaissance to modernity. In D. Howes (Ed.), *Empire of the senses: The sensual culture reader* (pp. 70–84). Oxford: Berg.

Classen, C., Howes, D., & Synnott, A. (1994). *Aroma: The cultural history of smell*. New York: Routledge.

Dewey, J. (1963). *Experience and education*. New York: Collier Books. (Original work published 1938)

Drobnick, J. (Ed.). (2006). *The smell culture reader*. Oxford: Berg.

Gieryn, T. F. (2006). City as truth-spot: Laboratories and field-sites in urban studies. *Social Studies of Science*, *36*(1), 5–38.

Honey, P., & Mumford, A. (1982). *Manual of learning styles*. London: P. Honey.

Howes, D. (2003). *Sensual relations: Engaging the senses in culture and social theory*. Ann Arbor, MI: University of Michigan Press.

Illeris, K. (2002). *The three dimensions of learning*. Roskilde: Roskilde University Press/Leicester: NIACE.

Illeris, K. (2003). Workplace learning and learning theory. *Journal of Workplace Learning*, *15*(4), 167–178.

Kolb, A. Y., & Kolb, D. A. (2005). Learning styles and learning spaces: Enhancing experiential learning in higher education. *Academy of Management Learning & Education*, *4*(2), 193–212.

Kolb, D. A. (1976). *The learning style inventory: Technical manual*. Boston, MA: McBer.

Kolb, D. A. (1981). Learning styles and disciplinary differences. In A. W. Chickering (Ed.), *The modern American college* (pp. 232–255). San Francisco, CA: Jossey-Bass.

Kolb, D. A. (1984). *Experiential learning: Experience as the source of learning and development*. Upper Saddle River, NJ: Prentice Hall.

Kolb, D. A. (2015) *Experiential learning: Experience as the source of learning and development* (2nd ed.). Upper Saddle River, NJ: Prentice Hall International.

Lewin, K. (1947). Frontiers in group dynamics: Concept, methods and reality in social science; social equilibria and social change. *Human Relations*, *1*(1), pp. 5–41.

Mavor, C. (2006). Odor di Femina. In Drobnick, J. (Ed.), *The smell culture reader* (pp. 277–288). Oxford: Berg.

Michelson, E. (1998). Re-membering: The return of the body to experiential learning. *Studies in Continuing Education*, *20*(2), 217–233.

Oxford Living Dictionaries. (2017). Architecture. Retrieved March 1, 2017 from https://en.oxforddictionaries.com/definition/architecture

Pavlov, I. (1903). The experimental psychology and psychopathology of animals. Paper presented at *The 14th International Medical Congress*, Madrid, Spain, April 23–30.

Piaget, J. (1970). *Science of education and the psychology of the child*. New York: Orion.

Robinson, S. K. (1980). Architects as tourists. *Journal of Architectural Education*, *33*, 27–29.

Sennett, R. (1994). *Flesh and stone: The body and the city in Western civilization*. London: W. W. Norton.

Skinner, B. F. (1935). Two types of conditioned reflex and a pseudo type. *Journal of General Psychology*, *12*, 66–77.

Watson, J. B. (1930). *Behaviorism* (rev. ed.). Chicago, IL: University of Chicago Press.

17

CULTURALIZING SCENT

Current Steps towards Integrating the Sense of Smell in Art and Design Education

Ashraf Osman, Claus Noppeney and Nada Endrissat

Introduction: Scent in Contemporary Culture

Compared to centuries past in the Western world, thanks to improved sanitation and hygiene, smell is no longer perceived as the problem it once was (Corbin, 1986), and is in fact barely noticeable in our lives: "In our postmodern world smell is often a notable (or, increasingly, scarcely noticed) absence. Odours are suppressed in public places, there are no smells on television, the world of computers is odourfree, and so on" (Classen, Howes, & Synnott, 1994, p. 203). In other words, the contemporary smellscape is at least as much a function of deodorization as it is of odorization. As a culture, we have managed not only to culturalize, but also commodify and industrialize our aversion to smell: from scent-free and unscented personal hygiene and household cleaning products to the sanitation infrastructure that makes our deodorized cities possible.

One facet of the antipathy towards scent in the general culture is its frequent dismissal by many arbiters of taste in art and design as mere gimmickry. For instance, *TIME* magazine named Smell-O-Vision, a theatrical scent system that released multiple smells during a film, developed by Swiss-American entrepreneur and fragrance enthusiast Hans Laube, as one of "The 50 Worst Inventions of All Time," labeling it an "ill-fated . . . gimmick" (Fletcher, 2010). However, as Gilbert (2008) states, referring to Smell-O-Vision and AromaRama: "The loud mockery of the pundits strikes me as a cheap shot . . . . If it's really such a bad idea, why does the public remain so fascinated by it?" (p. 147).

Another persistent symptom of scent's marginalization in our culture, specifically in art and design, has been the narrowness of means available to acquire any formal knowledge of it. Arguably, that has been largely the purview of perfumery, which is dominated as an educational path by the large fragrance companies, where such knowledge is conveyed largely in terms of professional training rather than a more overtly academic education. As such, the sense of smell has been mostly left out of the academic disciplines of art and design, and the role of scent in art and design education has been rather marginal and limited – if not non-existent.

What we propose here is a reconsideration of how, where and why scent, art and design should be integrated into higher education, based on observations of emerging trends in the practices of these fields. We begin by surveying trends in these three realms separately, even though the boundaries between them have begun to blur, especially with the rise of trans-disciplinary programs. Ultimately, it is our intent to facilitate collaboration and cross-fertilization between these fields in a cohesive, meaningful manner by sketching how smell could be integrated in art and design curricula, and conclude by summarizing some of the benefits that could be gained from this.

Culturalizing Scent

For a long time, the sense of smell has been the domain of two discrete areas of knowledge: physiology and perfumery. While the former is explicitly scientific/medical in focus, the focus of the latter is primarily vocational/technical. However, a strand of thoughtful reflection began to emerge around scent in the 20th Century that brought it out from between the scientific and the vocational and placed it increasingly in the cultural realm. Some of the earliest examples of that were the writings of German sociologist Georg Simmel, whose 1908 book *Soziologie* included the influential section, "The Sociology of the Senses." Another example comes from George Orwell's *The Road to Wigan Pier* of 1937, with its thoughtful critique of class-based prejudice as reflected through smell, "the real secret of class distinctions in the West," and encapsulated in the "four frightful words": "the lower classes smell" (Orwell, 1937, p. 112).

Part of the culturalization of scent, obviously, is this publication, as well as those others noted in the reference lists of the various chapters within it. However, while a comprehensive history of the culturalization of scent is beyond the scope of this chapter, we hope to highlight here a few notable examples of where this 'cultural turn' – to borrow a phrase from Fredric Jameson (1998) – has led, as well as some from our own practices.

Another notable facet of the culturalization of scent has been the trans-disciplinary academic institutions that emerged to fill in the gaps of studying scent in such a culturally oriented manner – efforts that would have otherwise fallen through the cracks of traditional academic disciplines and institutional structures. One such institution is the Centre for Sensory Studies (CSS) at Concordia University in Montreal, Canada. CSS is an inter-disciplinary collaboration platform established in 2011, based on a research team (CONSERT), co-founded by David Howes and Anthony Synnott in 1988. Amongst the cultural dimensions of CSS is a journal, *The Senses and Society*, started in 2006, and a website (www.sensorystudies.org) offering a research directory, books and events of note, as well as sensory curricula and syllabi, amongst other resources.

Another similar example is the Centre for the Study of the Senses (CenSes) at the University of London in the UK. CenSes facilitates collaborative sensory research across disciplines, connecting groups of researchers from different fields and sectors who can benefit from one another's activities. However, while these centers have done much to advance the cultural discourse around smell, it is worth noting that this has occurred within a larger discourse on the senses (albeit a more equitable one).

In this vein, with a more explicit emphasis on smell, we established the Scent Culture Institute (SCI) in 2014 as a hub for different kinds of inter-disciplinary activities in the realms of culture, business and society. SCI originated from the two research projects funded by the Swiss National Science Foundation, and based at Bern University of the Arts and the Business Department at Bern University of Applied Sciences. SCI's activities capitalize on the collaboration between disciplines such as business, management, art, design and cultural studies, through various activities, including research, consulting, education and professional practice.

Scent in Contemporary Art

The reflection of the cultural implications of smell, however, is only one facet of its culturalization. Another important facet is cultural production with scent (as opposed to its mere cosmetic use). That can be traced, in part, to Marcel Duchamp, who was one of the first to utilize scent as such in the visual or fine arts. Duchamp challenged the dominance of sight in art through his rejection of what he termed the 'retinal,' thus shifting emphasis to the conceptual and opening the door for the other senses (Arnason & Prather, 1998, p. 274).

In the first part of the 20th Century, some Surrealists and Futurists discussed using smell in their art, but few actual works were ever produced (Classen, 1998; Verbeek & van Campen, 2013). The second part of the century saw an increasing number of installations and performances involving odors, but as Shiner and Kriskovets (2007) argue, odor remained "an ancillary factor" in those works, as opposed to the "remarkable increase in the number of artworks that foreground scents and odors during recent years," and "make a statement about our experience of smell and its associations" (p. 273).

Nevertheless, another way to evaluate the emergence of olfactory art is the curatorial attention it is finally receiving. The year 2008 saw two groundbreaking exhibitions in this regard: at the Reg Vardy Gallery at the University of Sunderland, UK was an exhibit titled *If There Ever Was: An Exhibition of Extinct and Impossible Smells*, curated by Robert Blackson; and the Esther Klein Gallery at the Monell Center in Philadelphia, Pennsylvania had an exhibition called *Odor Limits*, curated by DisplayCult (Drobnick & Fisher, 2008). However, 2015 proved to be a truly momentous year for group olfactory art shows in Europe, including *There's Something in the Air! Scent in Art* at the Museum Villa Rot in Germany, *The Smell of War* in Belgium, *Voyage dans les sens* in Paris, *I'm Here but You've Gone* at the Fiorucci Art Trust and *Tate Sensorium* at the Tate Britain in London.

The most remarkable exhibition in 2015, however, was a monumental effort at the Museum Tinguely in Basel, Switzerland titled *Belle Haleine – The Scent of Art*. The largest of its kind to date, the exhibition comprised olfactory artworks by more than 40 international artists. Despite its claim that it is not a "historically comprehensive chronological collective exhibition" of olfactory art, *Belle Haleine* was by far the closest we've come to that yet (Museum Tinguely, 2015). The exhibition was accompanied by an inter-disciplinary symposium, which further highlighted the culturalization of scent.

The art world's biggest exhibition, the *Venice Art Biennale*, included several remarkable examples of olfactory art in 2015. The entrance of the Central Pavilion at the Giardini greeted one with the ominous smell of Oscar Murillo's oil-soaked black banners (*signaling devices in now bastard territory*). At the Arsenale, Tania Bruguera's *Untitled (Havana, 2000)* was recreated with the sickly sweet smell of molasses covering the ground, instead of the rotting sugarcane husks of the original. The Dutch Pavilion was filled with the smell of dried rose petals thanks to herman de vries' *108 pound rosa damascena*,[1] while the Finnish Pavilion smelled of tar in IC-98's timeless installation, *Hours, Years, Aeons*. The Swiss Pavilion had artist Pamela Rosenkranz collaborating with perfumers Dominique Ropion and Frédéric Malle on *Our Product*, which featured "a scent evoking the smell of fresh baby skin."[2]

This surge in olfactory art practices is getting recognition by further means of culturalization and institutionalization. One such instance is the Institute for Art and Olfaction in Los Angeles, which in 2015 launched the first award for olfactory art in the form of the Sadakichi Award for Experimental Use of Scent. The first winner of the award was the deserving project *Famous Deaths* by CMP Concept, a Dutch group (Art and Olfaction Awards, 2015).

Since the turn of the millennium, there has been a veritable boom in olfactory art production around the world. Artists of all nationalities are creating scent artwork that is as diverse as contemporary art can be, spanning its various modalities, be they installation or performance, political or environmental, institutional critique or street art. These examples show how artists increasingly recognize the medium of scent to express their ideas.

Scent in Design

The fragrance industry has traditionally divided its output into two main categories: fine fragrances (perfume) and functional fragrances (cosmetic and domestic products), with a rapidly growing category of other (nontraditional) applications. While simple symmetry may tempt one to equate fine fragrances with art and functional fragrances with design, the reality is that the whole fragrance industry is more closely aligned with design (be it fashion, product or industrial design) than to art, due to its association with use, function and commerce. As such, it is not surprising that scent has made more inroads in design than it has in art.

Fine and functional fragrances certainly have long had their place in design in ways we no longer even question, primarily under the broad rubric of branding, be it of fashion labels (Hermès, Chanel, Dior etc.), celebrities (Elizabeth Taylor, Beyoncé, P. Diddy etc.) or other lifestyle brands (Porsche, Vespa, *Playboy* etc.). However, the instances of scent in design that we discuss here do not fall under these main categories of fine or functional fragrances, but rather under that growing third category of 'other' applications.

It is important to note, however, that the focus on smell in design does *not* mean the addition of applied scent to all aspects of the design, but rather the awareness of smell as an inherent dimension of it. In many cases, this awareness may lead to the realization that the appropriate solution is the removal or masking of a smell. As such, the negotiation of odorizing and deodorizing is a skillset that should be part of every designer's toolbox, as that awareness in itself is increasingly an important factor in the success or failure of a design.

That said, the forays of scent into design can be understood under three main categories: product or industrial design, communication or interaction design and spatial or interior design. Product design includes instances where smell is a significant dimension of the product, be it inherent or applied. Communication design includes instances where smell is used to communicate, emphasize/enhance or bring attention to a message (often marketing or advertising-oriented). Spatial design includes instances where smell is manipulated to enhance the built environment, for example in retail or hospitality spaces.

A classic example of scent in product design is the smell of new cars, which is often an emotional incentive in the purchasing decision for buyers (Schifferstein & Hekkert, 2011). Similarly, many people cite smell as a reason for their attachment to paper books and their reluctance to switch to digital readers (Kraft, 2015). Related to this is another area that has experimented with applied scents: stationery products, including pens (such as Sanford's Mr. Sketch scented watercolor markers), pencils, crayons (such as Crayola's Magic Scent), erasers, stickers (such as Peaceable Kingdom's Scratch & Sniff series), adhesives (such as Pritt's celery-scented glue stick, to deter children from ingesting it) and notepads. Typically, these scents are added to make stationery for school use more appealing to children; but many are now added to humanize the antiseptic office experience, reduce stress and increase enjoyment of the workplace.

Another notable area for experimentation with nontraditional uses of scent in product design is the fashion industry. However, these are not the typical uses of fragrance as a branding accessory of fashion labels; rather, these applications fall under two categories. The first is that of added scents, embedded in the material of the clothing article itself, to create scented shoes (such as Reebok's Kool-Aid-scented footwear) or jeans (such as Naked Famous' Scratch-N-Sniff Denim), for instance. The second application is of smell-resistant or antibacterial products, often sports-related, to counter the smell of sweat produced through use. It is worth noting that such applications vary from the aesthetic to the functional, solving a problem of masking unpleasant scents, formerly dealt with at the level of the body and personal hygiene.

A related functional strategy in product design, which brings it close to communication design, is the use of smell (often unpleasant odors) to signal damage or unviability of an item (typically food). There have been other uses of product design that come close to communication design from a more aesthetic angle, from scented credit cards (by Al-Hilal Bank) and tourist guidebooks (the city of York in the UK) to business cards infused with individual pheromones.

More typical communication design applications of scent work in concert with more traditional media to emphasize their messages, as in the case of a bus shelter advert for baked potatoes reinforced with the fresh smell of these potatoes (Filippetti, 2012). Such multisensory multimedia campaigns have proven to be rather effective, as with the case of a campaign by Dunkin' Donuts in South Korea, where a jingle playing on public buses was accompanied with an atomizer releasing the aroma of coffee: "The campaign increased visits to Dunkin' Donuts outlets near bus stops by 16% and sales at those outlets by 29%" (Anonymous, 2015, p. 28).

Cases where things have not worked out quite so well, however, make an even more compelling case for an informed approach to such efforts, and hence the necessity for the proper education of their designers:

> A recent ad campaign for the liqueur "di Saronno" proposed to flood the London underground with a two week "scent-wafting" campaign, pumping the distinctive almond aroma of the drink into the ventilation system and directly appealing to consumer's taste buds. The campaign was pulled after only a day due to the coincidental release of an article the same day on the signs of terrorist activity in the popular broadsheet [newspaper] the *Daily Mail*. The article warned commuters to be cautious of the smell of almonds, especially whilst underground – an indicator of the presence of the dangerous gas cyanide!
>
> *Lim, 2014, p. 84*

These examples of communication design link to the third category of design uses of scent, which is spatial. Beyond the marketing and advertising focus of the above, there have been other uses for scent in the design of commercial spaces, especially in retail and hospitality contexts. In retail, the effects of scent could include an increase in turnover, the desire to shop and the dwell time within the retail environment (Gilbert, 2008). Similarly, hotels and other hospitality venues can enhance customer experience through scent. For instance, in Luzern, Switzerland, Hotel Montana offers guests a choice of ambient scents in their rooms. Hence, guests can tailor their olfactory environment to individual preferences, needs and desires. Similarly, healthcare environments, such as hospitals and senior care facilities, use environmental aroma technology to reduce stress, increase alertness and enhance comfort, pleasantness and perception (Ziegler, 2015).

As stated earlier, however, one can often learn as much from negative feedback as from positive; and people are increasingly expressing satisfaction or dissatisfaction with products, services and experiences based on scent. A cursory browse of reviews on online travel forums such as TripAdvisor reveals the deleterious effects for a hotel that comments about a disgusting smell in the hallway or a stinky sink can have, as opposed to one about the great smell of the lobby (TripAdvisor, 2011). Similarly, reviews on online shopping sites such as Amazon abound with comments about the inadvertently off-putting smell of a meat grinder or a chair (Amazon, 2014).

One can get a better idea, perhaps, of the future role of scent in design by doing a search for scent/smell/odor on design portfolio websites, such as Bēhance.[3] Projects range across the three categories outlined above: from product design (a scented ring that controls appetite, a food-smelling app, an olfactory game etc.) and communication design (olfactory design marketing campaign, advertising, art direction etc.) to spatial design (office way-finding, interactive home design concept etc.). One simply cannot turn up one's nose these days at the role of scent in design.

Scent in Higher Art and Design Education

As seen above, the increased culturalization of scent and its forays into contemporary art and design practices make it difficult for higher education institutions to continue ignoring the sense of smell as an area of cultural knowledge and production. But given that recent olfactory productions do not fit neatly into the traditional divisions of human knowledge, one notable way that scent has made its way into academia is via multi-/inter-/trans-disciplinary programs that have proliferated in higher education in recent years.

According to Evans (2010), such programs ensure that students "study concepts, issues and problems from more than one disciplinary perspective" and "recognize that knowledge can be constructed using more than one epistemological framework, even when working with the same or similar 'topics'" (p. 2). Trans-disciplinary programs engage students in probing "the idea that knowledge derived from the academic disciplines alone may be incomplete as a basis for addressing complex social, ecological, technical and other problems" and "the validity of various constructions of knowledge in the context of comprehending complex phenomena and addressing complex problems." They also engage students in "questioning the division of the knowledge world into discretely bounded disciplines," in "inquiry that makes use of knowledge generated both in- and outside the academy," and "academic work that both integrates and transcends disciplinary knowledge in an effort to both comprehend and directly address complex problems." Moreover, they "serve as a vehicle for students and faculty members to comprehend complex phenomena and problems and to engage in actions that address complex social, ecological, technical and other problems" (p. 2).

Thus, trans-disciplinary programs "increase the ability of members of society to engage in transformative capacities with complex issues in their work and their civic and social lives in an increasingly complex world." Such programs should, therefore, "generally be expected to demonstrate heightened abilities to engage with and evaluate knowledge drawn from multiple contexts both in- and outside the academy and to participate effectively in problem solving work". As such, this "should derive from the highly integrative and explicitly purposeful orientation of trans-disciplinary curricula" (Evans, 2010, p. 2).

In recent years, with the emergence of trans-disciplinary academic programs, scent started finding a footing in higher education, affording it the cross-disciplinary knowledge it requires. One of the first programs to explicitly integrate scent and design was that of Transdisciplinary Design at Parsons' New School for Design in New York. In 2010, in order to mark the establishment of this new Master of Fine Arts program, Parsons held an event titled *HEADSPACE: A Symposium on Scent as Design* in partnership with the Museum of Modern Art, International Flavors & Fragrances Inc., Coty Inc. and Seed. According to Parsons, this day-long event brought together "for the first time leading designers, scholars, scientists, artists, and perfumers to explore this new territory for design" (New School, 2010).

That experience did not translate, however, into a trend of integrating scent into art and design education. For that to happen, there needs to be a meeting in the middle of two currently disparate realms of higher education: art and design on the one hand, and scent or perfumery on the other. Despite the growing overlap between the worlds of perfumery, art and design, there remains very little structured cooperation between them.

In this respect, the diverse activities on multisensual design at the art university Schloss Giebichenstein at Halle (Germany) remain exceptional, even in its experimental stage. In context of the industrial design study program (1999–2000), students were introduced to the spectrum of the senses, including the principles of scent creation (Bork, 2002). Based on this input, the students developed and designed new interfaces to deliver scent, particularly in performative contexts (Lange, 2002; Röder, 2002). A key learning outcome from this experiment is that even a limited presence of the sense of smell in design education sparks off a substantial innovative potential.

Conclusion

Art and design education, in general, is dynamic and open to new fields, such as lighting design and scenography. It professionalizes and modernizes traditional vocations, creating fashion designers from tailors, product designers from carpenters and type designers for digital screens from typesetters. The same principle can be applied to the education and practice of perfumery to transform it into scent design.

The current situation of scent education remains highly technical and fragmented, dominated by the private institutions of large commercial fragrance companies. Based on these observations and our arguments so far, we strongly advocate an expansion of the traditional realm of perfumery to include the contemporary practices of art and design. This would shift the emphasis of most perfumery schools from traditional aesthetics, narrow technicality and market focus to emerging innovative artistic and design-focused uses of scent.

Those changes might include institutions offering young artists and designers the opportunity and preparation to make scent part of their artistic and design practices. Established institutions have traditionally provided their students with workshops to familiarize them with certain crafts. They could extend that to providing access to material and settings (labs) where artists and designers can familiarize themselves with – and experiment in – this dimension.

The basis for these practical program expansions would be the related but distinct area of intellectual exploration of scent described earlier in the section "Culturalizing Scent." As such, these recommendations for academic curricula in art and design education can be summed up as follows:

Scent Cultural Studies: Academic and Intellectual Training

This would explore the role that the sense of smell plays in culture and society at large. It is more theoretical, observational and analytical, but is, nevertheless, an essential grounding. It could include topics such as the following:

- scent and smell as the subjects of social science and cultural analysis, reflection and theory;
- education of the nose: education of the mind;
- fragrance criticism: scent and smell as critical reflection;
- exploratory investigations;
- sensory research methods (e.g. sensory ethnography);
- cultural practices involving smell;
- making sense of smell: writing and other sense-making methods.

Olfactory Design: Optional Creative Competences

Art and design schools have traditionally provided their students with workshops to familiarize their students with certain crafts. Not everyone needs to know how to do everything, but those interested should have sufficient access to acquire a level of familiarity with the medium that would enable them to think freely and creatively in it, as well as communicate clearly and effectively with the involved technicians. Thus, in addition to practical techniques and know-how, schools could extend that to providing access to material, tools and technologies (diffusion, analysis etc.) and settings (labs) where artists and designers can familiarize themselves and experiment in this dimension. This would include:

- creative olfactory competences for various design disciplines (fashion, interior design etc.);
- linking to other design disciplines;
- access to resources: work with materials;
- access to delivery technologies;

- understanding the interaction of the different senses;
- the 'sensual' education of the body;
- optional elective.

Scent Design: From Perfume Craft to Scent Design

Being part of an arts and design institution, perfumers would think of scent more in a wider aesthetic context, and would be enabled to integrate other practices taking place around them. As such, the focus shifts from narrow technical innovation to a larger cultural innovation:

- development of fragrances as a creative practice;
- scents as signifiers;
- fragrances in resonance with other cultural developments;
- perfume as an element of cultural practice.

Art and design schools nowadays are not only places for practical training, but also institutes for reflecting on culture and society. This role could be expanded to sensory studies by offering courses that guide students in thinking of their practices in the context of multisensory design that goes beyond traditional media. They could then explore the innovative incorporation of scent in the design of everyday objects and experiences, and the effective use of scent as an emotive and communicative medium in art.

Notes

1 de vries stylizes his name in lower case on his artwork, "to avoid hierarchy."
2 www.designboom.com/art/swiss-pavilion-venice-biennale-2015-pamela-rosenkranz-05-08-2015/ Retrieved March 1, 2017.
3 https://www.behance.net Retrieved March 1, 2017.

References

Amazon. (2014). Sayl chair by Herman Miller. Retrieved March 1, 2017 from www.amazon.com/SAYL-Chair-Herman-Miller-Adjustable/dp/B0055E1Q8Q

Anonymous (2015). The science of sensory marketing. *Harvard Business Review, 93*(3), 28–30.

Arnason, H. H., & Prather, M. F. (1998). *History of modern art: Painting, sculpture, architecture, photography* (4th ed.). New York: Harry N. Abrams.

Art and Olfaction Awards. (2015, April 11) Finalist: Famous Deaths. Retrieved March 1, 2017 from https://artandolfactionawards.wordpress.com/2015/04/11/finalist-famous-deaths/

Bork, K.-H. (2002). Designolfaktorik. In P. Luckner, B. Giebichenstein, & Hochschule für Kunst und Design (Eds.), *Multisensuelles Design: Eine Anthologie; Ergebnisse und fachwissenschaftlicher Kontext des Modellversuch im Hochschulbereich an der Burg Giebichenstein-Hochschule für Kunst und Design Halle* (pp. 265–296). Halle: Saale.

Classen, C. (1998). *The color of angels: Cosmology, gender and the aesthetic imagination.* New York: Routledge.

Classen, C., Howes, D., & Synnott, A. (Eds.) (1994). *Aroma: The cultural history of smell.* London: Routledge.

Corbin, A. (1986). *The foul and the fragrant: Odor and the French social imagination.* Leamington Spa: Berg.

Drobnick, J., & Fisher, J. (2008). Odor limits. *The Senses & Society, 3*(3), 349–358.

Evans, T. (2010). Purposes of multidisciplinary, interdisciplinary and transdisciplinary programs. Retrieved March 1, 2017 from https://www.academia.edu/2315140/Purposes_of_Multidisciplinary_Interdisciplinary_and_Transdisciplinary_Programs_April_2010

Filippetti, J. (2012, February 7). Multisensory bus shelter ad smells of baked potatoes. *designboom.* Retrieved March 1, 2017 from www.designboom.com/design/multisensory-bus-shelter-ad-smells-of-baked-potatoes

Fletcher, D. (2010, May 27). The 50 worst inventions of all time. *TIME*. Retrieved March 1, 2017 from http://newsfeed.time.com/2010/05/27/the-50-worst-inventions-of-all-time/

Gilbert, A. (2008). *What the nose knows: The science of scent in everyday life*. New York: Crown.

Jameson, F. (1998). *The cultural turn: Selected writings on the postmodern 1983–1998*. London: Verso.

Kraft, A. (2015, December 14). Books vs. e-books: The science behind the best way to read. *CBS News*. Retrieved March 1, 2017 from www.cbsnews.com/news/kindle-nook-e-reader-books-the-best-way-to-read/

Lange, S. (2002). Sniffman und Debut de Parfum. In P. Luckner, B. Giebichenstein, & Hochschule für Kunst und Design (Eds.), *Multisensuelles Design: Eine Anthologie; Ergebnisse und fachwissenschaftlicher Kontext des Modellversuch im Hochschulbereich an der Burg Giebichenstein-Hochschule für Kunst und Design Halle* (pp. 429–450). Halle: Saale.

Lim, C. J. (2014). *Food city*. New York: Routledge.

Museum Tinguely. (2015). *Belle Haleine – The Scent of Art*. Retrieved March 1, 2017 from www.tinguely.ch/en/ausstellungen_events/ausstellungen/2015/Belle-Haleine.html

New School. (2010). Leading designers, scientists, artists and perfumers transcend boundaries to explore scent as new territory for design [Press release]. Retrieved March 1, 2017 from www.newschool.edu/pressroom/pressreleases/2010/headspace.aspx

Orwell, G. (1937) *The road to Wigan Pier*. London: Victor Gollancz.

Röder, G. (2002). Der Einsatz von Duftstoffen im Multimediabereich. In P. Luckner, B. Giebichenstein, & Hochschule für Kunst und Design (Eds.). *Multisensuelles Design: Eine Anthologie; Ergebnisse und fachwissenschaftlicher Kontext des Modellversuch im Hochschulbereich an der Burg Giebichenstein-Hochschule für Kunst und Design Halle* (pp. 491–498). Halle: Saale.

Schifferstein, H. N. J., & Hekkert, P. (2011). Multisensory aesthetics in product design. In F. Bacci & D. Melcher (Eds.), *Art and the senses* (pp. 543–569). Oxford: Oxford University Press.

Shiner, L., & Kriskovets, Y. (2007). The aesthetics of smelly art. *Journal of Aesthetics and Art Criticism*, *65*(3), 273–286.

Simmel, G. (1908). *Soziologie: Untersuchungen über die Formen der Vergesellschaftung*. Leipzig: Duncker & Humblot.

TripAdvisor. (2011). "Beautiful! What's that amazing scent when you walk in the lobby? Friendly staff." Review of ARIA Resort & Casino. Retrieved March 1, 2017 from www.tripadvisor.com/ShowUserReviews-g45963-d91925-r106119531-ARIA_Resort_Casino-Las_Vegas_Nevada.html

Verbeek, C., & van Campen, C. (2013). Inhaling memories: Smell and taste memories in art, science, and practice. *The Senses and Society*, *8*(2), 133–148.

Ziegler, U. (2015). Multi-sensory design as a health resource: Customizable, individualized, and stress-regulating spaces. *Design Issues*, *31*(1), 53–62.

18

OLFACTORY EDUCATION IN ART AND DESIGN

Joanna Norman and Nicola Pozzani

Introduction

The introduction of smell can help stimulate innovation in all areas of art, design and creation. The designer's 'cognition,' defined by Oxford Living Dictionaries (2016a) as "The mental action or process of acquiring knowledge and understanding through thought, experience, and the senses," can be informed by activating the sense of smell and how it interacts with the other senses. Experiential ways of teaching and learning to use smell to connect practice to multisensory, life-focused design were developed by the authors to encourage artists and designers to include scent more effectively in their projects. A key advantage of this arises from the relationship of smell with memory and emotions (Curtis & Williams, 2001) and how important this is for artists and designers to reflect these in their work.

To illustrate olfactory education in these areas, case study examples are taken from both Joanna Norman's teaching of the module 'Sensory Design' to postgraduate MA Fashion students at Kingston University (KU), and Nicola Pozzani's BA Art and Design students at Bern University of the Arts (BUA), in how to incorporate fragrance in their work. The 'Sensory Design' module was devised at KU, specifically focusing on the sense of smell to complement the students' existing understanding of sensory awareness. The BUA course was developed to encourage students to experience cross-disciplinary stimuli, reflecting the emerging 'olfactory art' scene. These, combined with a case study from *Design with Scents*, an interactive five-day course on designing a 'scent for space,' led by both Nicola and Joanna jointly, examines multisensory techniques used for fragrance development and communication. This course was developed because the authors perceived a lack of olfactory education existing within art and design. This was confirmed by the international audience of students and professionals from the creative disciplines who have attended the workshop and have found such education hard to find. The knowledge gained from these case studies provides practical guidance regarding the different materials and methods recommended to include olfaction more effectively within art and design teaching.

Case Study 1: Postgraduate Fashion Design and Fragrance

Designers are at home with visual, tactile and auditory stimuli. The use of color, texture and sound is essential to their work. By including smell, their sensory practice becomes more complete and inquisitive. The link between the two luxury industries of fashion and fragrance is well established on a commercial level. Since Paul Poiret and Coco Chanel initiated fashion house-inspired fragrances in 1911 and 1921 respectively, the majority of fashion designers launch an accompanying perfume and beauty line. *The Fashion Book* (Phaidon, 1998) remarked in its glossary definition of haute couture: "In terms of revenue, haute couture no longer pays its way but is still shown twice a year (in January and July) as an exhibition of luxurious fashion for the purpose of selling the couture houses' perfume" (p. 506).

Moving to our first example, the aim of the postgraduate 'Sensory Design' module was to explore fragrance as key to the actual fashion creation process. At Kingston University, within the MA Fashion course, Joanna developed this module, which introduces the concept of olfaction as inspiration for the fashion designer. Instead of developing a fragrance which represents the fashion brand or garment, the designer develops a collection or installation which is an interpretation of the fragrance itself. By incorporating scent at the design stage, fragrance inspires the creative process rather than becoming an optional accessory.

The students are asked to identify areas of their research that will benefit from the inclusion of smell, whether it be an object, a place, a space, a body or a person. Their brief is for the design development to capture the mood or essence of a scent, or to incorporate scent into its physicality. The objective is to engage the senses more fully. They are also asked to consider the technology of delivery of how they intend to include the scent with their work. Will it be sprayed onto their garment or installation using a traditional spray (alcoholic or non-alcoholic), or incorporated directly into other materials or substrates such as fabric, plastic, wood or candle wax? The students may need to evaluate the feasibility and practicality of their design idea, researching processes and manufacturers which specialize in perfuming different materials.

The teaching and learning strategy includes tutorials and workshops given by perfume specialists. The focus is on developing a descriptive fragrance language by examining the fragrance structure and the emotional triggers it evokes. The ethereal nature of smell makes it hard to describe. Often people know whether they like or dislike a smell, but find it hard to articulate their reasoning. Qualitative market research focus groups which Joanna ran when working at leading fragrance supplier IFF[1] for products such as fabric conditioners and shampoos, as well as fine fragrances, found that consumers often ran out of words after using the words 'fresh' and 'sweet.' 'Fresh' corresponded with a consumer who liked the smell, whilst 'sweet' was usually found to have a negative connotation, meaning it was over-sweet or cloying. Perfumers have developed a vocabulary to communicate the technical structure of odors, often borrowing from other senses. For example, the smell of a single ingredient is referred to as a 'note,' taken from the musical world, and the smell of grass is referred to as 'green,' borrowed from vision and color. Whilst it is important to have an understanding of these descriptive terms, it is also vital to explore the sense of smell with regard to our emotional and sexual responses. This is because the nervous impulses from the sense of smell are the only ones to flow directly to the limbic system of the brain: "The limbic system is concerned with our feelings, moods, emotions and sexual behavior and with memory. It also exerts modifying influences on behavior in relation to past experience" (Curtis & Williams, 2001, p. 37). Thus the olfactory system is truly a direct extension of the brain, and is more closely associated with our feelings, emotions and behaviors than any of the other senses.

The students are encouraged to smell a variety of raw materials. A useful tool to aid descriptive terminology is by using opposite terms.[2] For example, on a technical level: do you find the smell hot or cold, wet or dry, earthy or clinical? On an emotional level: how does it make you feel – energetic or relaxed, adventurous or pensive, romantic or sporty? An interactive exploration of perfumery practices on how fragrance ingredients combine to form new accords and build fragrance families develops this further. In this specific exercise, the focus is on the categories which underpin

TABLE 18.1 Typical constructions of traditional fragrance families within perfumery

Fragrance family	*Typically contains the following raw materials:*
Citrus Cologne	Bergamot, neroli, petitgrain, lavender
Floral	Rose, jasmine
Chypre	Bergamot, *Cistus labdanum*, oakmoss, patchouli
Fougère	Citrus notes, lavender, coumarin, woody notes, musk
Oriental	Vanilla, incense

most commercial fragrances, namely: Citrus Cologne, Floral, Chypre, Fougère and Oriental families to develop a context. Each perfume house has its own interpretation of these families. Curtis and Williams (2001) describe them in more technical depth in *An Introduction to Perfumery*, and Fairley and McKay (2014) describe them in an easy-to-read guide in *The Perfume Bible* with many examples of commercial fragrances in each family; Joanna has summarized them in Table 18.1. Ultimately, the aim is to extend the vocabulary from words to design practice, whether that develops into a garment, an accessory or an installation.

The students are encouraged to paint or draw using color to illustrate their interpretation of the odor of both raw materials and fine fragrances. Perfume specialists are involved in running one-day workshops, including in this case Givaudan, a leading supplier of fragrances, who provided materials. Sense, a UK national charity for the deaf/blind, gives an opportunity to ascertain how much those who are visually and hearing impaired rely on their other senses of touch, taste and smell. Together, these two organizations add a real-life experience of working with fragrances and how crucial they are in our daily lives. For example, Sense described how a deaf/blind child at the end of the school day is given a handkerchief smelling of his/her mother to know that it is time to put his/her coat on and get ready to leave.

The first three years the project was run let the students find their own fragrance as inspiration, but during the last two years, the students were offered a choice of fragrances as their starting point. Both methods led to a wide variety of final outcomes. The latter allowed the students to concentrate on the smell rather than the 'hunt' for the smell, which takes more time. Each year, the winning projects were exhibited in public as part of their end-of-year show and collated within a 'Digital Book' entitled *Sensory Design Project*. In 2010, the first year that this activity was developed, Shriya Kariappa produced a digital smell concept as part of the Facebook platform. The following year, Fay Gascoigne (2011) won the overall prize for best design with a lavender-inspired garment which evolved from working with soldiers and memory. These design students found their own fragrance or smell to inspire their practice. In Shriya's case, she connected with a company who were investing heavily in social media and the youth market, the focus being on describing scents on the Internet, whilst in Fay's project, the lavender was the smell used in scented letters sent to their loved ones. Lavender evoked loving memories symbolic of home life and support. In some cases, the soldiers were grateful for the added closeness the smell brought them; for others, they requested not to receive scented letters since the smell evoked memories too hard to bear.

Other projects have included Laura Buechner's *Eve on the Ladder to Heaven* (2012). Her premise was that sperm cells possess the same type of odor receptors that allow the nose to smell, suggesting sperm navigate towards a fertile egg by detecting its scent. She claims: "The chemo attractant for human spermatoza is bourgeonal which is an aromatic aldehyde used in perfumery and has a fragrance reminiscent of lily of the valley" (Buechner, 2012). Female sexual organs and muguet served as inspirational source for her end-of-year catwalk design.

Elizabeth Gilbey (2012) drew inspiration from the role of scent in choosing sexual partners used by the Amazonian Desana tribe, which mirrored the choosing of partners to a lesser extent in current society. Her project *Tribal Redolence* used "hand crafted, scented beads placed over the areas of the body where temperature is highest to allow for scent based exploration of the human form" (Gilbey, 2012). To create the tribal theme, the garments were hand-crafted using techniques including macramé and leather engraving. This texture allowed for a tactile experience combined with scent and a visually attractive accessory resulting in multisensory jewelry pieces.

Recently (2013–2017), the module has been run such that the students are presented with a collection of fragrances before they initiate their projects. These fragrances were provided by Givaudan, and each student chose their preferred scent for inspiration. No name or description of the scent was given until the project is complete. This method has shown communal links in interpretation between the perfume creator and the designer; for example, one project claimed to be inspired by the citrus elements of a coded fragrance, and developed this into a powerful, futuristic textile construction involving sensory leather sculpted around the body. Interestingly, the fragrance was revealed as being called *Kinetic Energy* by the perfumer.

Working with select fragrances seems also to have engendered a more inclusive use of involving the scent with others: for instance, an art project was developed involving the surrounding community to transform the streets using a chalk-based scented spray paint. The aim was for the public to work together to convey their message on local walls, creating their own scent tag.

Specific examples include Sarah Hellen (2014), who was inspired by *Scent No. 1* (see Figure 18.1) to produce a white knitted garment full of dense texture to produce a cocooning feel of protection and warmth set to the sound of music. For her research, she ran sensory workshops in a care home centered on fragrance and art, whilst Niro Wang (2014) developed a tactile fabric vocabulary (see Figure 18.2) to interpret different smells via the sense of touch. His garments were created from a paper-like fabric, also in white, removing any visual clues. The introduction of smell became central to creating their fashion design development at the same time as arousing awareness of the interaction of all the senses.

The students' design outcomes were truly diverse, including improving our well-being, enhancing people and products, initiating sensual arousal, creating tactile, 'talking' clothing and

FIGURE 18.1 *Scent No 1* by Sarah Hellen, 2014.

Image courtesy of Ezzidin Alwan, Kingston University London.

FIGURE 18.2 *Tactile Fabric Vocabulary* by Niro Wang, 2014.

Image courtesy of Ezzidin Alwan, Kingston University London.

communicating smell via social media. The project delivered areas of exceptional insight, and explored personal boundaries from base animal instincts to a refined sense of luxury. All the students said that they had a greater awareness of the sense of smell and how it affected their emotions in both a positive and negative sense, and that it would influence their practice in the future. A good example of this is from Shriya, who, five years on, runs a design consultancy in India called IdeaCulture. She says she makes a conscious effort to try to engage the sense of smell, especially when working in the retail design and service experience.

Case Study 2: The Introduction of Olfaction across All Art and Design Programs

The first of its kind for academia, *Olfactory Art & Design* is part of the study offer of the Y Institute Centre for Trans-disciplinary Tuition of Hochschule der Kuenste, Bern (HKB), Bern University of the Arts, and has been running since 2014. Their website claims: "The Y Institute sees itself as a place of communication between various forms of realisation and representation."[3] *Olfactory Art & Design* is aimed at students undertaking degrees in the following departments: Design and Fine Arts, Conservation-Restoration, Music, Opera/Theatre and the Swiss Literature Institute. The aim is for students to experience fragrance as an art form, whilst benefiting from cross-disciplinary knowledge shared between olfaction and the other 'traditional' artistic disciplines taught at HKB.

The idea of the course was born out of an encounter between Nicola Pozzani and Prof. Claus Noppeney of HKB, at the Berlin Perfume Laboratory of Perfumers and Artists, Christophe Laudamiel and Christoph Hornetz, also known as 'Les Christophs.' With them, Nicola was animating the *Duft Visons* perfume workshops, presenting fragrance as an art form to the Berlin audience. The encounter resulted in HKB inviting fragrance into their building as part of their curriculum.

The *Olfactory Art & Design* course introduces fragrance as a new medium. It provides an experimental platform for fragrance education in art and design, given the increasing use of scents and the lack of available education in these fields. During the week-long, highly practical, 30-hour workshop, students produce an *Olfactory Art & Design* project as an end product of a cross-sensory study program.

The olfactory journey begins by exploring the links between the students' areas of expertise and a palette of 40 perfume ingredients kindly provided by Les Christophs, along with perfume compositions, such as harmonious combinations of perfume ingredients. Every year, the course is adapted according to the students' backgrounds, and it is Nicola's job to develop a connection with them, understanding which elements can be taken into the classroom to facilitate the conversation: a grand piano, a palette of artificial sounds, watercolor paints, poems and so on.

Nicola's teaching approach is very much based around an induced 'synesthesia,' defined as "The production of a sense impression relating to one sense or part of the body by stimulation of another sense or part of the body" (Oxford Living Dictionaries, 2016b), a process which he studied at the Universita' dell'Immagine of Milan, Italy, combined with teaching techniques learnt through his Teaching English as a Foreign Language qualification and experience, adapted to the teaching of fragrance.

Although the use of scent in marketing and design is mentioned, given the students' backgrounds, the core focus is on art. Whilst training regarding the sense of smell, the course raises students' awareness about the work of some of the leading artists using scents in their practice. Students are exposed to an exceptionally close way in which to 'look at' this, thanks to artists' very kind engagement, for which both Nicola and students are especially grateful. Laudamiel and Hornetz have, for instance, shared examples of what they call generically 'scent sculptures,'[4] the pioneering fragrance artwork which they have exhibited at art galleries and museums around the world.

In order to illustrate the curatorial and commercial opportunity of exhibiting and selling scent artwork, Andreas Hermann, Director of Mianki Gallery Berlin, kindly talked to the students about putting together Laudamiel's shows at Mianki: *Emotions* (2014, with light artist Jakob Kupfer) and *Reconstruction* (2015). Furthermore, students have been introduced to *Per l'eternita*, the scent installation by Italian artist Luca Vitone with perfumer Maria Candida Gentile (Italian Pavilion, *Venice Art Biennale 2013*). Gentile gave the students a heartfelt account of the artistic process behind their work, the first to enter the *Venice Art Biennale*. In 2015, students were taken on a visit to Museum Tinguely in Basel, where *Belle Haleine – The Scent of Art*, the world's first major exhibition around the role of scent in art, took place, further acknowledging scent in art.

After studying fragrance ingredients, compositions and examples of fragrance artworks, students finally venture into creating their own fragrance artwork. Their job is that of choosing a theme explored in their area of study and translating that into scent. Their *Olfactory Art & Design* project, which they complete working in groups, is presented on the final day: every group shares with peers and tutor the artistic process, cultural context and the composed scent, along with its possible applications. Nicola's role is to guide them throughout the design of the idea to fragrance formulation and evaluation, finally commenting with them on coherence of idea, group work and scent.

Students' interest in the olfactory exercise has been very high, as measured with feedback forms, and has produced some fine work. The class of 2014 developed several group projects, including *the smell of the apocalypse*, the result of an inter-disciplinary collaboration with the *Verhalten bei Weltuntergang* (How to act when the end of the world is near) course, led by Dr. Florian Werner. Focusing on the study of the apocalypse, the fragrance which the olfactory students produced was the answer to the question, "What does the whore of Babylon smell like?" which Werner's students asked them.

In 2015, The Neighbours Ballet group decided to work on the theme of 'dance,' taking as inspiration opposed dance movements, aiming for their fragrance to embody balance as a result of opposition. This started by comparing *Triadisches Ballett* (Triadic Ballet) by Oskar Schlemmer (Germany, 1922) from the Bauhaus, a robotic ballet dance in three acts, where each act expresses different movements and comes with a different color, to its opposite, the soft movements coming from images of contemporary classical ballet dance. This was translated into a fragrance which fully embraced the concept of opposition, with its harmonized juxtaposition of molecules of a very different 'movement' or weight.

Whilst working on the *Unknown* project, other students were particularly impressed by the challenge of translating a powerful 'immaterial' idea with the materiality of fragrance molecules. This project expressed the excitement and surprise of suddenly experiencing unknown culture, whereby a piece of Arabic music by Ziadh Rahbani, unknown to the group, was repeatedly played, and translated as a result into an erupting combination of visual and scent elements.

Another group of students developed a project called *Madness*, taking its inspiration from the novel *On the Road* by American writer Jack Kerouac, a work from the Beat generation which chronicles the author's roaming throughout America. Students created first a short film and consequently a scent composition of intentional 'disturbing qualities,' as explained by them, reflecting the lives of the protagonists spent between drugs and poetry.

All in all, feedback forms collected at the end of the course and the high demand for it (it was oversubscribed twice) demonstrate its relevance as part of the academic offer. This became clearer during the second year, as some of the students who took the course were working on olfaction as part of their thesis in Fine Art and Art Conservation.

Case Study 3: *Design with Scents*, Kingston University London, July 2012

Design with Scents was devised as the first international workshop for the design of places through the sense of smell (Design with Scents, 2016). It was aimed at designers of all disciplines to discover how scent is a powerful communication tool at work and at home. Through a five-day interactive course, participants learnt how to enhance their projects (both academic and practitioner), to communicate an idea to an audience with fragrance. They developed new design skills, exploring the sense of smell and experimenting with blending oils whilst working on a scent design solution.

Led by Nicola Pozzani, founder of the fragrance agency S Sense, with guest speakers Jo Norman and John Ayres from the fragrance and training consultancy Pandora Ltd, *Design with Scents* has become a multisensory training program for the scent enthusiast. The first workshop was held in 2012 at Kingston University London, and had participants from China, Germany, France, Belgium, Saudi Arabia and the UK. Their backgrounds covered sensory marketing, the urban environment, tobacco flavoring, jewelry design, communication, cosmetics and event design. Academics, professionals and perfume lovers were all welcome.

Fragrance oil kits (see Figure 18.3) and sensory materials including a range of textiles of different colors and textures, foods such as marshmallows and ice cream, along with paints, were provided for individual use. The olfactory journey began with the holistic perfume experience, discovering how scents can affect emotions. Synesthesia was used to study perfume as a creative language. Through interactive sessions, the students discovered the scents that form the perfumer's palette.

The course included how scents link with social history, fashion and culture, and we were delighted to have Dr. Victoria Henshaw present her PhD research *Smell in the City*, which helped raise awareness of the relationship between scents and the urban environment. As the workshop progressed, students formed 'Scent Agencies,' analyzing a specific place and developing scent

FIGURE 18.3 Dr. Victoria Henshaw and Prask Sutton, *Design with Scents*, 2012.

Image courtesy of Ezzidin Alwan, Kingston University London.

design solutions. Guided by John Ayres, the students experienced how to translate their group projects into scent through fragrance creation.

On the final day, the results were presented whilst peers, course leader and guest speakers provided feedback on the whole creative process. Working on retail, residential and shared spaces projects, each team showed concepts and fragrance solutions. The combination of experience and interest from the students contributed to the success of the course, the learning process, the exchange of knowledge and passion in the creation of a scent design network.

The workshop has since become an annual one-week short course, currently held at London College of Fashion, University of the Arts, London. The alumni have formed an online and offline community of individuals working on scent projects around the world. They are now also guest speakers scoping their individual disciplines and sharing their olfactory interests and inspiration with the next incoming group of students. Dr. Victoria Henshaw summed up her week immersed in scents as having "learnt more in the one week than in any other single week of my education."

Conclusion

Nicola and Joanna's experiences from running these projects have shown there is a clear thirst for information and a keenness to experiment with smells and odors. Outcomes from these courses have included a basic knowledge of the physiology of smell, an understanding of fragrance creation, and most importantly, a discovery of the emotional links that come from 'listening' to your nose, with resulting creative output. The results seen here have been wide-ranging, visceral and enlightening, leading to an increased perception of what was previously taken for granted. The smell or affects of smell of a project are now being taken into consideration at the initial stages of design. Developing a comprehensive language has proved key in conveying images of smell. The artists and designers have demonstrated they can communicate using a wide variety of media other than the traditional use of words, and indeed use smells and fragrances as part of a palette of tools to both generate and inspire their work.

A particular challenge in providing olfactory education is sourcing specialist knowledge and fragrance materials. Where they are not available themselves, Nicola and Joanna would suggest partnerships with a perfumer and/or perfumer suppliers who may provide such support. In the future, we see more collaboration between academia and industry, perhaps in the form of olfactory departments within either the science or art and design faculties.

Acknowledgments

All the students whom Joanna cites kindly gave their permission to quote from their individual *Sensory Design* projects.

Notes

1 International Flavours and Fragrances (IFF) is a global supplier of fragrances and flavors. Joanna worked for IFF in all aspects of fragrance project development from 1982 to 2000.

2 Used with students on the International Centre for Aroma Trades Studies Diploma, a distance learning course in perfumery education and training for the aroma trades industry (www.icatsaromaeducation.com/ Retrieved March 1, 2017).

3 https://www.hkb.bfh.ch/en/hkb/about-us/divisions/y-institute/ Retrieved May 27, 2017.

4 A term which they have branded for their specific use and according to their peculiar quality criteria under the trademark Air Sculpture®. An Air Sculpture® creation by Les Christophs should convey that a scent is meticulously built like a sculpture; created, composed, worked by the artists and experienced by the sniffers all in three dimensions. It is specifically designed for the air and diffused in the museum/art space

through the use of specific diffusing technologies which display the precise sensations of volumes, textures and other special olfactory effects to allow the brain to recreate further a three-dimensional emotion.

References

Buechner, L. (2012). *Sensory design project*, MA Fashion. Kingston University.

Curtis, T., & Williams, D. G. (2001). *An introduction to perfumery* (2nd ed.). Weymouth: Micelle Press.

Design with Scents. (2016). *Official Website*. Retrieved March 1, 2017 from www.designwithscents.com/

Fairley J., & McKay L. (2014). *The perfume bible*. London: Kyle Books.

Gascoigne, F. (2011). *Sensory design project*, MA Fashion. Kingston University.

Gilbey, E. (2012). *Sensory design project*, MA Fashion. Kingston University.

Hellen, S. (2014). *Sensory design project*, MA Fashion. Kingston University.

Kariappa, S. (2010). *Sensory design project*, MA Creative Economy. Kingston University.

Oxford Living Dictionaries. (2016a). Cognition. Retrieved March 1, 2017 from https://en.oxforddictionaries.com/definition/cognition

Oxford Living Dictionaries. (2016b). Synaesthesia. Retrieved March 1, 2017 from: https://en.oxforddictionaries.com/definition/synaesthesia

Phaidon. (1998). *The fashion book*. London: Phaidon Press.

Wang, N. (2014). *Sensory design project*, MA Fashion. Kingston University.

19

SITUATING OLFACTORY LITERACIES

An Intersensory Pedagogy by Design

Melanie McBride and Jason Nolan

Introduction

> Classrooms, like wine, are known by their smell and tactile qualities as well as by their sight. As Jackson (1968) remarked, the smells of chalkdust and stale milk are almost certain signs of an elementary school classroom.
>
> *Eisner, 1991, p. 68*

Eisner and Jackson's recognition that smell is often the first, and most powerful, sign of the social and cultural life of a place calls attention to a way of knowing few of us are ever explicitly taught. As schoolchildren, we may learn the differences between colors, shapes and musical notes, yet there is no such 'alphabet for the nose' (Tolaas, 2011). Reflecting on this missing, and often hidden, curriculum, our chapter asks how it is possible to train built environment professionals, planners and designers, as well as public users of designed spaces, to communicate their olfactory knowledge, preferences or needs in the absence of shared literacies.

Dismissed as "vestigial, a relic of our mammalian ancestors who got around on all fours" (Agapakis & Tolaas 2012, p. 569), olfactory knowledge is not so much forgotten as it is "systematically suppressed" by the same structures that repress other 'essences' "considered threatening to the social order" (Classen, Howes & Synnott, 1994, p. 4). These socio-cultural dimensions of smell are especially consequential for urban planners and designers of public space, given the role of odor in assigning and maintaining gendered, racialized and classist social categories of discrimination (Corbin, 1986; Henshaw, 2014; Rhys-Taylor, 2013). In her research into the UK's urban smellscapes, Henshaw (2014) observes how smells associated with cultural diversity, such as "international food odours" (p. 96), are often deemed "out of place" (p. 97) in xenophobic disputes over heritage.

While there are many specialized tools available for detecting and analyzing environmental odors (Henshaw, 2014), these instruments do not 'capture' the more subjective dimensions of smell as a socio-culturally 'situated' (Haraway, 1988) or 'lived' (Van Manen, 1997) knowledge. This distinction also reflects, as de Castell and Jenson (2004) observe, how tools are often "decontextualized, either hidden or abstracted, from their particular socio-cultural locations" (p. 652). We argue that in order to meaningfully situate smell, we must begin from the 'margins' (hooks, 1994), rather than the center, of our ocularcentric 'sensory order' (Howes, 2005).

If we are to realize Victoria Henshaw's vision of "designing more humanistic places for the future" (2014, p. 224), this necessarily involves an explicit commitment to diverse ways of knowing, being and sensing. Building on our ongoing research of situated and marginalized 'knowledges' and literacies, our chapter contributes a pedagogically oriented framework of perspectives, principles and considerations intended to scaffold designers, planners and built environment professionals in equitably and meaningfully communicating and designing with smell. Beginning with arts educator Elliot Eisner's (1991) idea of 'educational connoisseurship,' we examine the 'hidden curriculum' (Jackson, 1968) of traditional olfactory knowledge domains and their investments in text-bound modes of literacy (Nolan & McBride, 2015). We then consider the participatory and do-it-yourself (DIY) practices of olfactory artist-practitioners as a promising alternative to traditional, mastery-oriented domains of olfactory expertise. Our chapter concludes with critical considerations for learning, communicating and designing with smell in practice.

The Art of Appreciation: Learning as Connoisseurship

According to Elliot Eisner (1991), education can be regarded as a form of connoisseurship similar to that of wine appreciation. Eisner defines educational connoisseurship as an "art of appreciation" that discerns the "qualities" as well as the environmental and socio-cultural "conditions that give rise to these qualities" (p. 65). Eisner distinguishes his critical use of appreciation from its conventional association with 'liking' through his example of 'appreciating' the "subtle but significant forms of sexism and racism found in books, newspapers, laws, and other seemingly neutral [learning] materials" (p. 69). Discerning dimensions of experience and perception that are often systematically rendered 'invisible' is not a matter of intuition, but rather a deliberate practice. As Eisner notes, "learning to see what we have learned not to notice remains one of the most critical and difficult tasks of educational connoisseurs" (p. 77).

The art of appreciating qualities that may be 'hiding in plain sight' informs Philip Jackson's (1968) concept of the 'hidden curriculum,' which refers to institutionally valued, but concealed, power relations, beliefs, norms and dispositions required for success in traditional structures of education. From this standpoint, the pre-eminence of language over the physical knowledge of the body is among the most definitive expressions of the hidden curriculum (Nolan & McBride, 2014, 2015). This relation, of the word over the body, is also, paradoxically, characteristic of Western structures of olfactory knowledge and expertise.

Olfactory Knowledge: What We Talk about When We Talk about Smell

While linguistics researchers have established the centrality of language in the domain of olfactory expertise (Croijmans & Majid, 2015), we know very little about the ways non-experts communicate about smell without access to training or specialized vocabulary. In their review of the literature of olfactory expertise, Royet, Plailly, Saive, Veyrac and Delon-Martin (2013) find that the domain of olfactory knowledge "remains poorly understood, mostly likely because experts in odor, such as perfumers, sommeliers, and oenologists, are much rarer than experts in other modalities, such as musicians" (p. 1). The authors also note that the existing literature does not clearly articulate to what degree olfactory expertise is the result of training versus 'lived' exposure to sensorially rich environments and experiences.

More recent work suggests that specialist training in aroma descriptors has limited application outside of the given contexts in which they are used. In their empirical study of expert and non-expert olfactory language use, psycholinguists Croijmans and Majid (2015) find that while wine and coffee experts have more opportunities than non-experts to acquire specialized vocabulary, they were no better than untrained individuals in their ability to identify and name smells. As celebrated wine critic and writer Jancis Robinson (2000) suggests, this may be because "there is no way of extracting and comparing the sensory impression each of us associates with a given word because the business of tasting is so essentially hidden and subjective" (p. 34). These differences are most often framed as a problem of language, inspiring endless schemes and systems of sensory classification that typically privilege Western categories and examples.

In their study of olfactory language development among the Jahai speakers of the Malay Peninsula, Majid and Burenhult (2014) identify the "dangers of extrapolating from English speakers to the rest of humanity" (p. 270) when it comes to categorizing and naming smells. Croijmans and Majid (2015) note that, in contrast with the Jahai, who learn a language of odors in childhood, Western olfactory experts acquire their specialist descriptors "late in life and long after any critical period for language learning" (p. 487). These findings highlight the limitations of language as a foundation for olfactory knowledge and communication, and also raise questions about where, when and how we acquire odor literacies, beyond the word.

Physical Knowledge: Embodying Olfactory Literacies

In the field of early childhood education, sensory learning is traditionally valued for its developmental benefits. However, many schools and institutions throughout North America are increasingly adopting 'scent-free' and 'scent-reduced' policies, despite a lack of scientific evidence of their necessity (Senger, 2011). This is especially concerning for educators, as early childhood is often our last opportunity to autonomously explore and construct our own ways of sensing through direct physical inquiry of our immediate environment (Kamii, 1991). As a consequence of the dominant 'ages and stages' model of development, 'social knowledge' in the form of language begins to abstract and distance us from more embodied ways of understanding and communicating (Kamii, 1991; Nolan & McBride, 2015), such as smelling, tasting or touching as 'significant' ways of knowing.

Considering smell's primal role as our "chemosensory custodian of survival" (Martin, 2013, p. 3), it is curious that we do not place a greater premium on teaching children empirical skills of detecting, identifying and discriminating between odors as a form of physical knowledge. Elaborating on Henshaw's (2014) contention that "built environment professionals do not consider the physicality of the body within the environment" (p. 4), and her emphasis on the role of 'bodily states' in general (Henshaw, 2014), we argue that physical knowledge constitutes a neglected priority for learning and designing with smell. Indeed, as proponents of the multimodal 'turn' (Kress, 2013; Jewitt, 2009) suggest, there is a need for 'non-canonical' forms of literacy, learning and representation that move 'beyond language' (Jewitt, 2009).

In response to these challenges, we next turn to practices of olfactory artists who are contributing accessible, inclusive and intersensory approaches to designing with smell that also have implications for learning and literacy.

Smelling in Practice: An Intersensory Turn

Some of the most innovative interventions of our ocularcentric and logocentric knowledge paradigms are being undertaken by contemporary olfactory artists who have reconfigured olfactory 'art' beyond bounds of traditional perfumery to ask new questions about the social and cultural life of smell. By and large, these artists embrace smell's more transgressive, ephemeral and playful qualities as an opportunity to challenge conventional beliefs, values and assumptions about odors. As artist-practitioners Agapakis and Tolaas (2012) note, "a deeper understanding of odors and olfaction allows us to tolerate smells, to value their importance and their power, and to smell their nuanced and subjective beauty" (p. 574). For many olfactory artists, the everyday world is their atelier.

In the realm of olfactory art as a mode of 'place-making,' UK artist-practitioner Kate McLean's (2015b) urban smell maps extend Henshaw's (2014) urban smellscapes research through techniques of fine art and applied design. McLean invites smellwalk participants to visually represent their affective responses to urban smellscapes through a combination of visual, text-based, sculptural and volatile media. For example, McLean's *Anosmics, Hyposmics and Normosmics: Sharing Olfactory Experience* workshop (McLean, 2015a) engaged sensory-diverse participants in a 'buddy smellwalk' that encouraged them to share their associations, memories, and experiences of smellscapes through watercolors and 'smell notes.' One of the participants even created her own custom smell wheel to express the more muted smellscapes of anosmia and parosmia (McLean, 2015a). Unlike mastery-focused domains of olfactory expertise such as perfumery or wine tasting, projects like McLean's are intended for ordinary citizens whose olfactory 'meaning-making' is grounded in their own bodily, lived and ecological experiences.

In her research into McLean's smellwalking and smell mapping workshop in Marseille (McLean, 2015b), author McBride observed how participants made use of their whole bodies, including other material affordances such as architectural features, to access a perceived smell source. As a component of her fieldwork in Provence, France, McBride has documented the role of intersensory ambiances, including sound (McBride, 2015a), to situate smell as a relational and contingent part of the total social, cultural and environmental ecology to engage with smell as a whole-body experience. In Figure 19.1, author McBride applies the practices of smellwalking, crouching on the sidewalk in order to smell the fragrance emitted from a pipe at the base of a retail store in Aix-en-Provence, France. This illustrates how the embodied physical practices of smellwalking are contingent on the 'ecological affordances' (Gibson, 2013) of the environment in which they are perceived. This is an example of 'making' sense as a deliberate, physically embodied and ecologically contingent activity that puts smells into context.

FIGURE 19.1 Author McBride smelling perfume vent, Aix-en-Provence, France.

Image courtesy of Carolyne Kauser-Abbott.

McBride and her colleague, acoustic ethnographer Milena Droumeva, revisited approaches for an intersensory workshop, *Making Sense, Sensing Space* (Droumeva, 2015; McBride 2015b), which involved listening, smelling, touching and tasting as mode of sensory inquiry into the ambiances of urban environments in downtown Toronto. Drawing on a combination of practices from smellwalking and acoustic ecology, workshop participants used video, audio, still images and drawing to document and mediate their experiences of varied stimuli, with an emphasis on and attention to their own bodily responses in relation to the total environment (Droumeva, 2015).

In an effort to apply these principles in practice, McBride collaborated with human–computer interaction researchers to create an original aroma composition for an intersensory museum interaction involving a replica 16th-Century prayer nut (see Figure 19.2) (Harley et al., 2016). As these historical artifacts are traditionally presented behind glass, this tangible intersensory interaction is designed to provide museum visitors with access to sensory stimuli otherwise unavailable in traditional museums. Using DIY instructions from the website of smell artist Maki Ueda, who combines "chemical and kitchen skills in order to extract the scents of daily life, including foods, ambient aromas, and bodily scents" (Ueda, 2015), McBride translated historical lists of ingredients, such as lavender, rose petals, saffron and cinnamon, into two custom scents using raw materials such as spices, resins and essential oils (Harley et al., 2016).

FIGURE 19.2 Prayer nut.

Image courtesy of Jamie Kwan.

FIGURE 19.3 DIY smellbox.

Image: authors' own.

Figure 19.3 shows a custom DIY aroma kit created by author McBride that features everyday aroma materials that are locally sourced. As a former public school teacher, McBride was inspired to locate materials that are accessible to the average public school teacher who may have only a few days to plan an activity on a very low budget. These examples illustrate how, in the absence of a common curriculum for designing with smell, designers, researchers, educators and artist-practitioners can establish a more accessible model for olfactory knowledge through the open documentation, and also attribution, of works that contribute to a highly emergent context of practice that is different from that of fragrance industry professionals or technicians.

The following considerations and questions, drawn from examples developed in this chapter, are intended to scaffold artist-practitioners, design educators and built environment professionals in establishing their own inclusive and humanistic principles and practices for designing with smell.

Situating Smell: Considerations for Intersensory Design

1. Sensory Standpoint: Whose 'Senses' Are We Talking About?

Before we can design for others, we must first examine the sources of our own beliefs, assumptions and values of smell and the senses. Identifying our 'sensory standpoint' involves carefully considering the influence of these often unexamined assumptions and beliefs on our chosen philosophies and approaches. For example, one might ask: How do my own assumptions or investments in particular perspectives or theories of smell and the senses reflect the interests of dominant, versus non-dominant, groups or individuals? How do my orienting metaphors, or tropes, of smell implicitly communicate and reinforce particular beliefs and assumptions about smell and the senses?

2. Sensing Inclusion: Design from the Margins

To establish truly inclusive and 'humanistic' design principles for smell, our approaches need to reflect clear commitments to equity, diversity and inclusion. Given the role of smell in marginalizing non-dominant individuals, groups and ways of sensing, this consideration asks: To what degree will non-dominant groups see their identities, needs, and lived experiences reflected in our design schemes and approaches? Have inclusion and accessibility guidelines been developed in collaboration with the non-dominant groups or individuals whose interests we claim to represent?

3. Making Sense of Smell: Do-it-yourself Learning Inquiry

Reflecting on the often do-it-yourself approaches of olfactory artists, we suggest that DIY approaches can provide an initial and self-directed route to olfactory learning for designers who may not have the resources to participate in programs or workshops involving costly travel expenses. As with the sensory standpoint, this practice-focused consideration encourages the development of one's own custom, intersensory knowledge base through self-directed inquiry. For example: one might begin a relationship with smell through the daily practice and documentation of odor encounters using traditional or digital media.

4. Smelling in the Present Tense: Embodying Sensory Knowledge

While smell is often celebrated as a kind of Proustian time machine, this is a misleading claim that not only neglects how olfactory triggered memory actually works (Gilbert, 2008), but also reduces smell to a 'playback device' (Ingold, 2011). We argue that practices such as smellwalking function to put smelling into the 'present tense' by relating with the sensations of odors as a profoundly physical, rather than purely interior, way of knowing. Extending this logic, this principle asks: To what degree do our chosen design methodologies reflect current knowledge of the functional and physiological process of olfaction? Furthermore, how do our design principles engage with the 'intersensory' dimensions and ambiances of our immediate environment at the level of varied and sometimes conflicting sensations and bodily states?

5. The Internet of Smell: Networked Learning and Sharing

The ubiquity of networked digital and mobile communications and media calls for new ways of representing and communicating smell. We suggest that designers, planners and practitioners explore distributed knowledge sources through online communities of practice and participatory social media platforms such as YouTube, Instructables, Twitter and Facebook. While the use of mobile or personal media can distract from focused practices of smelling, it is also a crucial means of documenting sensory encounters. This principle encourages designers and planners to consider the strategic use of mobile and networked media as a means of encouraging citizen participation through user-generated data collection and digital applications that promote documenting and sharing olfactory experiences.

Conclusion: Smelling in the Present Tense

In our efforts to situate socio-cultural, physically embodied and intersensory contexts of olfactory learning and design, we have proposed a more humanistic, inclusive and accessible foundation for designing with smell that educates from the 'margins' (hooks, 1994), rather than the center, of our 'sensory order' (Howes, 2005). More critically, our emphasis on the physical act of smelling emphasizes the corporeality of sensations. Unlike expert domains of olfactory knowledge, which function to qualify the mastery of specialized practices and terminology, our emphasis on self-directed inquiry implies that designers, practitioners and the public are capable of constructing our own 'situated knowledges' (Haraway, 1988). Echoing Rancière's critical treatise on education, *The Ignorant Schoolmaster* (1991), educational researchers de Castell and Jenson (2004) ask why it is still so 'radical' to believe that "intelligence is not something that one individual has more or less of and that, in fact, everyone is equally intelligent, given how widely distributed is humanity's capacity to focus its intellect, effort, and attention on accomplishing its various goals" (p. 395). We argue that learning to appreciate, in Eisner's critical sense of the term, how the low significance of smell rearticulates other forms of marginalization involves a 'de-schooling' (Illich, 1973) of traditional modes of literacy to discover curricula we really need. To sense what we have been conditioned not to sense through the ordinary act of smelling is, finally, our most extraordinary encounter with knowledge as a living and responsive ecology. Most plainly, the physical act of putting our noses to the ground brings us back down to earth and into our bodies as a primary way of knowing.

References

Agapakis, C. M., & Tolaas, S. (2012). Smelling in multiple dimensions. *Current Opinion in Chemical Biology*, *16*(5), 569–575.

Classen, C., Howes, D., & Synnott, A. (1994). *Aroma: The cultural history of smell*. London: Routledge.

Corbin, A. (1986). *The foul and the fragrant: Odor and the French social imagination*. Cambridge, MA: Harvard University Press.

Croijmans, I., & Majid, A. (2015). Odor naming is difficult, even for wine and coffee experts. In D. C. Noelle, R. Dale, A. S. Warlaumont, J. Yoshimi, T. Matlock, C. D. Jennings, & P. P. Maglio (Eds.), *Proceedings of the 37th Annual Meeting of the Cognitive Science Society (CogSci 2015)* (pp. 483–488). Austin, TX: Cognitive Science Society.

de Castell, S., & Jenson, J. (2004). Paying attention to attention: New economies for learning. *Educational Theory*, *54*(4), 381–397.

Droumeva, M. (2015, December 1). Making sense – intersensory workshop [Blog post]. Retrieved March 1, 2017 from http://natuaural.com/?p=1477

Eisner, E. (1991). *The enlightened eye: Qualitative inquiry and the enhancement of educational practice*. New York: Macmillan.

Gibson, J. J. (2013). *The ecological approach to visual perception*. New York: Psychology Press.

Gilbert, A. (2008). *What the nose knows: The science of scent in everyday life*. New York: Random House.

Haraway, D. (1988). Situated knowledges: The science question in feminism and the privilege of partial perspective. *Feminist Studies*, *14*(3), 575–599.

Harley, D., McBride, M., Chu, J. H., Kwan, J., Nolan, J., & Mazalek, A. (2016). Sensing context: Reflexive design principles for intersensory museum interactions. *Proceedings of the 20th Annual Conference of Museums and the Web*, Los Angeles, California. Retrieved May 27, 2017 from http://mw2016.museumsandtheweb.com/paper/sensing-context-reflexive-design-principles-for-inter-sensory-museum-interactions/

Henshaw, V. (2014). *Urban smellscapes: Understanding and designing city smell environments*. London: Routledge.

hooks, b. (1994). *Teaching to transgress: Education as the practice of freedom*. London: Routledge.

Howes, D. (2005). *Empire of the senses: The sensual culture reader*. Oxford: Berg.

Illich, I. (1973). *Deschooling society*. New York: Harper and Row.

Ingold, T. (2011). *Being alive: Essays on movement, knowledge and description*. London: Routledge.

Jackson, P. (1968). *Life in classrooms*. New York: Teacher's College Press.

Jewitt, C. (Ed.). (2009). *The Routledge handbook of multimodal analysis*. London: Routledge.

Kamii, C. (1991). Toward autonomy: The importance of critical thinking and choice making. *School Psychology Review*, *20*(3), 382–388.

Kress, G. (2013). *Multimodality: A social semiotic approach to contemporary communication*. London: Routledge.

Majid, A., & Burenhult, N. (2014). Odors are expressible in language, as long as you speak the right language. *Cognition, 130*(2), 266–270.

Martin, G. N. (2013). *The neuropsychology of smell and taste*. London: Psychology Press.

McBride, M. (2015a). Lavender-Puyloubier-France. Retrieved March 1, 2017 from https://soundcloud.com/multisensorymel/lavender-puyoubier-france

McBride, M. (2015b, November 3). Making sense, sensing space: An inter-sensory walk & workshop [Blog post]. Retrieved March 1, 2017 from https://relab.blog.ryerson.ca/2015/11/03/making-sense-sensing-space-an-inter-sensory-walk-workshop/

McLean, K. (2015a, January 20). Anosmics, hyposmics and normosmics: Sharing olfactory experience [Blog post]. Retrieved March 1, 2017 from http://sensorymaps.blogspot.ca/2016/01/anosmics-hyposmics-and-normosmics.html

McLean, K. (2015b). Smellscape mapping Marseille. Retrieved March 1, 2017 from http://sensorymaps.com/portfolio/smellscape-mapping-marseille/

Nolan, J., & McBride, M. (2014). Beyond gamification: Reconceptualizing game-based learning in early childhood environments. *Information, Communication & Society, 17*(5), 1–15.

Nolan, J., & McBride, M. (2015). Embodied semiosis: Autistic "stimming" as sensory praxis. In P. Trifonas (Ed.), *The international handbook of semiotics* (pp. 1069–1078). Berlin: Springer.

Rancière, J. (1991). *The ignorant schoolmaster: Five lessons in intellectual emancipation* (K. Ross, Trans.). Stanford, CA: Stanford University Press.

Rhys-Taylor, A. (2013). The essences of multiculture: A sensory exploration of an inner-city street market. *Identities, 20*(4), 393–406.

Robinson, J. (2000). *How to taste: A guide to enjoying wine*. New York: Simon & Schuster.

Royet, J. P., Plailly, J., Saive, A. L., Veyrac, A., & Delon-Martin, C. (2013). The impact of expertise in olfaction. *Frontiers in Psychology, 4*(928), 1–11.

Senger, E. (2011). Scent-free policies generally unjustified. *Canadian Medical Association Journal, 183*(6), E315–E316.

Tolaas, S. (2011). An alphabet for the nose. *Journal for Artistic Research*, 7. Retrieved March 1, 2017 from https://www.researchcatalogue.net/view/?weave=1036

Ueda, M. (2015). Bio. *Atelier Maki Ueda*. Retrieved March 1, 2017 from www.ueda.nl/index.php?option=com_content&view=category&id=94&Itemid=757

Van Manen, M. (1997). *Researching lived experience*. London, Ontario: Althouse.

PART VI

Historic and Theatrical Smellscapes

INTRODUCTION

Jonathan Reinarz

This part deals broadly with smell in the arts, past and present. It begins, however, with the future. The Futurists, as their name denotes, were an avant-garde movement that, among other things, repudiated the past and imitation in artistic performance, frequently rebelling against harmony and good taste. As one might imagine, their works reserved a place for scents, primarily those that invoked feelings of disgust. Like other scents, however, theirs were ephemeral, which inspired Caro Verbeek (Chapter 20) to explore the possibility of reviving some of those employed in Futurist works, including scented poetry recitals, aromatic toys, and perfumed sculptures.

One of the first members of this artistic movement to contemplate the realm of scent was Fedele Azari, a decorated pilot of World War I, better known for his aerial performances. In his work *Flora Futurista* (1924), Azari experimented with smells, predicting the death of flowers and natural odors, or at least heralding the arrival of abstract scents, such as those invented by French perfumers Coty and Chanel (although they do not seem to have discussed Chanel's scents). One of the challenges of Verbeek's project is to consider how Azari's scents would have been perceived by interwar audiences, who inhabited very different sensory worlds, certainly distant from ours, having only just encountered less concrete fragrances, such as *Chanel No. 5*, a perfume with which today's consumers have long been familiar.

The search for such a formative experiential moment is what Verbeek describes as her pursuit for the 'period nose.' One might similarly consider the way in which an 18th-Century public first responded to the taste of other novel consumer goods, including chocolate or coffee. Today's shoppers, accustomed to a broad palette of tastes and scents, might easily underestimate the impact these once made on a somewhat naive public that was far less seasoned when it came to the senses. Textual sources are therefore drawn upon by Verbeek in an effort to recapture these primal sensual moments. In another experiment, Verbeek seeks to reconstruct the Futurists' *Tactile Dinner Party*, to memorable effect. Whether or not she is ever able to capture the elusive fragrance of a Futurist perfume, it seems likely that such a scent would have had an extremely volatile top note. Either way, Futurist works, which Verbeek has successfully re-enacted, continue to have the power to shock.

Stuart Eve (Chapter 21) is similarly chasing after lost scents in a fragrant heritage project that aims to excavate the lost scents of heritage sites. If the potent scents employed by Futurists eventually lost their pungency with the passage of time, more subtle fragrances from even earlier eras leave even fewer traces in the archaeological record. Eve nevertheless recognizes the power of smells to invoke intense feelings and meanings, and describes the use of an innovative portable device, which might easily introduce scents in situ and forever change the way in which historical venues are experienced. The process begins with digital maps, in this case the fields of Waterloo, the location of the final battle of the Napoleonic Wars, the terrain of which has retained many of its features. In designing scents to correspond with this decisive European conflict, Eve has identified base note smells that characterized this region over the *longue durée* surrounding the clash of British, Prussian and French forces in June 1815. These odors were then overlaid with top notes of cannon fire and the stench of rotting flesh, which briefly punctuated the atmosphere of this timeworn landscape. Working with war veterans, Eve is concerned with attaining something authentic, given previous dubious attempts to perfume museum exhibitions, but is equally sensitive to the ethics of experimenting with smell. Unlike the Futurist artists explored in Caro Verbeek's chapter, Eve wishes to avoid adding to the trauma endured by those who experienced past conflicts. Thus, although scents are regarded as a crucial part of reconstituting the past in its totality, the dangers of total recall prompted by an accurate smellscape are equally worthy of consideration.

The second half of this part comprises two chapters that examine the use of scent on stage. McGinley and McGinley (Chapter 22), like Eve, discuss the power of scent to draw audiences fully into theatrical performances. In particular, they explore a singular experimental theater piece performed in the Soap Factory in Minneapolis, Minnesota in 2014. Just as scent is regarded by Stuart Eve as a means of allowing visitors to Waterloo to immerse themselves fully in the field of battle, smell in this instance blurs the boundaries between stage and spectators, drawing theatergoers into a fictional performance space. As in the fields outside Waterloo, this is done by way of technology, but only after the McGinleys fully consider the science of smelling. In this case, a performance of *Treasure Island* was enhanced by way of fragrant interludes, introducing audience members to ambient scents redolent of a pirate ship, an old tavern, sea travel and a jungle environment. At times, scents piped into the seating gallery shocked unsuspecting audience members, but were intended to add meaning and depth to scenes and storylines. Unlike the interwar period, when seemingly less respected audiences were assaulted by experimental Futurist theater, today's audiences can expect scent warnings to accompany all other advertising and information about shows. Scent-free performances are also offered for less adventurous showgoers.

Chapter 23, Anna D'Errico's "The Nose Onstage," unmistakably places scent center stage. Dealing with some of the same issues raised in the McGinleys' chapter concerning scents and the theater, it also returns readers to issues discussed at the outset of this part. In particular, the chapter reflects on the strong reactions that odors can trigger. While recognizing the ability of smells to conjure mood and ambience, not only among audience members, but also performers, D'Errico equally notes the difficulties of working with scent in theater. Not merely ephemeral, scents, and their meanings, vary with each nose. Art in this case is clearly in the nose of the beholder. Fragrances are also hard to control in the playhouse or concert hall. As in a previous age of Smell-O-Vision, scents reach some seats sooner than others, and do not always vacate the auditorium, or oditorium, at the end of a scene. Neither can audience members easily 'look' away from undesired odors, for scents enter the person through the nostrils without permission. And while a distinct scent might be relatively innocuous in the opinion of one spectator, it may trigger very different, even violent, responses in others. Thus, it would seem that caution should be applied to the use of scents in a theatrical context.

Some of the plot lines in this part will be familiar to those versed in the science of scent scholarship. For example, the authors here are as concerned with the physiology of smelling as the use of scents in theater. D'Errico also considers scientific theories of smelling before outlining the histories of her unique performances. She describes the potential of some scents to over-stimulate the olfactory epithelium, and advises others avoid harsher scents, such as ammonia. With some substances possessing the ability to inflame nasal mucosa and trigger headache and nausea, one might as easily wonder how these materials were overlooked by the Futurists. In any case, the differences between the 1930s and our era rapidly become apparent. That said, the language employed by today's scent scholars appears as playful as one might have expected from the proponents of Futurism. Thus, rather than refer to her museum audience, Verbeek speaks of an 'odience.' When describing her installation, she notes the responses of 'onsmellers,' rather than onlookers. Such wordplay has no doubt become familiar to most scholars working in sensory studies today.

D'Errico also considers far more than the impact of scents on actors alone. She also reflects on the smells emitted by audience members, and ponders how these too might influence the experience of theatergoing. Perhaps the simple issue of the cost of theatergoing today, where audiences are drawn largely from 'perfumed' socio-economic backgrounds, let alone the expenses associated with the Japanese art of *koh-do*, have removed us from the more democratic experience of early modern theater. Such ideas should remind some readers of Sally Banes' work on olfactory performances, which D'Errico summarizes before considering the interaction of scents and performance through an examination of three particular performance pieces: Michel Roudnitska's *Pierre de Soleil*, *Les Parfums de l'Âme* and *Duft*, a piece designed by D'Errico herself. The first of these is inspired by a poem underscored by an olfactory score, while the second takes odor as its starting point, from which characters and setting gradually emerge. D'Errico's production, inspired by Nietzsche's *Thus Spoke Zarathustra*, uses scents to signal changes in an imaginary metamorphosis. In all of these, like the chapters in this part, scent plays the title role. Like any novice performer thrust into a leading part, there is still room for improvement. But, to end where we began, as far as debuts go, scent in historical and theatrical performances appears to have a promising future.

20

INHALING FUTURISM

On the Use of Olfaction in Futurism and Olfactory (Re)constructions

Caro Verbeek

Introduction: The Use of Olfaction in Futurism

> The so-called sweet perfumes of flowers prove insufficient to our nostrils that demand more violent olfactory sensations . . . . Hence we begin a new field of investigation and artistic creation.
>
> *Azari, 1924a, p. 1*

This quote could come from one of the many contemporary olfactory artists, such as Sissel Tolaas, Peter de Cupere or Clara Ursitti, who are interested in olfactory taboos (like body odors), and in the more unpleasant side of the dichotomy of foul and fragrant, as opposed to perfumery. Instead, its source is from the Italian Futurist artist Fedele Azari. His elaborate publication, entitled *La Flora Futurista ed Equivalenti Plastici di Odori Artificiali* (The Futurist Flora and the Plastic Equivalents of Artificial Odors), is the first Futurist manifesto that discusses smell as a direct medium (see Azari, 1924a).[1] It contains all the Futurist traits: the typical bold and expressive use of language, the role of education, an emphasis on synesthesia, the claim of originality, the shock element, and the interest in bodily or lower senses. A manifesto on touch, *Il Tattilismo* (Tactilism), had already appeared (Marinetti, 1921), and a text addressing taste, *La Cucina Futurista* (The Futurist Cookbook), was to follow (Marinetti & Fillìa,[2] 1932).

Futurism provides many more examples that refer to smell, or directly address the sense of smell through performances, artists' books, and even toys with an aromatic dimension. Founded by the fascist poet and recital artist F. T. Marinetti (1876–1944), Futurism was born as a literary movement. Therefore, there are numerous textual sources that enable us to study and analyze not just which substances the artists may have used, but also the role olfaction played for them. In the following paragraphs, I will discuss one of the most olfaction-oriented manifestos, and in addition give examples of Futurist aromatic objects, some of which still exist. I will conclude by discussing the value and meaning of Futurist sensory reconstructions, and their presentation to a wider audience (or rather *odience*). How should we (re)present and interpret these reconstructed historical scents?

The Futurist Flora and the Plastic Equivalents of Artificial Odors

In his groundbreaking manifesto *La Flora Futurista ed Equivalenti Plastici di Odori Artificiali*, Azari (1924a) was the first Futurist to elaborate on the possibilities of actual smells as an artistic means. Like all Futurists, Azari believed in the supremacy of mankind, the synthetic and the machine over nature, ideally resulting in a manufactured *flora futurista*. Azari elaborated on the absolute need for the replacement of natural flowers, considered hectically sentimental and bourgeois, by synthetic ones. This, ideally, did not just include their appearance, but also their other essential aspect – perfume:

> Ascertained by this time that the flowers provided to us by nature do not interest us any more, we futurists to brighten, animate and decorate our paintings and our environments, have started the creation of a plastic[3] flower.
>
> Most original
> Absolutely invented
> Most colorful
> Most aromatic
>
> *p. 2*

and

> We keep plastically constructing our flower and violently colouring and perfuming her with the most intense perfumes.
>
> *p. 2*

The idea of these man-made flowers was realized by one of the most elusive Futurist figures, Giacomo Balla, who created numerous giant wooden and brightly colored, very angular specimens that contrasted with natural plants in as many ways as possible. Whether these were ever accompanied by scents is uncertain. By now in-odorate, it remains a question whether this is due to the volatile nature of scent, or its complete absence from the start. But Azari (1924a) left us with clues on specific perfumes used to add scent to Futurist sculptures, as he mentions some of the *odori artificiali* (artificial odors) by name: "I have realised a few plastic coloured interpretations of synthetic perfumes that are most in vogue ([l']Origan, C[h]ypre, Contessa Azzurra, etc.)" (p. 2). Fortunately, the first two of these perfumes were reproduced by the Osmothèque (a smell conservatory and perfume archive in Versailles, France), carefully reconstructed and kept in the right conditions. This makes it possible to not just study their cultural backgrounds, but excitingly, also, their inherent aroma.

Chypre, composed in 1917, is a famous composition created by the renowned French perfumer François Coty.[4] The composition, now no longer available in stores, is often described as the first step to abstract perfumery. It consisted of synthetic ingredients that 'amplify' or highlight other ingredients, generating extreme contrasts between top, heart and base notes (Stamelman, 2006, p. 186).[5] Furthermore, Coty was one of the few contemporary perfumers to use raw materials without waxes. As a result, his compositions were much stronger, or 'loud,' as if someone had turned up the volume. Naturally, this appealed to the Futurists, who glorified noise and speed, synesthetically expressed by bright colors and dramatic contrasts.

Strikingly, or maybe rather logically, the most famous perfumer and perfume theoretician of all time, Edmond Roudnitska, also sees correspondences between visual and olfactory art: he describes *l'Origan*, which came into existence in 1905 and which was also created by Coty, as "a translation of fauvist sensibility," borrowing all its "vividness, violence and audacity" (Stamelman, 2006, p. 186). Perfumers are often aware of art and art history. But besides knowing which perfumes were fashionable, there is little indication that Futurists knew about perfume theory or history.

It is almost certain that Azari did not just talk about perfumes, but actually experimented with them, as proven by a letter sent to his friend and colleague, the Futurist Fortunato Depero, just a few days prior to the first presentation of his manifesto on synthetic flowers during a Futurist conference. In the postscript, we read:

> I need to remind you to bring me: one bottle of the perfume "dans la nuit" by Worth (the flat one, not the spherical one, 12 cm in height, costing appr. 100 fr.) and a bottle of "tabac blond" by Caron (the perfume, not the eau de toilette). Don't forget!!!
>
> *Azari, 1924b, p. 8* verso

Tabac Blond (composed in 1919) was one of the revolutionary postwar perfumes designed to appeal to those independent European women wearing trousers and smoking cigarettes. The powdery leathery aroma would blend in with the smell of tobacco, now part of their unusual and new olfactory identity. It was the first in a class of masculine non-floral perfumes (Stamelman, 2006). Now it is a much more common note. Back then, it was revolutionary.

Apparently, it is possible to retrieve and recreate perfumes once used by the Futurists, for example by archival research. But then another question arises. Once obtained, how to present, diffuse and communicate smells for a wider audience to consume? Can we smell what a contemporary perceiver once smelled? To answer this question, a reconstruction of *Tabac Blond* and other perfumes were on olfactory display during the exhibition *Something in the Air – Scent in Art*, which took place at Villa Rot, Germany from March to August, 2015 (Verbeek, 2015). The show dealt with the role of smell in both contemporary and classic modern art, and featured about 15 installations by artists from 11 different nations, such as Peter De Cupere (Belgium), Maki Ueda (Japan) and Luca Vitone (Italy). The most important and recurring themes within olfactory art now, and in the past, such as memory, synesthesia and olfactory mapping, were installed throughout 12 rooms (including the historical bathroom of the former villa). One of the rooms was dedicated to the first half of the 20th Century. Photographs of Surrealist and Futurist exhibitions, catalogues, original manifestos and other texts adorned the walls and filled the showcases there. Glass laboratory vials containing scents provided the onlookers and *onsmellers* with olfactory cues of fragrant events that took place almost a century ago (see Figure 20.1).

It is almost impossible to imagine a smell just by its description. Only by inhaling its smoky perfume might the museum visitor understand how the leathery, misty aroma of *Tabac Blond* would have blended in with the smell of burning cigarettes, which were more woody and opiate at the time. To meet with these types of fashionable smokeables, the dry-down (base notes) reveals cedar and clove (Herman, 2013).

But we are not even close to an historical sensation yet. For by solely smelling this perfume, a museum visitor would still not understand why it was experienced as novel, or why the Futurists were interested in this composition in particular. To fully grasp this, we need to address our noses even more by getting to know (aromatically) the perfumes that were created before, during and after a given period in history, and by doing so creating a *period nose* that can obtain contextual information olfactorily. *Tabac Blond*'s strong (loud), dark and sultry tobacco-leather bouquet juxtaposes the modest and innocent florals of the (by then) more old-fashioned perfumes (Herman, 2013). When compared, it is not hard to realize the revolutionary changes perfume underwent. Today, many perfumes – Byredo's niche perfume *Baudelaire*, for example (meant for men, but worn by the author of this chapter) – contain tobacco notes. Consequently, these are not experienced as shocking or innovative since we are used to them already. We need to be aware of these changes in context at all times.

Another important part of developing a period nose is training a capacity for distinguishing the *formal* characteristics of a perfume (similar to color, contrast and perspective in a visual composition), for they exist independently from a cultural-historical context. These are the volume (meaning

FIGURE 20.1 Futurist exhibits at the *Something in the Air – Scent in Art* exhibition, Villa Rot, Germany. Image: author's own.

that it reaches far, leaving a long sillage, or trail), strength and the powdery quality of *Tabac Blond*, which perfectly coincide with the two perfumes by Coty from more or less the same era discussed earlier. As stated previously, the Futurists were interested in superlatives, synthetic artificial ingredients, and exaggerations; criteria which all the perfumes mentioned by Azari certainly met. One can read about these qualities extensively, but only learn to recognize them by actually smelling (compare reading about blue without ever having seen the color). Based on these arguments, it is safe to assume that an historical notion obtained from textual sources, in combination with the sense of smell, can clearly help us – the contemporary perceivers – get more insight into what it might have been like to inhale Caron's perfume in the late 1920s, and truly gain a deeper understanding of our evaporated past.

La Cucina Futurista

After Futurism's tactile and synesthetic explorations based on smell, Marinetti and Fillìa (1932) proposed a revolution in food and dining habits in *La Cucina Futurista* (The Futurist Cookbook). As the Futurists aimed to merge life and art, the leader of Futurism sought to revitalize one of the oldest and most firmly rooted Italian traditions linked not only to gustation, but also to touch, hearing, sight and olfaction. In fact, there are few activities as multisensory as the act of eating.

The recipes in *The Futurist Cookbook* consisted of minute descriptions of the preparation and visual appearance of food, the desired physical properties of servants (sex and ethnicity), music, choreography, clothing, temperature, tactile impressions such as sandpaper next to the plates, and last but not least, the ambient diffusion of perfumes and other odors, preferably with an airplane propeller. These smells varied from aluminum and the smell of battle to lavender, carnations and vanilla. Because the vocabulary referring to smell and taste is almost non-existent, the literary educated artists came up with new words to describe taste and smell concepts within the text, such as *conprofumo* (a term that indicates the olfactory affinity of a given perfume with the flavor of a given food), *disprofumo* (a term that indicates the complementary nature of a given perfume with the flavor of a given food), and *polibibita* (a drink with several flavors, as in a cocktail) (Marinetti & Fillìa, 1932; Agapakis & Tolaas, 2012).

One of the most sensual performances envisaged by Futurism took the form of *A Tactile Dinner Party*, in which none of the senses was left out. This was imagined and described by Marinetti and Fillìa (1932). According to their text, guests were supposed to wear pajamas covered with different tactile materials such as sponge, sandpaper and silk. Each had to choose a dinner partner in complete darkness, according to tactile impression and inspiration. Once seated, diners would see waiters dancing with grand geometric movements and serving a 'polyrhythmic salad' of music boxes filled with lettuce leaves, dates and grapes. Between one dish and the next, guests had to "let their fingertips feast uninterruptedly on their neighbour's pyjamas" (Marinetti & Fillìa, 2014, p. 171). The second course, called 'magic food,' consisted of caramels stuffed with different ingredients like garlic, raw meat or, if you were more fortunate, candied fruits. The final dish included cooked and uncooked vegetables put in front of the participants. Without the use of their hands, diners were forced to bury their faces in the food, and to display uncivilized, or at least uncommon, behavior, but with a sensory goal. In this manner, the tongue as well as the lips and cheeks were in full contact with the textures of the greens, and hence the name of this course – 'tactile vegetable garden.' Each time someone dared to raise his/her head the waiters sprayed his/her face with perfumes of lavender and eau de Cologne (Marinetti & Fillìa, 1932). It is a well-known fact that the Futurists were intrigued by multisensory approaches and that they did not shy away from shock elements. But what exactly were the Futurists trying to achieve with this bombardment of sensory output?

During the course *The Other Senses*, taught annually at the Royal Academy of Art, The Hague, a group of students re-enacted this historical multisensory event of *A Tactile Dinner Party* in an attempt to find out at least what the effect would be nowadays. Whereas three 'waiters' prepared the performance by studying the available text, the diners – in this case, two male fellow students chosen at random – were unaware of what was about to happen, just like the original guests. During the entire re-enactment, the remaining students danced in geometric patterns to music composed especially for the event. The performers instructed the diners from time to time when they did not follow the rules, saying things such as: "Keep touching the objects on each other's garments when you are not eating!" The diners obediently followed these demands, causing cheers and laughter among those present. The sight of two people constantly touching each other's clothing had a somewhat sensual connotation, as one cannot avoid touching the other person's body. Whether the Futurists meant this to be humorous or sensual was not clear, but the absurdity of the situation during the re-enactment was undeniable (see Figure 20.2).

The 'magic food,' or balls of caramel filled with garlic or candied fruit, turned out to be either absolutely disgusting or pleasant. As the teacher, I felt I should try one, not wanting to appear a 'killjoy.' But in spite of my curiosity, I had to get rid of the 'tasty treat' after a few minutes of trying to swallow it. Reading about the dish definitely did not have this effect, but in reality, the combination seemed uncommon. For the 'tactile vegetable garden,' the students picked broccoli and served it both cooked and raw. The fact that the diners were not allowed to use their hands for this proved quite a challenge, as broccoli fell out of their mouths and some traces were left on their faces. Since the cooked (soft) and raw (hard) broccoli hardly displayed any visual differences, the students hypothesized that the tactile surprise effect must have been the reason for the Futurists' decision to serve raw and cooked vegetables.

For the perfumes, the executors used essential oil of lavender and the classic *Eau de Cologne 4711* (see below), mixed with alcohol to enhance the volatility. And indeed, every time a diner raised

FIGURE 20.2 Students re-enacting *Tactile Dinner Party*.

Image: author's own.

his/her head, he/she was treated with a splash in the face, which made the diners close their eyes and mouths. The perfume that slowly impregnated the air of the classroom was considered agreeable. So why did the Futurists favor these exact perfumes? Lavender water (*acqua di lavanda*) and eau de Cologne were both specific types of fragrances in those days. For example, famous perfume houses such as Coty and Rigaud issued their own eau de Cologne at the time of the Futurist movement, named after the medicinal water composed by the Italian barber Gian Paolo Feminis and marketed in Cologne in 1709 (Stamelman, 2006, pp. 53–54). The *Eau de Cologne 4711* (the address of this brand's store in Cologne is Glockengasse 4711) used in our re-enactment of *A Tactile Dinner Party* is allegedly very close to the original so-called 'miracle water.' Containing a large percentage of alcohol and essential oils such as bergamot, rosemary and lavender, it has strong antiseptic qualities.

By contrast, lavender water often consisted of many more ingredients, such as musk, rose and citrus fruits. Lavender grows in the Italian Alps, and was considered a typical national smell during the 1930s. Magazines of the time advertised it as a refreshing odor, suitable for outdoor activities such as tennis. In such instances, perfumes were typically sprayed onto handkerchiefs, clothes and the forehead, and used to refresh oneself. Nowadays, we would not necessarily associate lavender with sports in this manner. This emphasizes how we may have different associations with perfumes and scents that were once part of contemporary daily life. Thus, *Eau de Cologne 4711* could remind us of our grandmother, and lavender of a holiday in southern France. Accordingly, it is important to realize that it takes more than the reproduction of an historical substance to enhance an historical sensation.

Besides the performances, with their fleeting and dynamic nature, the Futurists wrote about objects with smells. Giacomo Balla and Fortunato Depero speak of toys "with the most intense and exciting colors, sounds and smells" in *Ricostruzione Futurista dell'Universo* (Futurist Reconstruction of the Universe) (Balla & Depero, 1915). Just like Balla's flowers, it was never proven scientifically if any of the toys were ever enriched with an olfactory dimension. Or might inherent instead of additional smells have provided for an aromatic quality? One of the materials used by Depero to construct toys was buxus, or boxwood, which is soft, easy to model, and reeks of cat urine when freshly cut. Although the Futurists fancied the shock element and promoted the foul smells of machines and war, it is not very likely they tried to stimulate the young through the smell of a feline bodily secretion. Many other toys were made of metal components. Although metal is odorless, when in contact with the skin it catalyzes reactions among the organic molecules that coat our bodies, emitting a typical metallic musty smell.

There is another category of objects made out of metal. Several books by Futurist artists were fashioned in tin for its multisensory qualities. The first metal artist's book, *Parole in Libertà Futuriste: Olfattive Tattili Termiche* (Futurist Words in Freedom: Olfactory, Tactile, Thermal), was a collaboration between Marinetti (words) and Tullio D'Albisola (images) (Marinetti & D'Albisola, 1932). Both binding and pages are made of lithographs printed on metal. No less than two of the poems focus on olfaction: "Paesaggio d'odori del mio cane-lupo" (Landscape of odors from my dog-wolf) and "Ritratto olfattivo di una donna" (Olfactory portrait of a woman).

The aromatic dimension of these poems is threefold. On a *material* level there is the inherent smell of the metal. The hand turning the page will give off an odor. Second, on a *textual* level there are numerous references to olfaction, such as descriptions of odors and their shapes and movements, expressed in the development of shapes, such as "3 6 nove spirali di odori di vaniglia" (3 6 nine spirals of scents of vanilla) and "questo soave agilissimo volume ovoidale di profumi freschi" (this soft most flexible egg-shaped volume of fresh perfumes). And there are newly invented words like *caldagrodolce*: a contraction of *caldo*, *agro* and *dolce*, meaning 'warm,' 'bitter' and 'sweet' (Marinetti & D'Albisola, 1932). The third is the most esoteric, least graspable level, yet it is most akin to Futurism, and modernism as a whole. The style, layout and typography are supposed (and believed) to evoke multisensory perceptions (and in this specific case, olfactory perception) synesthetically through the 'vibrations' Marinetti's words were supposed to generate. Thus, Marinetti in his 1913 manifesto, *Distruzione della Sintassi–Immaginazione senza Fili–Parole in Libertà* (Destruction of Syntax–Imagination without Strings–Words-in-Freedom) stated that:

> my Adrianople Siege-Orchestra and my Battle + Weight + Smell required many onomatopoetic harmonies. Always with the aim of giving the greatest number of vibrations . . . .

and

> On several parallel lines, the poet will throw out several chains of color, sound, smell, noise, weight, thickness, analogy. One of these lines might, for instance, be olfactory, another musical, another pictorial.

These 'vibrations' deserve some explanation. In the same year, Carlo Carrà (1913) wrote that "sounds, noises and smells are nothing but different forms and intensities of vibration." This reflected a contemporary and widespread paradigm based on the conviction that all sensory stimuli originate from the same vibrational source that was believed to be omnipresent (Devos, 2013). This idea lies at the core of the contemporary concept of synesthesia or the merging of senses. In *Traité du Verbe*, the symbolist poet René Ghil (1886) speaks of 'audition colorée,' or colored hearing (p. 34). He and many of his fellow artists believed that certain auditory perceptions simultaneously evoked color sensations. He argued that poetic words unite images (or meanings/ideas behind words) and sounds (the phonetic dimension). Words that are in harmony, on the level of both sound and meaning, reflect the perfect universal order (Devos, 2013). Similarly, Marinetti and the other Futurists often employed onomatopoeia, which refers to the property of a word to phonetically imitate, resemble or suggest the source of the sound it describes. Onomatopoeic words thus resonate with other sensory realms by their common vibrational ground.

Making Sense of Historical Scents

In this chapter, I have discussed how to reconstruct, present and interpret our olfactory past by using the sense of smell as a methodological tool, in addition to text and images: in one case by diffusing an historical perfume (*Tabac Blond*) in a museum and providing the *onsmeller* with the appropriate textual and visual context, and in another by re-enacting a documented performance (*Tactile Dinner Party*). Smith (2007) states that sensory history only becomes history when a particular sense is analyzed and put in the right historical context. Scents are neither universal nor trans-historical, he argues. Not only the object, in this case a smell, but also the perceiving subject changes constantly over time and place. Therefore, he warns historians not to overvalue sensory reconstructions. Smith grants language (and language alone) the power to overcome such subjective interpretation. But this should not lead to completely neglecting the sensory modalities constituting the original context or situation as granting us valuable information or evoking historical sensations. Academics could benefit from using as many strings of information as possible. As we have seen in the given examples, text and sensory input can reinforce each other. In the case of smell, it can even be dangerous and misleading to treat language as a substitute or isolated methodological tool. It is true that we have poor vocabulary when it comes to smell (Agapakis & Tolaas, 2012). But pursuing the quest for a proper descriptive language is not the (only) answer: smell is not connected to language in the way sight is, but the problem is that it is *thought* to be so. As sensory anthropologist David Howes (2002) eloquently puts it: "one is not aware of this huge difference in smelling and seeing related to thinking and language" (p. 67). This reflects Arnheim's (1969) assertion that "one can indulge in smells and tastes, but one can hardly think in them" (p. 18).

Indeed, 'olfactorizing' or imagining scents by verbal cues is only possible for 30% of the Western population (Köster, 2002). Olfactory memory, unlike our visual memory, is passive and incidental (Van Campen, 2014). Smell memories occur when we *re-encounter* a scent from the past, not when we try to *imagine* a scent from the past, at least for most of us. In a famous episode, Proust exactly describes what is so characteristic of an olfactorily (or at least chemically) triggered memory, in this case internally or retronasally experienced through the mouth (Verbeek & Van Campen, 2013). Marcel, the main character of *In Remembrance of Things Past*, suddenly feels as though he is transported back in time, remembering the days he spent in Combray, while soaking a madeleine in lime tea in the exact same way he did a few years before:

> . . . weary after a dull day with the prospect of a depressing morrow, I raised to my lips a spoonful of the tea in which I had soaked a morsel of the cake. No sooner had the warm liquid, and the crumbs with it, touched my palate than a shudder ran through my whole body . . . .
>
> *Van Campen, 2014, p. 13*

Marcel describes the magic of the moment and its emotional impact, for the memory of the days he spent during his early childhood was long lost. He starts contemplating that neither the word 'madeleine' nor the sight of the cake were capable of evoking this vivid memory of 'things past,' but senses such as taste could (Van Campen, 2014).

Even in cases where we encounter smells we have never perceived before, their physical presence can still convey knowledge. Annick Le Guerer (2002) explains it quite accessibly: ". . . just because the sense of smell is coined as anti-intellectual, doesn't mean we cannot benefit from its information intellectually. Smell is revelatory not just of substances, but also of moods, climates, and even existential states" (p. 11).

Olfaction is a subtle tool of knowledge that allows for an intuitive and pre-linguistic understanding. It enables us to have a profound relationship with our environment. Hippocrates recommended that a doctor is "a man with an open nose" (Le Guerer, 2002, p. 11). Photographs, lacking actual plasticity and conveying only one sensory dimension, are rarely questioned as means of representation or documentation of the multisensory and three-dimensional world of the past.

They do not just document, but additionally *shape* collective visual memory (Erll & Rigney, 2009). Scents could and should be treated in the same way. Past smells can become part of a shared memory beyond the idiosyncratic and individual: they may be trans-historical in the sense that they enable us to develop a period nose that recognizes a perfume's socio-historical context aromatically.

If we can somehow secure fleeting art historical objects, by documenting, replicating and presenting their volatile sensory heritage, it is up to future generations to (re)contextualize and interpret them, both academically and for a wider audience, and to overcome the supposed subjectivity Smith (2007) considers perilous. The pages of the olfactory history of art may one day transcend electronic or ink and paper representations, with texts being accompanied not just by visual illustrations, but ones which address the other senses, including olfaction.

Acknowledgments

This research is part of the PhD *In Search of Scents Lost – (Re-)constructing the Aromatic Heritage of the Avant-garde*. It is financially supported by Vrije Universiteit Amsterdam, the Netherlands Organisation for Scientific Research and the Royal Academy of Art, The Hague, in collaboration with International Flavours & Fragrances Inc., the Rijksmuseum, Amsterdam and several heritage institutions.

Notes

1 The manifesto *Arte degli Odori* (Art of Scents) by Ennio Valentinelli was written earlier by the Italian Futurist Ennio Valentinelli around 1916, but was not published.
2 Fillìa (1904–1936) was the name adopted by Luigi Colombo, an Italian artist associated with the second generation of Futurism.
3 'Plastic' means three-dimensional.
4 François Coty (1874–1934) was a French perfumer and businessman. He was a founder of the fascist organization Solidarité Française. The company he founded in 1904 is now Coty Inc., based in New York City (Toledano & Coty, 2009).
5 Perfumes are composed according to the principle of variation in volatility. The top notes are most volatile, followed by heart and base notes.

References

Agapakis, C. M., & Tolaas, S. (2012). Smelling in multiple dimensions. *Current Opinion in Chemical Biology*, *16*(5), 569–575.

Arnheim, R. (1969). *Visual thinking*. Berkeley, CA: University of California Press.

Azari, F. (1924a). *La flora futurista ed equivalenti plastici di odori artificiali: Manifesto futurista*. Rome: Direzione del Movimento Futurista.

Azari, F. (1924b). Letter by Fedele Azari to Fortunato Depero. MART (Museo d'Arte Moderna e Contemporanea di Trento e Rovereto) archive 900, Reference Dep. 3.1.7.5

Balla, G., & Depero, F. (1915). *The Futurist reconstruction of the universe*. Retrieved May 23, 2017 from www.unknown.nu/futurism/reconstruction.html

Carrà, C. (1913). *The painting of sounds, noises and smells*. Retrieved March 1, 2017 from www.unknown.nu/futurism/paintsound.html

Devos, P. P. D. (2013). *Talend lichaam. De visuele en haptische waarneming in de avant-gardepoëzie van Huidobro en Péret*. Doctoral thesis. Groningen: University of Groningen.

Erll, A., & Rigney, A. (Eds.). (2009). *Mediation, remediation, and the dynamics of cultural memory*. Berlin: Walter de Gruyter.

Ghil, R. (1886). *Traité du verbe*. Paris: Chez Giraud.

Herman, B. (2013). *Scent and subversion: Decoding a century of provocative perfume*. Guilford, CT: Lyons Press.

Howes, D. (2002). Nose-wise: Olfactory metaphors in mind. In C. Rouby, B. Schaal, D. Dubois, R. Gervais, & A. Holly (Eds.), *Olfaction, taste, and cognition* (pp. 67–81). Cambridge: Cambridge University Press.

Köster, E. P. (2002). The specific characteristics of the sense of smell. In C. Rouby, B. Schaal, D. Dubois, R. Gervais, & A. Holly (Eds.), *Olfaction, taste, and cognition* (pp. 27–44). Cambridge: Cambridge University Press.

Le Guerer, A. (2002). Olfaction and cognition: A philosophical and psychoanalytic view. In C. Rouby, B. Schaal, D. Dubois, R. Gervais, & A. Holly (Eds.), *Olfaction, taste, and cognition* (pp. 3–15). Cambridge: Cambridge University Press.

Marinetti, F. T. (1913). *Destruction of syntax–Imagination without strings–Words-in-freedom*. Retrieved March 1, 2017 from www.unknown.nu/futurism/destruction.html

Marinetti, F. T. (1921). *Il tattilismo*. Retrieved March 1, 2017 from www.futurismo.altervista.org/manifesti/tattilismo.htm

Marinetti, F. T., & D'Albisola, T. (1932). *Parole in libertà futuriste olfattive tattili termiche*. Rome: Edizioni Futuriste di Poesia (Lito-Latta). Retrieved March 1, 2017 from http://brbl-dl.library.yale.edu/vufind/Record/3548809

Marinetti, F. T., & Fillìa (1932). *La cucina futurista*. Milan: Sonzogno.

Marinetti, F. T., & Fillìa. (2014). *The futurist cookbook* (S. Brill, Trans.). London: Penguin Classics.

Smith, M. (2007). Producing sense, consuming sense, making sense: Perils and prospects for sensory history. *Journal of Social History*, 40(4), 841–858.

Stamelman, R. (2006). *Perfume: Joy, scandal, sin – a cultural history of fragrance from 1750 to the present*. New York: Rizzoli.

Toledano, R. B., & Coty, E. Z. (2009). *François Coty: Fragrance, power, money*. Gretna, LA: Pelican Publishing.

Van Campen, C. (2014). *The Proust effect: The sense as doorways to lost memories*. Oxford: Oxford University Press.

Verbeek, C. (2015). *Es liegt was in der Luft – Duft in der Kunst/Something in the air – scent in art*. Burgrieden: Hoenes Stiftung.

Verbeek, C., & Van Campen, C. (2013). Inhaling memories: Smell and taste memories in art, science and practice. *The Senses and Society*, *8*(2), 133–148.

21

A DEAD MAN'S NOSE

Using Smell to Explore the Battlefield of Waterloo

Stuart Eve

Introduction

As an archaeologist, I am constantly searching for clues about what it was like to live in the past. By excavating what remains of archaeological sites, we can see and touch the artefacts. We can also smell them, but in most cases many years in the soil fades and changes the smell,[1] leaving archaeological sites strangely olfactorily sterile. Chemical analysis of pottery vessels that used to contain food or unguent residues can give us clues to their contents, but the smell of these is almost always absent. There have been many attempts to recreate the look and feel of archaeological sites, and using modern technology, we are now used to *seeing* three-dimensional reconstructions of an archaeological site (for examples, see Renfrew, 1997). We are even able in some cases to *hear* what certain places sounded like (Pentcheva, 2011), but it is very rare for us to be able to smell them.

Sensory psychologist Avery Gilbert has written extensively about the power of smell to augment a feeling of place and space. When discussing the references to smell in James Joyce's *Ulysses*, Gilbert (2008) writes:

> A reader [of *Ulysses*] may be able to reimagine a familiar smell, but for one he doesn't know, he's left to guess. To re-experience the smells of times gone by, one needs the actual stuff; without it, written references and therefore literature eventually lose their power.
>
> *p. 206*

Gilbert argues that without the real bodily experience of the smells of Bloom's gorgonzola sandwiches or the salty odor of Dublin Bay, we are in fact not able to truly understand *Ulysses*, or the smell of times gone by.

During the interpretation and discussion of a site, archaeologists often attempt to evoke an atmosphere by vividly describing sounds or smells, and yet it is harder for readers to summon up the smell of something than it is to close their eyes and see the picture of the landscape being described by the words. Yet (as discussed in other chapters in this volume) smell has the ability to invoke intense feelings and emotions, such as disgust, attraction, fear or comfort. In some cases when used in a heritage context, smell has been deliberately toned down, for fear of creating too strong a reaction. A classic example of this is the *Tyrannosaurus rex* model in the Natural History Museum in London, which was intended to be accompanied by the stench of rotting flesh and blood. The smell was considered so bad that it was more likely to put people off coming to the exhibit, and therefore, the smell of a swamp was chosen instead (BBC, 2001). Smell has also been shown to aid in the recall of certain contextual facts: using the famous smells of the Jorvik Centre in York, Aggleton and Waskett (1999) have shown that visitors to the museum remember the exhibit more clearly if they are exposed to the same smells at a later time.

The recreation of archaeological smell has so far been restricted mainly to exhibits in museums, and is very rarely used to actually investigate an archaeological site, especially not in an *in situ* exploratory and embodied way. Within this chapter, I present a device that can be used to experiment with smells at archaeological sites and show how this could change the way we think about and experience such sites.

Experiencing Smell

Whilst the corporeal or chemical reaction to a stimulus such as a smell may be similar for many people, the perceptual or cognitive reaction may be very different. Of the Jorvik Centre, Kevin Walsh (1990) writes:

> . . . the contextualising of smells from the Viking period and placement in a twentieth-century tourist attraction seems highly dubious as each person visiting the centre will have a different perception or attitude towards a smell and it is quite likely that these will be very different from those held by the people who originally produced and lived with the smells.
>
> *p. 287*

This view is also supported in more recent literature on the subject, which challenges the popular and commonly held belief that before modern hygiene and sanitation, the past stank. We cannot deny that to our noses the past would have smelled bad; however, smell is inherently cultural – we quickly adapt to smells, and rather than always being unpleasant, they become familiar. As Morley (2014) states:

> the characteristic smell of a Roman city – a mixture of urine, shit, decay, smoke incense, cooked meat and boiled cabbage – would be a stench to us, because that is our characteristic response to most of these odours. For the regular inhabitants of ancient cities, it was the smell of home – and perhaps even civilization, as opposed to the more backward countryside where those smells were less intense, and barbaric regions where many of them might be absent altogether.
>
> *p. 119*

If people do become habituated to the smells of everyday life, then it could be argued that by recreating and exploring archaeological smells, we are not going to experience the site as the people in the past did. However, even when we look at a virtual reality reconstruction, we are not *really* seeing the site as it was experienced by past people; instead, it is giving us an impression of a site that may spark ideas or open avenues of investigation that were not apparent before. By recreating past smells and experiencing them *in situ*, we are challenging our perceptions and considering the place in a new way, both spatially (i.e. how the smells change as we move through the site) and temporally (i.e. how differently the place may have smelt in the past).

Constructing and Delivering the Smells of the Past

As previously discussed, recreations of historical smells very often take place within museum exhibits or galleries. As a field archaeologist, I am interested in experiencing the smells *in situ*, as

I walk around, excavate in and explore a site. Dale Air, international suppliers of a multitude of different scents and the company used by most of the major museums that incorporate smells, have a vast range of different smell dispensers (Dale Air, 2016). However, most of these dispensers are static and non-portable, only dispense one type of scent, and/or have no way of being remotely controlled. The challenge, therefore, was to find a way to create something that was portable enough to be worn by the archaeologist or visitor in the field, could be programmed to release a scent when instructed depending on the user's spatial location, and preferably, that would allow more than one smell to be released.

The first step was to create a digital map of the site, using a Geographical Information System (GIS).[2] This digital map is used to situate the user, and also to pre-program the areas in which certain smells should be emitted. The construction of this 'smell map' will be explained in further detail below when discussing the case study (see also Kate McLean's artistic practice reported in Chapter 7). The smell map forms the basis of an application which, when deployed on a mobile device (smartphone or tablet), uses the Global Positioning System (GPS) location of the user to begin the emission of the relevant smell. Due to the need to combine a physical/mechanical action (the release of a scent) with the computer-based information (from the GIS and GPS data), it is necessary to use a microcontroller board to act as the bridge between the two mediums. The recent rise in 'maker culture' means it has become increasingly easy to create with and program micro-controllers, and they have been used in the fabrication of many different gadgets, from a set of simple flashing light-emitting diodes to a self-watering gardening system that sends a message via Twitter if your plants get too dry (Gordon, 2013). I used an Arduino Uno (Arduino, 2017) board as the basis for the system. The Arduino boards allow a relatively low-cost and simple plug-and-play-type approach to electronics. They reduce the need for soldering or the creation of complex circuits, and use a simple coding language (Processing) to perform the actions.

As can be seen from Figure 21.1, the Dead Man's Nose[3] consists of an Arduino board, battery, a BluetoothLE chip, and four small fans contained within small boxes. The boxes contain miniature drawers in which a scent can be placed on a small piece of cotton wool, which is then dispersed when the fans blow. The fan boxes also have small pieces of Velcro on the back, allowing them to be attached either to clothing or to a board that can be worn around the neck. A separate mobile device runs an application built using the Unity3D gaming engine that reads the GPS position of the user and compares it to the smell map previously created. Each of the areas

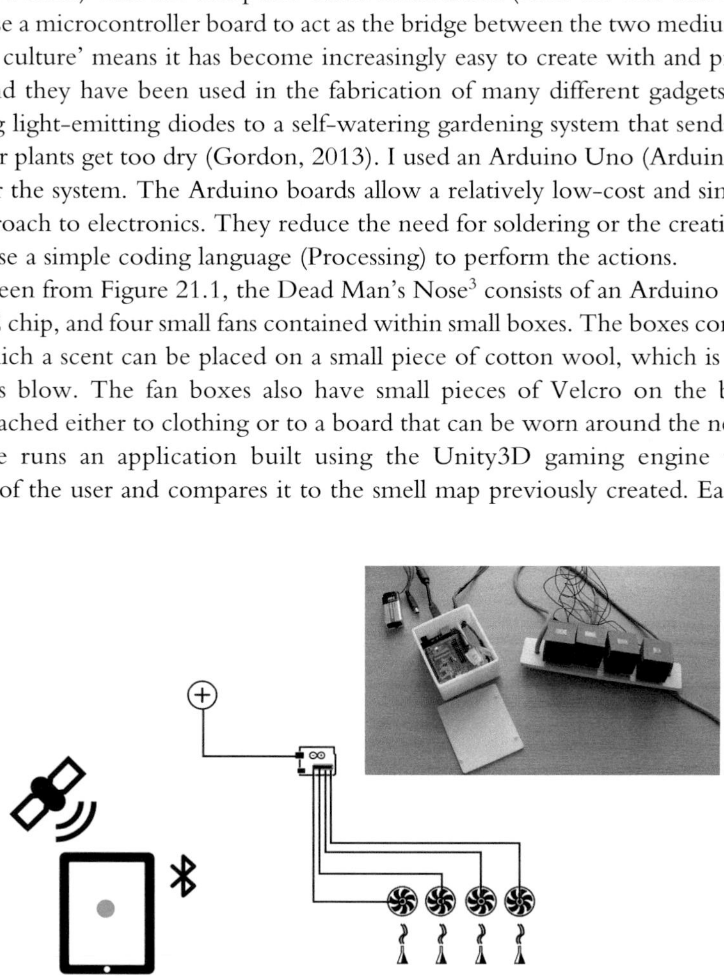

FIGURE 21.1 The Dead Man's Nose.

Image: author's own.

in the smell map is configured to correspond to one of the scents in the four boxes. When the user walks into one of the smellzones, a message is sent from the mobile application via Bluetooth to the Arduino, which in turn provides power to the corresponding fan (or fans) and releases the smell.

Using the Dead Man's Nose at Waterloo

The Dead Man's Nose can be used in countless different situations and archaeological sites (for an example of its use on Bodmin Moor, southwest England, see Eve, 2015). For the purposes of this chapter, I will explore the use of the device to recreate the smells of the battlefield of Waterloo, Belgium (see Figure 21.2). The Battle of Waterloo, fought on June 18, 1815, was the final battle in the Napoleonic Wars, and represented the beginning of a period of relative peace within Europe. Nearly 120,000 soldiers took to the field from most of the major nations of Europe, resulting in 65,000 being either wounded or killed. During 2015, the multi-national *Waterloo Uncovered* project (Waterloo Uncovered, 2014) began a five-year campaign to excavate and record the historic battlefield. The project is unusual in that as well as professional archaeologists, the team is also made up of wounded army veterans from modern conflicts. It has been shown by previous projects, such as Operation Nightingale (Defence Archaeology Group, 2015), that archaeology can aid in the rehabilitation and skills development of injured soldiers.

Battlefield archaeology is unique in that most battles take place over a short amount of time, sometimes a week, sometimes only a few hours. As archaeologists, we are used to excavating landscapes that have been utilized over millennia, and picking the archaeology of a single day apart

FIGURE 21.2 Waterloo in context.

Image: author's own, using data from www.openstreetmap.org.

from the centuries of other activity on the site is both challenging and also incredibly rewarding. Yet we also need to remember that because the area of the battlefield has usually had previous and subsequent usage, then this should be respected in terms of the smellscape of the battlefield as well. To that end, the Dead Man's Nose needs to not only represent the smells of the battlefield on the day of the battle, but also have a temporal depth that respects the possible smells of the landscape before and after the battle.

The *Waterloo Uncovered* project has concentrated its first year of excavations on the chateau complex at Hougoumont. Hougoumont was a key part of the allied defense against Napoleonic forces during the battle and was fiercely fought over throughout the day. Prior to the battle, it was a working farm, with a tower-type chateau and associated formal gardens. Following the battle, Hougoumont gradually returned to being a working farm – although the formal gardens never quite recovered from the actions of the troops stationed there. As can be seen from Figure 21.3, for the purposes of the spatial aspect of the application I have divided Hougoumont into four separate areas: the gardens, the farmyard, the woodland and the orchard. I have also divided the temporal aspect into four separate time periods: June 17, 1815 (the day before the battle), June 18, 1815 (the day of the battle), June 19–21, 1815 (the period directly following the battle, which involved cleaning-up, burial/cremations of bodies etc.) and late 1815 (when the farm reverted to its former usage). Table 21.1 shows a breakdown of the probable smells in each area during the different time periods. In this case, I have designated foreground smells[4] – that is, examples of the probable overriding smells during that period – but it should be noted that these will of course have been mixed with other less distinct background smells as well. I obtained the smells from Dale Air (see Table 21.2), matching the desired smells as closely as possible to smells from their range.

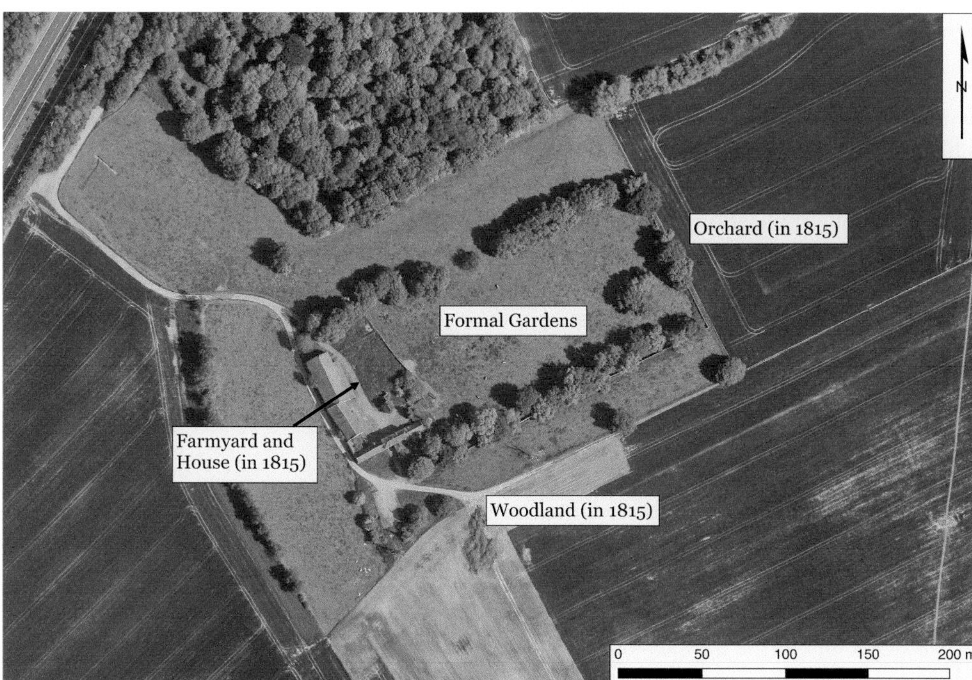

FIGURE 21.3 Aerial view of Hougoumont.

Image courtesy of Service Public de Wallonie.

TABLE 21.1 The probable smells in each area and time period

	Gardens	*Farmyard*	*Woodland*	*Orchard*
June 17, 1815	Flowers, clipped box	Farmyard smells	Summer trees	Fruit trees
June 18, 1815	Gunpowder, crushed flowers	Smoke, burning buildings/flesh	Cannon fire, gunpowder, blood, crushed plants	Cannon fire, gunpowder, blood, crushed plants
June 19–21, 1815	Soil, decaying bodies	Burnt buildings/ flesh, smoke	Soil, decaying bodies, cremations	Soil, decaying bodies
Late 1815	Flowers, grass	Farmyard smells	Trees and fallow field	Fruit trees

TABLE 21.2 The Dale Air scent corresponding to each smell

Type of smell	*Dale Air scent*
Flowers, clipped box, crushed flowers, crushed plants, grass	*Fresh Air*
Gunpowder	*Ship's Cannon*
Soil, decaying bodies	*Cannibal's Cave*
Farmyard smells	*Farmyard*
Smoke, burning/burnt buildings/flesh, cremations	*Burning Witch*
Summer trees, trees and fallow field	*Leaves*
Cannon fire, blood	*Man-O-War*
Fruit trees	*Pear*

It would be theoretically possible to create custom smells, but Dale Air have a wide range and supply the smells in small tester phials which are ideal for the Dead Man's Nose.

Once the areas and temporal conditions are configured, it is a matter of connecting the Dead Man's Nose to the application, loading the smellboxes with the correct smells (as instructed by the application), attaching the smellboxes to your person (usually by hanging the boxes around your neck) and walking out into Hougoumont. The application has a simple interface that allows users to set the desired time period to explore, and then, as they physically walk around the site and enter the various different smellzones, the correct fans begin to rotate and the smell is wafted to the nose.

Mixing Temporal and Spatial Scales

As smells have both a spatial and temporal dimension, the application is designed to create a temporally changing smellscape, which updates as the user moves through the real landscape. However, it is also possible to experiment with the scales of the temporal and spatial interactions. For instance, in order to get an idea of how an area smelled throughout the various time periods, the user could configure the application to release all of the smells from all time periods configured for that location at once, providing the user with an atemporal smellscape. Equally, it is possible to configure the application to release all of the smells for a single temporal period at once (e.g. all of the smells of June 18, 1815) regardless of the user spatial location – creating a smellscape of the entire battlefield, perhaps whilst sitting at a desk.

Implications of the Dead Man's Nose

Currently, no official user testing or survey of the usage of the Dead Man's Nose has been undertaken at Waterloo – beyond *ad hoc* usage in the field. There are a number of reasons for this, the first being the prototype nature of the application. At this stage, I have concentrated on developing the technology to work, therefore the user interface is not fully completed. The second reason is ethical, and brings to the fore the emotive nature of smells. In many cases, smells can be a powerful trigger of panic attacks, especially among those who suffer from combat stress or post-traumatic stress disorder (PTSD). Studies undertaken with Cambodian refugees who escaped from the Pol Pot regime have shown that olfactory-triggered panic attacks can be caused by exposure to a number of different odors, including car exhaust fumes, cigarette smoke and cooking smells (Hinton, Pich, Chhean & Pollack, 2004). The *Waterloo Uncovered* project team includes a number of army veterans suffering from PTSD, therefore exposure to smells may well have been unnecessarily traumatic if not correctly handled. To that end, a formalized testing phase is planned for the 2018 season, in conjunction with the Ministry of Defence's Defence Medical Services, to assess the ethics and efficacy of experimenting with smell – and perhaps to begin using the device as part of the veterans' therapy program.

Conclusion

Throughout this chapter, I have presented a prototype smell device that can be used on archaeological or heritage sites, to recreate a past smellscape. By making the device portable, it is possible to walk around the existing site and experience past smells *in situ*. The use of the device at Waterloo as part of an excavation that included army veterans has highlighted the emotive power of smell and brought to the fore some of the ethical issues that are involved when experimenting or designing with smell. The Dead Man's Nose is a relatively low-cost device, and the expandability and endless possible configurations mean that the device and application have great future potential for enriching the exploration, interpretation and enjoyment of archaeological sites.

By recreating the smells of the past, we are not necessarily experiencing them or interpreting them as past people would have done; instead, they can be used as an aid to challenge the way we think about a site or landscape. In the case of Hougoumont at Waterloo – now a bucolic landscape of wheat fields and gently rolling hills – beyond the various memorial stones, there is little hint on the ground of the slaughter and chaos that took place 200 years ago. However, by being directly assaulted with the acrid smell of battle, burning buildings and human anguish, some of that landscape's hidden history is revealed. Walking through this historical smellscape, we gain a greater understanding of our place within the story of that foreign field.

Notes

1 However, some olive oil factories in Sicily retain their smell 2000 years later (Koloski-Ostrow, 2014, p. 103).
2 For further discussion of the use (and some inherent problems) of GIS within archaeology, see Conolly and Lake (2006).
3 This is named after the M. R. James short story "Dead Men's Eyes," in which a pair of old binoculars allows the protagonist to see into the past.
4 Here I am adapting Schafer's (1977) identification of three different types of sound: the undifferentiated background sounds, such as the sound of the sea to a sailor (*keynotes*), sounds that are made to draw attention, such as the crying of a child (*foreground sounds*), and finally, sounds that have a particular significance to a community, such as church bells ringing (*soundmarks*).

References

Aggleton, J. P., & Waskett, L. (1999). The ability of odours to serve as state-dependent cues for real-world memories: Can Viking smells aid the recall of Viking experiences? *British Journal of Psychology*, *90*(1), 1–7.

Arduino. (2017). Arduino – boards. Retrieved March 1, 2017 from http://arduino.cc/en/main/boards

BBC. (2001, February 6). Roaring T rex robot unveiled. Retrieved March 1, 2017 from http://news.bbc.co.uk/1/hi/sci/tech/1155517.stm

Conolly, J., & Lake, M. (2006). *Geographical information systems in archaeology*. Cambridge: Cambridge University Press.

Dale Air. (2016). Dispensers. Retrieved March 1, 2017 from www.daleair.com/dispensers/machines

Defence Archaeology Group. (2015). *Defence Archeology Group*. Retrieved May 29, 2017 from www.dag.org.uk

Eve, S. (2015, March 11). Finding out what the past smelled like. *The Atlantic*. Retrieved March 1, 2017 from www.theatlantic.com/technology/archive/2015/03/finding-out-what-the-past-smelled-like/387352/

Gilbert, A. (2008). *What the nose knows: The science of scent in everyday life*. New York: Crown Publishing.

Gordon, W. (2013, July 3). Build an Arduino-powered, tweeting, self-watering garden system. Retrieved March 1, 2017 from http://lifehacker.com/5579086/build-an-arduino+powered-tweeting-self+watering-garden-system

Hinton, D., Pich, V., Chhean, D., & Pollack, M. (2004). Olfactory-triggered panic attacks among Khmer refugees: A contextual approach. *Transcultural Psychiatry*, *41*(2), 155–199.

Koloski-Ostrow, A. O. (2014). Roman urban smells: Archaeological evidence. In M. Bradley (Ed.), *Smell and the ancient senses* (1st ed., pp. 90–109). London: Routledge.

Morley, N. (2014). Urban smells and Roman noses. In M. Bradley (Ed.), *Smell and the ancient senses* (1st ed., pp. 110–119). London: Routledge.

Pentcheva, B. V. (2011). Hagia Sophia and multisensory aesthetics. *Gesta*, *50*(2), 93–111.

Renfrew, C. (1997). *Virtual archaeology*. London: Thames & Hudson.

Schafer, R. M. (1977). *The tuning of the world: Toward a theory of soundscape design*. New York: Alfred A. Knopf.

Walsh, K. (1990). The post-modern threat to the past. In I. Bapty & T. Yates (Eds.), *Archaeology after Structuralism: Post-Structuralism and the practice of archaeology* (pp. 278–293). London: Routledge.

Waterloo Uncovered. (2014). Retrieved March 1, 2017 from www.waterloouncovered.com/

22

OLFACTORY DESIGN ELEMENTS IN THEATER

The Practical Considerations

Michael McGinley and Charles McGinley

Introduction

Theatrical productions include lighting and sound to enhance the theatergoing experience by immersing the audience's senses in 'the scene.' Periodically, artists have experimented with scents in theater productions. In order to incorporate scent, the theatrical designers must ask themselves how they should integrate odors into a scene in analogous ways to how they consider lighting, sound and other elements of stage design.

Designers must consider how this integration into the design is accomplished in a stylistic and creative way. Is scent added to spotlight a specific moment or event in the production? Are scents added to provide an ambience to the scene? The audience already can be immersed into a scene with the right combinations of lighting, sounds and set design. Our strong memory and emotional connection to odors can pull the audience to a scene in very powerful ways. The currently unique aspect of theatrical scent design can add significant artistic value to the production.

In this chapter, discussion of theatrical scent design will provide examples from award-winning work conducted at The Soap Factory Art Gallery (hereafter The Soap Factory), a non-profit laboratory for experimentation in contemporary theater and art, housed in a historic warehouse in Minneapolis, Minnesota. Before exploring the approaches to olfactory design in theater, it is helpful to understand some of the basic principles of odor perception. Scent is meaningless without an audience to smell it and experience it. Knowing how we perceive odor and subsequently process it, both physiologically and mentally, is therefore vital to the successful use of scent in theater.

Olfactory Perception

A scent is made up of one or many chemical compounds. Each compound is individually an odorant. The odor is the perception in the brain as we detect and possibly recognize a stimulus. The odor can be a perception of one compound with a specific odor character, or multiple compounds combining for a complex mix of characters.

Anatomy

We first sniff and bring chemical odorants into our nose to pass by the olfactory epithelium and dissolve (transfer) into mucus at a rate dependent on their water-solubility and other mass transfer factors. The olfactory cells then receive the chemical odorant, and a response is created which is dependent on the mass concentration, or the number of molecules present. Each reception creates an electrical response in the olfactory nerves. A summation of these electrical signals leads to an 'action potential.' If this action potential has high enough amplitude (threshold potential), then a signal is propagated along the nerve, through the ethmoidal bone between the nasal cavity and the brain compartment, where it synapses with the olfactory bulb (Ohloff, Pickenhagen & Kraft, 2012).

All olfactory signals meet in the brain at the olfactory bulb, where the information is distributed to two other sections of the brain (Hawkes & Doty, 2009). One major pathway of information is to the limbic system, which processes emotion and memory response of odors. This limbic system pathway tends to yield involuntary responses, and is what gives odors their strong emotional connection. A brief smell of pine can bring us the warm feelings of Christmastime with family. That 'old building' odor can immediately bring feelings of anxiety from elementary school or college. Tobacco smoke can place us in our grandfather's arms.

The second major information pathway is to the frontal cortex. This is where conscious awareness takes place as information is processed with other sensations and is compared with cumulative life experiences for the individual to potentially recognize the odor and make some decision about the experience. For example, a person may consider whether the odor is harmful and they want to remove themselves from the location, or conversely, if the smell makes them feel good and they want to continue to enjoy it. Odors are therefore an initial signal as our nose antenna receives, and our brain interprets, our surroundings. The strong memories and feelings an odor brings us influence these interpretations and decisions.

Parameters of Odor

Fundamental concepts of odor perception include the 'power law,' thresholds, geometric progressions, logarithmic transformations, dilution ratios, and dose–response functions (McGinley, McGinley & Mann, 2000). Understanding these concepts sparks a greater appreciation of how chemical odorants lead to our ultimate observation of odors and how we receive and react to the odor experience. Odor is a psychophysical phenomenon. Psychophysics involves the response of an organism to changes in the environment as perceived by its senses (Stevens, 1960). Examples of the psychophysical phenomenon include how the human body perceives sound loudness, light brightness or odor strength.

In the 19th Century, E. H. Weber proposed that the amount of increase in a physical stimulus, to be perceived differently, was a constant ratio (Weber's Law) (Boring, 1942). This relationship can be expressed as:

$$\Delta I/I = \Delta C/C = k$$

where:

I is the stimulus intensity,
C is the measurable amount or concentration of stimulus, and
k is a constant that is different for every sensory property and specific stimulus.

As an example, this expression means that there would be the same perceived increase in sweetness intensity when changing a concentration of sugar in water from 10% to 11% as when changing the concentration from 20% to 22%.

In 1860, G. T. Fechner expressed Weber's Law somewhat differently by plotting the perceived intensity versus the stimulus magnitude on a semi-log scale (Fechner, 1966). Fechner's Law was expressed as:

$$I = k \log C + b$$

where:

I is the stimulus intensity,
C is the measurable amount or concentration of stimulus, and
k and **b** are constants.

In the 1950s and 1960s, through his work at Harvard University, S. S. Stevens proposed that apparent odor intensity grows as a power function of the stimulus odorant. Stevens showed that this 'power law' is captured in the following equation (Stevens, 1957, 1962):

$$I = k\,C^n$$

where:

I is the odor intensity,
C is the mass concentration of odorant (e.g. mg/m^3), and
k and **n** are constants for specific odorant or odorant mixture.

As shown by the example in Figure 22.1, this equation is a straight line when plotted on a log–log scale. The x-axis is the logarithm of mass concentration (mg/m^3) of the odorant (the dose). The y-axis is the logarithm of the perceived odor intensity, which could be measured with various scales (the response). The upward slope of the graph illustrates that the odor intensity of the

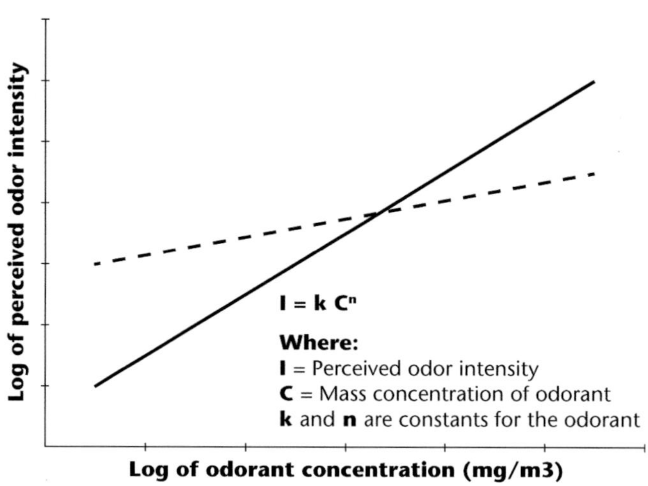

FIGURE 22.1 The 'power law' of two different odors.

Note: The solid line is a grassy meadow odor that increases in intensity at a high rate with increasing concentration. The dashed line is a candy odor that is initially more intense, but increases in intensity at a slower rate.

Image: authors' own.

odorants increase as the mass concentration increases (dose–response relationship). The slope of the power law is <1 for odors, since it takes a larger and larger increase in concentration to maintain a constant increase in perceived intensity.

In Figure 22.1, the potential difference between two odors can be seen. The solid line is a grassy meadow scent, which has a steep slope increasing in intensity with increasing concentration. The dashed line is a candy scent that is initially more intense than the grassy meadow at the same concentration; however, the flatter slope shows that as the concentration of the candy odor increases, the intensity increases at a lower rate than the grassy meadow odor. These two odors will therefore behave differently if dispensed in the same way in the same space.

Some researchers argue whether Steven's Law applies in all cases, or if the Weber-Fechner Law is more accurate or appropriate in specific instances or with specific levels of odors, such as concentrations closer to threshold (Cain & Moskowitz, 1974; Chen, Bundy & Hoff, 1999). The key point is that odor perception will increase with the odorant concentrations, and understanding the relationship of this dose–response function can help to set odor levels in a specific situation.

As the odor intensity approaches zero, another significant odor parameter is reached: the odor threshold. As can be seen in Figure 22.1, while the green meadow scent is more intense at higher concentration, as the lines extend to the left towards a perceived intensity of zero, the dashed line (candy odor) has a much lower odor threshold concentration than the solid line (green meadow); thus, as the concentration is lowered, the green meadow will be first to reach an intensity of zero. It would therefore take much less candy odor to be just perceived than the green meadow odor. The character of a scent or smell can also be affected by the concentration of its odorant mixtures or combinations. The perfume industry commonly discusses such combinations in terms of top, middle and base notes of odor (Turin & Sanchez, 2008). The top notes tend to be the character that is most prominent and obvious. For example, cleaning products often use a lemon or orange odor as a clear and obvious top note, signaling that the room is clean. The middle notes tend to be noticed after the initial observation of top notes. The middle notes are the main body or 'the heart' of a scent, and tend to be mellow odors. The base notes are the underlying odor that is meant to remain longest. As the scent concentration increases, the top note tends to overpower and become more intense (Cain & Drexler, 1974). Conversely, at low scent concentration, the top note is less impacting, and makes way for the base note to make a more lasting impression.

These odor parameters can be understood through experimentation with scents. However, in real life, perceptions could vary depending on the presence of other stimuli as well as the context of the environment. A distracting sensation may influence the cognitive ability of the brain to notice another sensation. In the context of theater, loud sound effects may require a higher concentration of odorants for scents to be noticed. Or inversely, a well-designed set could tune the audience in and make the scents more recognizable due to a contextual link. This potential variance must be considered. Each instance is unique, and adjustments may be required to get the odor level 'just right' – that is, to exactly meet the director's objective to influence the audience. In any art installation or theatrical presentation, the odors will need to be tested and these relationships will need to be evaluated to consider the impact on the audience of the intended application of the scents.

Olfactory Fatigue

Olfactory fatigue is a short-term adaptation to odors. As one is exposed to a constant presentation of an odor over time, it is a common experience that the odor will decrease in perceived intensity (Cain & Engen, 1969). An example most people have experienced is entering a home or business and noticing a distinct odor, but then after a short period of time, that odor goes into the background and is no longer perceived. Our olfactory system saves us from information overload by decreasing perceptions of what it deems unimportant in order to be prepared for anything new or changing. In this process, over time, as the odor perception is decreasing, it is possible the character of the odor could also shift slightly. For example, a base note may become more prominent as the perception of the top note decreases. Additionally, exposure to multiple cross-modal stimuli could also lead to another type of adaptation or fatigue. Loud sounds or bright lights may distract an individual, and thus decrease his/her perception of the scent. The art and design process can use this to advantage by optimizing the use of scents with consideration of when and how fatigue would set in. For example, in a theatrical scene, the perception of scent may be decreased as the action in a scene increases, thus the scent delivery can be adjusted accordingly.

Olfactory Design in Theater

These concepts of odor perception are important to consider when designing an olfactory theater experience. One can now consider how to utilize these elements by successfully adding scent in various ways to various types of scenes. The scents can immerse the audience in the reality of the scene through integration of scent with the complete production design, and thus add production value through realized artistic device and public interest.

Overall, ambience and spotlighting are the two main types of scent presentation for scenes. As a true spotlight focuses light upon a character, scent spotlighting involves highlighting a specific moment or event. Ambience is a presentation of the background odors in a scene or a moment/event of longer time scale. Spotlighting tends to be more challenging to control in terms of timing of the odor presentation at specific instances and then subsequent removal of the odor. A simple example of spotlighting is the addition of the smell of orange as an actor cuts into the fruit. In the *Haunted Basement* productions at The Soap Factory, scent spotlighting is used most often to emphasize an actor in a scene: for example, a sinister scent of animal odor from a character that only becomes noticeable in their presence. Ambience scent is more commonly used throughout scenes to help place the audience within the scene. An example here is baby powder odors that place us in a suburban family home nursery. Another example is musty, woody and perfume scent notes for an old Victorian home.

Each of these types of scenting is designed to stimulate the audience and immerse them into the reality a performance is attempting to convey. A stage is prepared with the scenes of a neighborhood party in a 1970s home. The room is dimly lit with period lamps, and the flicker of the TV is seen across the stage. Actors are dressed in bell-bottom jeans and dresses reminiscent of the disco era. Classic rock plays in the background, and banter of the guests can be heard. All of these various design elements bring us back to the 1970s. The scent of cooking food from the kitchen and alcohol and cigarette smoke from the guests break down the 'fourth wall' and allow the audience to truly feel they are in the room with these partiers. Our strong memory and emotional link to odors allow the audience to mentally transport themselves into the scene. Scent somehow makes it more real, and brings the stage and seating into the same existing space.

Scenting Devices

Selection of scent delivery modes depends on the theater architecture. Three common delivery approaches include: (1) into facility ventilation systems, (2) from mezzanine/catwalk/pipe hanger systems above the audience and (3) at seat level. Delivery of a vapor into the theater heating ventilation and air conditioning system requires that this system is isolated from other parts of the facility, as well as being accessible for the installation and control of the scenting device.

FIGURE 22.2 Ventilation system controls for the 2014 production of *Treasure Island* at the Theatre in the Round in Minneapolis, Minnesota.

Image: author's own.

Dispensing the scent above the audience near lighting instruments also requires easy access for installation and system control. Presentation of scent at seating level is commonly much more accessible; however, proximity to the audience requires more consideration of a system design that is unobtrusive.

Generally, there are two physical types of scenting systems: (1) mechanically aerosolizing and (2) passive diffusion. Each of these modes of delivery requires consideration of air movement to transport the scent. Many commercial devices could be used to aerosolize a scent into the air, either directly into an air duct or dispersed near room air supply ducts. It is not critical to utilize facility air systems, as it is possible that natural air movement in the theater will disperse the scent; however, conditions during testing in an empty theater can differ from when an audience is present. Passive diffusers include pads or other sorbent materials with scent applied or scented solids. Mechanical ventilation (e.g. fans and air ducts) can increase the evaporation and also spread the scent through the theater space. The selection of the physical scenting system will depend on the scent properties (e.g. physical state, odor intensity characteristics, etc.).

Figure 22.2 displays an example of a system utilized for a production of *Treasure Island* at the Theatre in the Round in Minneapolis, Minnesota in April 2014. Commercially available respiratory therapy nebulizers were adapted to dispense the scent into the ventilation ducting, and were operated by stage technician using compressed air-controlled valves. Labels can be seen for the four main ambience scents of the pirate ship, the old tavern, sea breezes and Skull Island jungle.

The Scent Experience

Scent design should always add artistic value to a performance. In some cases, the addition of scent may create a buzz or public interest for a show; however, when it is time to see the production, the scents should be integrated in a performance with the same care as lighting and sound. 3D movies have always been a device to sell a movie to the public with a technology that literally adds dimension to the experience. Early 3D movies were filled with gimmicks and sight gags with objects thrown directly at the screen. While everyone loves a good sight gag, more recently 3D technology is being used successfully with an emphasis on adding depth and beauty to a movie. Likewise, in adding scent to the theater, people may appreciate the introduction of a gross or

FIGURE 22.3 Scene from *Haunted Basement* (2010), 'Clown Stage.'

Image: author's own.

otherwise intriguing scent; however, the truly successful use of scent design will add depth to scenes.

Scent has immense artistic value when it can be manipulated as an artistic device in overall production design. A similar example with sound is the use of slow music playing from a radio on stage while a chaotic scene of a fight is taking place. Regarding scent, Noah Bremer, director of the *Haunted Basement* at The Soap Factory, stated: "We don't always want to have a scene where somebody is being butchered and there is the smell of blood. Maybe we want the smell of lilacs. It's weird and disorienting" (The Soap Factory, 2013). The point here is that scent can be used creatively and counter-intuitively to play on the audience's emotions and expectations – to surprise and disturb. An example of such olfactory stimulation was found at the 2010 *Haunted Basement* production (see Figure 22.3). Patrons found themselves in a dark room with the only sensory cue the scent of popcorn and bubblegum. Suddenly, spotlights illuminated them to reveal that it was they who were 'on stage' in front of an audience of demented clowns!

As with the use of strobe lighting and gunshots, scent warnings should be displayed in the production program and posted in the lobby. A scent-free alternative performance can also be presented.

Conclusions

Olfactory design in theater provides amazing potential to immerse the audience further into a production. Lighting and sound are commonly designed with artistic elements to add to a production. Directors and stage designers can add elements of scent in similar ways.

With an understanding of how we perceive odors, design teams can work to present theatrical scents in an efficient and practical way. Scents can spotlight a specific character or moment, or they can create an ambience to fit the scene. A variety of scent devices can be used to vaporize the odorants into the air, with use of facility air systems or designed mechanical systems to disperse the scents to the audience.

Olfactory design will continue to advance, with creative artists experimenting and taking risks. While scent gimmicks can be fun for the audience, scent use should impact the general production value, adding depth and creative impact. This could lead to a future where playwrights provide specific olfactory instruction and directors utilize what will then be common artistic olfactory devices.

References

Boring, E. G. (1942). *Sensation and perception in the history of experimental psychology*. New York: Appleton-Century-Croft.

Cain, W. S., & Drexler, M. (1974). Scope and evaluation of odor counteraction and masking. *Annals of the New York Academy of Sciences*, *237*(1), 427–439.

Cain, W. S., & Engen, T. (1969). Olfactory adaptation and the scaling of odor intensity. In C. Pfaffmann (Ed.), *Olfaction and taste III* (pp. 127–141). New York: Rockefeller University.

Cain, W. S., & Moskowitz, H. R. (1974). Psychophysical scaling of odor. In A. Turk, J. W. Johnston, Jr., & D. G. Moulton (Eds.), *Human responses to environmental odors* (pp. 1–32). New York: Academic Press.

Chen, Y., Bundy, D. S., & Hoff, S. J. (1999). Development of the relationship between odor intensity and threshold dilution ratio for swine units. *Journal of the Air & Waste Management Association, 49*(9), 1082–1088.

Fechner, G. T. (1966). *The elements of psychophysics* (H. E. Adler, Trans.). New York: Rinehart and Winston. (Original German work published 1860)

Hawkes, C. H., & Doty, R. L. (2009). *The neurology of olfaction*. New York: Cambridge University Press.

McGinley, M. A., McGinley, C. M., & Mann, J. (2000). Olfactomatics: Applied mathematics for odor testing. Proceedings of *WEF Odor/VOC Specialty Conference*, Cincinnati, Ohio, April 16–19.

Ohloff, G., Pickenhagen, W., & Kraft, P. (2012). Scent and chemistry: The molecular world of odors. *Chemické Listy*, *106*, 685–692.

Stevens, S. S. (1957). On the psychophysical law. *Psychophysical Review*, *64*(3), 153–181.

Stevens, S. S. (1960). The psychophysics of sensory function. *American Scientist*, *48*(2), 226–253.

Stevens, S. S. (1962). The surprising simplicity of sensory metrics. *American Psychologist*, *17*(1), 29–39.

The Soap Factory. (2013, October). *Haunted Basement production program*. Minneapolis, Minnesota.

Turin, L., & Sanchez, T. (2008). *Perfumes: The guide*. New York: Viking Press.

23

THE NOSE ONSTAGE

Olfactory Perceptions and Theatrical Dimension

Anna D'Errico

> If no one asks me, I know; if they ask and I try to explain, I do not know.
>
> *Saint Augustine,* Confessions, *Book XI, ch. XIV*

Saint Augustine here was talking about time, but I believe his words apply to odor perception too. In fact, both time and odor impressions are ephemeral. This fleeting attribute is also common to the performing arts: with each event being unique, anchored to the here and now. But how in practice can odor interact with a theatrical space, with actors and the public?

Odor can induce strong reactions in the human body, catalyzing emotions and feelings in actors and spectators. This brings us to the question of why, as Paquet (2004) has suggested, is the contemporary theater anosmic? To draw the line between potentialities and difficulties of handling smell and engaging olfaction in theatrical space, it is worth thinking about how this sense works and what implications this has for human perception. Therefore, this chapter begins by briefly outlining human neuroanatomy and some human olfactory capabilities. Following this, I will outline specific showcases, highlighting different approaches to the use of odor in performative spaces. According to Banes' (2001) classification for olfactory performances, overall these cases have at least two levels of application and interpretation in terms of odor design: first, mainly illustrative, meaning the use of specific odors to depict ambiences, moods and characters; second, to frame and contextualize a piece or a performance. Moreover, I question the implications of odor use for actors and performers – in particular, how an odor impacts acting and actors' feelings, and how this can be integrated into actorial training and into the composition process.

Anatomical and Physiological Characteristics of Human Smell: Implications for the Performing Arts

Many sensory stimuli can trigger our memories, but amazingly, the most powerful of them is one we usually forget. Olfaction, the sense of smell, has the unique feature of being directly connected with the deep and old part of our brain – or the so-called 'reptilian brain.' This cerebral area is also involved in the processing of emotions and memory (Shepherd, Chen, & Greer, 2004; Gottfried, 2011). The result is that odor can be a powerful trigger for recalling old memories and emotional states, or for learning.

In humans, as in all other mammals, the olfactory system is organized in three main stages: the main olfactory epithelium, in the nose, and the olfactory bulb and olfactory cortex, both in the brain (Shepherd et al., 2004). Olfactory stimuli reach our receptors in the nose with discontinuity because of respiration. The sense of smell is connected with the act of breathing, for anatomical reasons: air passes into the nose during inhalation and carries odor molecules which can diffuse in the mucus and thereby come into contact with the olfactory sensory neurons. Even without active sniffing, we cannot avoid odor stimuli, simply because we need to breathe. This aspect leads us already to an important consideration for the use of smell within a space, theatrical or otherwise: odor comes inside us without our permission, it is invasive, and thus it can assume a brutal connotation.

Sited within the olfactory epithelium are specialized neurons expressing olfactory receptors that can bind odor molecules. The interaction between odorant and olfactory receptor determine what is termed a molecular transduction cascade within cells that translates chemical information into an electrical signal – used by neurons to carry information – which is transmitted to the brain (Su, Menuz & Carlson, 2009). Humans have a repertoire of around 400 functional olfactory receptor genes expressed in almost 12 million olfactory neurons per epithelium, left and right. Each person expresses variability in his/her olfactory receptors (Mainland et al., 2014) and therefore has a slightly different odor-sensitivity. In a way, therefore, it could be said that we all have a different nose. This genetic variability is another element to take into account when we consider the use of odor for any purpose, as it means that one person might perceive a given odor differently to another.

The way an odorant can be decoded in the brain is still unclear, but the accepted model among the scientific community is a combinatorial code: odor molecules can bind to different receptors, and those different receptors can recognize different molecules with different affinity in a combinatorial fashion (Malnic, Hirono, Sato & Buck, 1999; Su et al., 2009). The sensitivity and selectivity of the receptors is variable too; some are broadly tuned and can respond to many odors, others are very selective. In the olfactory epithelium are also present trigeminal terminations. They have important functions as the sensory input triggering protective reflexes – respiratory, secretory and cardiovascular – in response to high concentrations of irritating vapors. In such instances, trigeminal stimuli produce psychophysical sensations ranging from pungency to burning and pain, and thereby engender aversion to certain odors like ammonia, which can cause irritation, or the cooling sensation felt with menthol odors. The physiological reactions subsequent to trigeminal over-stimulation and aversion include swelling of the nasal mucosa and vasomotor dysfunction at the level of the larynx, which can give rise to muscular paralysis with coughing and vocal congestion (Cometto-Muñiz & Cain, 1991; Dalton, 2003; Smeets and Dalton, 2005). Over-stimulation could also cause headache and nausea due to vagal reflex. When selecting a palette of odors for an olfactory 'score' within a theatrical performance, we should therefore bear in mind to avoid those odors that could over-stimulate in this fashion, as they could negatively affect both the audience and the performers.

Olfactory Dimensions in Theater

Western theater, from Greek and Roman theater to the 19th Century and after, has always been very smelly. Scents were part of the *mise-en-scène* to different degrees, not least in respect of the background odors emitting from audience members all sat closely together and, in certain periods of history, often in poor hygienic conditions. In terms of the performance itself, playwrights have often used smell strategies to surprise or scandalize their audiences and the public, or make a play more realistic. However, olfactory communication in theatrical space has been always considered an extra addition to the other sensory channels. This may explain why, as Banes (2001) notes, we do not have any semiotic analysis, critical understanding or systematic description of the use of smell in performative space. Therefore, smell design in theatrical space often involves many different 'experiments,' without a proper analysis of the use and meanings of odor in such contexts. We can probably include this consideration in the explanation why the modern Western theater, despite several examples of the use of odor onstage, appears anosmic.

Since olfaction is a powerful sense loaded with many psychological and cultural implications, its use in performances and theater can become a precious tool in the hands of a playwright and/or director who is conscious of this. Banes (2001) highlights six different ways odor can be used in theater, dance and performing arts: (1) to illustrate words, characters, places and actions; (2) to evoke a mood or ambience; (3) to complement or contrast with other sensory channels; (4) to summon specific memories; (5) to frame a performance as ritual; and (6) to serve as a 'distancing device' (the odor in this case can call attention to itself as a theatrical effect with, for example, the function of deliberately undermining what is happening onstage). This taxonomy for theatrical aroma design focuses on the representational function the odors within the performance are supposed to convey. The categories are not mutually exclusive, and smell onstage can perform more of those functions at once. Indeed, the two last categories have more to do with the framing and contextualization issues than with representational or illustrative purposes.[1]

In the following sections, I will consider different cases of the use of odors in a performative context. In *Pierre de Soleil*, the use of odor in the performance is meant to evoke a specific mood and ambience, complementing other audio-visual input; in *Les Parfums de l'Âme*, the main characters are defined and illustrated through specific odors; in *Duft, an Olfactory Metamorphosis*, odors define specific ambiences and mood. The line that connects these cases is the use of olfaction as an aesthetical element which, extrapolated from the specific contexts of different cultures and performative designs, maintains an ability to bring new or additional meanings to performances. However, before examining these cases we begin by discussing the Japanese *koh-do* ceremony of incenses, which shows how the sense of smell can be pivotal for aesthetical experiences and art appreciation.

Koh-do Ceremony

The *koh-do*, literally 'listening to incense,'[2] offers a good example of use of smell in space, not as complement or support to other sensory expression, but as art *per se*. The Japanese taste for burning incense has a long history, reported in the first Japanese chronicle *Nihonshoki* in 595 CE (Aston, 1971), and consolidated in the Nara period (710–792 CE). The use of precious and aromatic wood for incense became a distinctive sign of elegance and wealth among the aristocrats of the Heian period (795–1185 CE), but it was only during the Edo period (1603–1867) that a rigorous classification of incense types according to their odor quality and provenance developed (Shikibu, 2010).

Why is the *koh-do* so special? It is a unique example of art centered on odor appreciation, which became a divertissement and aesthetical expression within Japanese culture. In this respect, philosopher Chantal Jaquet (2010) identifies that the evanescent attribute of odor reflects a culture of impermanence and detachment from physical things in accordance with Buddhist practice and well suited to a country so often exposed to natural disasters like tsunamis and earthquakes, which add instability to life and living. Moreover there are two cultural elements driving a development of such aesthetics in the *koh-do* tradition. First, Buddhism does not give any restrictive prescription for incense, thus its use for recreational purposes was not perceived as a profanation. Second, we should note a different taste in odor use, and in its purpose, in Japanese culture. In Western cultures, the use of fragrances was first meant to cover body odors, and became later a tool for seduction. By contrast, Japanese culture does not like strong scents at all, and instead of putting eau de Cologne and perfumes on their body, the Japanese prefer burning aromas, which can be more faint and subtle than a perfume. This habit allows the detachment of smell appreciation from any sexual connotation, bringing the experience of an odor to a different level.

The appreciation of incense is a pleasure *per se* and a divertissement enclosed in a traditional ceremony. A popular version of the *koh-do* ceremony is the *kumikoh*, with two or more different incenses. *Kumikoh* represents an intellectual diversion in the appreciation of incense, involving a repertoire of incense games, often in connection with a literary pursuit (Morita, 1992). In the context of theatrical performances, smell can be used in a similar manner to this, in that it potentially interacts with the audience and all the elements on stage, including the actors, to create an experiential effect. The result of this interaction depends principally on the way the olfactory encounter has been designed. Moreover, because smell perception is strictly connected with a person's attention, an odor can interfere both positively and negatively with the audience experience. In short, having an odor released during a performance could have two opposite effects: it could enhance the attention of the audience member, adding to their appreciation of the performance, or distract him/her from the scene on stage.

Using Smell in a Multisensorial Performance: *Pierre de Soleil* (Sun Stone) by Michel Roudnitska

Among the first to integrate olfactory performances with computer graphics, Michel Roudnitska gives us a modern interpretation of 'total spectacle,' as he calls it. In *Pierre de Soleil*,[3] a project inspired by the poem of the same title by Nobel Prize winner Octavio Paz, he translates into sounds, images and smell personal suggestions and feelings arising from the poem. The performance, presented for the first time in 1998 for the second symposium of *New Loving Art* in Paris, develops around the concept of multisensoriality. It involves a choreography of sounds and animated video projected onto the background of a 5 m-wide screen and the bodies of performers, while an olfactory score underlines the performance. Avoiding literal illustration, out of the lines of the original poem Roudnitska created seven olfactory ambiences with emotional resonance: (1) Live waters: from the river to the sea; (2) Sunlight atmosphere: sensual and fruiting summer; (3) Humus of the rain forest: solitude; (4) Smoke of wood fire: dry stone; (5) Cognac aroma: joy to be alive; (6) Monastic evocation: incense, myrrh, lotus, flower; (7) Love perfume and rose petals.

As Roudnitska points out, one of the difficulties of such a performance is maintaining a balance between the different sensory channels, and harmonizing the stimuli without creating a sensory overload. This process needs to take into account some practical issues, in terms of the timing required from the various source stimuli to be released – sounds, video projections and odor release systems. Not least, the sounds and movements of a performance are usually the first to be perceived by the auditory and visual senses, taking a few milliseconds to be processed in the brain. Smells, by contrast, are typically the last stimulants to be consciously perceived (Gottfried, 2011).

Pierre de Soleil reflected a period during the 1990s in which there was growing interest in the use of smell in performative space (Banes, 2001; Paquet, 2004). On the other hand, we should notice the differences between this production, which is openly experimental and inspired by the idea of synesthesia, and olfactive theater. In *Pierre de Soleil*, the work starts from the ideal translation of a poem in the form of sensory impressions, expressed as images and perfumed ambiences, and then integrated with choreography. Olfactive theater arguably requires a different approach, involving an olfactory plot within the dramaturgic design; we now move on to examine this in more detail.

Les Parfums de l'Âme, an Example of Olfactive Theater

The French playwright Violaine de Carné with her company Le TIR et la Lyre is developing a bottom-up approach in terms of olfaction, with the writing process starting from the odor. Conscious of the power of smell in relation to memories, she uses it as a tool to investigate human emotions. Starting from an odor, she creates characters, ambiences or specific places defined through the odor. And in *Les Parfums de l'Âme*, she creates a non-space, an undefined atmosphere that reflects the ephemeral attribute of smell and desires, and where each character is defined by specific odors, which helps us to get into their personal drama. The piece opens in the hall of a mysterious and futuristic research institute where six persons are waiting, each with a different provenance and background, but with one element in common: they all have lost a person close to them. The institute specializes in recreating smell, so, starting from personal objects like clothing, the researchers can give back to the 'patient' a perfume of their lost beloved. Unfortunately, smelling such odor arouses in the protagonists vivid memories, making them feel worse, so proceeding with a new life becomes impossible. To get over this mortal limbo, it is realized that, instead of recreating the odor of a lost person, it would be better to make a sort of olfactory testament. Thus, the researcher can acquire a person's olfactory memories and recreate a smell record of his/her life, so it can be donated to the future generation.

The reception of the audience in the theater was suffused with incense coupled with the background sound of a requiem. During the performance, 12 odors in total (aside from incense) were diffused in alcohol through a compressor and released into the theater audience via Sigmacom machines. Each odor was used to define a character or a specific memory from a character, and its diffusion was paired with them throughout the piece. One of the key features here is the use of odor not just as illustrative or contextualizing tool, according to Banes' (2001) classification, but as the driving force of the full creation. The work is therefore an exploration of the sense of smell conducted by the actors, the perfumer, the perfume-releaser and the author, together in concert.

For this production, Violaine de Carné worked with the independent perfumer Laurence Fanuel, joining a research project on the aesthetics of olfaction and its application in contemporary arts promoted by the philosopher Chantal Jaquet (l'Université Paris-Sorbonne) and the neuro-biologists Roland Salesse (Institut National de la Recherche Agronomique, Paris) and Didier Trotier (Centre National de la Recherche Scientifique, Paris). To this end, after five of the representations of *Les Parfums de l'Âme*, 319 assessment questionnaires from the audience and 35 individual interviews were collected. The overall evaluation of the *piece* was positive, with 77% of the people satisfied about the odor experience and the odor release system (Domisseck & Salesse, 2015).

The Odor as a 'Theatrical Object' for a Dancetheater Performance

Based on a similar approach to *Les Parfums de l'Âme*, I developed *Duft, an Olfactory Metamorphosis*, a dancetheater *solo* performance, presented in the form of work in progress at the sixth edition of the *Smell Festival* in May 2015 in Bologna, Italy, and as an open-air version in June 2015, during a private artistic event in Bellinzona – Martella Island, Switzerland, hosted by Nephos Swiss Fog.[4]

The general plot develops in three parts around an imaginary metamorphosis evoking the three metamorphoses from *Thus Spake Zarathustra*, the late 19th-Century philosophical novel by Friedrich Nietzsche (1999), where "the spirit becomes a camel, and the camel a lion, and lion at last a child" (see Mitcheson, 2013, p. 153). In the novel, the 'spirit' or 'soul' has to become a camel – carrier of heavy weights through the desert – in order to abandon pride and arrogance. The camel has then to become a lion – king of animals – in order to get freedom back and to turn the order (a 'super-ego' order, I would suggest) 'you must' into 'I want,' creating new value. To accomplish the metamorphosis, the soul has to change one more time, from lion into a child – an expression of innocence, oblivion and a new start – in order to reconquer the spirit's inner world.

The performance's composite theme anchors on three perfumed ambiences respectively, and the dramaturgical rhythm follows a line of free associations inspired by Nietzsche's tale and by the odors, without any literal translation from the novel. The choice of the three-odor theme was based on my personal interest, and on associations I made between the tale and odors. This was also the starting point for the composition. The three odor scores were developed with the performance following these main associations. 'Metamorphosis I' was captured by an iris/chypre odor, with the iris poudré note recalling the ambience of an old loft full of dust and antiques. The camel carrying heavy weights is associated with heavy thoughts arising from the fight of spirit or soul against individual vanity. 'Metamorphosis II' was a rose/tuberose odor. At this stage, a whiff of vanity is still there, the floral aroma reflects the vanity game of a lion which looks at himself and gradually develops his own will – with rebellion against conformism and a search for new value. 'Metamorphosis III' was captured with the odor *Animalia*. Here, animal-like notes reflect a stage where the soul is still troubled by the implications of its own change. The 'innocence' of a child or young boy is reflected in an animalic and natural odor note that leads the spirit to a new and more conscious discovery of the body.

The performance reflects an idealistic and surreal metamorphosis from a hetaeric and vacuous sense of purity and beauty to a more authentic and well-grounded self-presence and rediscovery of the sense of smell. In practice, the work for this composition of odor and theater began by examining actors' physiological and affective reactions to the individual odors. Actors' resultant improvisations and free associations were subsequently used as material for the final composition of the performance. The general concept is a performance that can become site-specific and can be adapted to different places according to a general idea of metamorphosis. Thus, it is not only the theme of the performance that is inspired by a metamorphosis, but also the performance *per se* changes by certain degrees depending on the space where it takes place. Each performance therefore becomes a unique event.

For the performance in May 2015, each audience member received a libretto composed of three chapters, each of which referred to one metamorphosis or part of the performance, and containing (instead of the traditional written plot) a black feather carrying the corresponding odor. The feathers make the odor release practical, inexpensive and suitable for an itinerant adaptation to different spaces (even if this was not the case here). Moreover, it allows individual members of the public to smell the feathers in the order intended by the performance, or to tailor the nature of their own olfactory input by referring to the feathers in alternative sequences. Indeed, aside from the general theme of the performance and a short line about each metamorphosis within the libretto describing the ambience, I decided to provide the audience with no additional didascalic explanations, thereby leaving each audience member free to live the performance in a personal and intuitive manner. This seemed better suited to the ephemeral nature of smell and its singular features of communicating without words.

The release of this performance at Martella Island offers an example of the use of smell in open-air performance spaces. In this case, the performance was integrated with a fog installation, and the odors released through the fog (see Figures 23.1 and 23.2).

The perfumed fog is a project born from a collaboration between the company Nephos Swiss Fog, which develops and constructs mist-propagation systems for industry and scenography, and the independent perfumer Alessandro Gualtieri. They developed a 'PJ (perfume jockey) console,' which allows for the mixing of odors and their subsequent release into the fog (see Figure 23.3). A preliminary test with an alcohol and water dilution showed a better and more consistent diffusion than with water dilution alone.

FIGURE 23.1 *Duft, an Olfactory Metamorphosis* – in open air with perfumed fog.

Image courtesy of Sabrina Montiglia.

FIGURE 23.2 *Duft, an Olfactory Metamorphosis* – in open air with perfumed fog.

Image courtesy of Sabrina Montiglia.

FIGURE 23.3 Perfume jockey at work for the perfumed fog installation.

Image courtesy of Sabrina Montiglia.

The advantage of this system is that it gives a reliable release of odor into the air, but it presents some technical difficulties due to the fact that fog as a medium carries the odor towards the ground, partially neutralizing the smell effect. Therefore, a second set of tests was run to calibrate the odor diffusion, its period of persistence in space, and to optimize the system for sequential odor releases. Currently, the PJ console can mix four odors in different proportions and release them reliably. This system of odor diffusion, coupled with the idea of an artistic experiment, and the perfumed fog installation, where three different olfactory ambiences created by Alessandro Gualtieri were released according to the action, together became part of the performance.

Conclusion

The aesthetics of olfaction brings us to a multifaceted field when we consider theatrical and performing arts: we are confronted with an integrated sensory input where visual and aural systems traditionally play the main role, and where the odor intermingles at different levels. At first glance, the sense of smell integrates with other sensory channels, constructing or deconstructing the main plot and/or specific ambiences and characters. But this is only the tip of the iceberg: as we have seen in this chapter, odor use in theatrical space has physiological and psychological implications in terms of experiences induced in actors and audience members. Odor can enhance or detach from realism, it can assume a strong symbolic or cultural connotation exaggerating or denouncing social stereotypes. Thus, the attribution of sense and creation of meaning through the use of aroma can become a powerful tool for authors and performers. The aesthetics of olfaction expressed in the Japanese art of incense offers us a unique example of how a performative art based on smell, the most neglected of our sensory systems, is possible. This observation might lead us, I suggest, to a use of odor as a 'theatrical object,' and not just as an illustrative or symbolic complement for a performance. We can see the potential of this approach in *Les Parfums de l'Âme* and in *Duft, an Olfactory Metamorphosis*. The 'awakening' of the nose during the training, for actors and performers, and in the composition process, can open a window onto new forms of creativity and consciousness, leading to an awakening of the audience too. The sense of smell is the most archaic of our senses, and in theater we have just consciously started to use it.

Notes

1 In Peter Brook's *Mahabharata* (1985), the burning incense throughout the full play can be considered as illustrative of the Indian context, but it also acts on a different level, shaping the full experience of the audience with a ritual connotation.

2 The word *koh-do* is probably derived from *mon-koh*, "the Japanese pronunciation of the Chinese expression, *wenxiang*" (Morita, 1992, p. 15), meaning 'listening to incense,' where *koh* means 'perfumed.'

3 *Pierre de Soleil*, first creation June 21, 1998, Paris. Choreography: Patricia Olive and Yann Lheureux; voice: Jean-Claude Marol; image creation, olfactory score and realization: Michel Roudnitska; technique of olfactory diffusion: Sigmacom; production: Olfactive Vision.

4 www.nephos.ch/ Retrieved March 1, 2017.

References

Aston, W. G. (Trans.). (1971). *Nihongi: Chronicles of Japan from the earliest times to A.D. 697*. Tokyo: Tuttle.

Banes, S. (2001). Olfactory performances. *TDR/The Drama Review, 45*(1): 68–76.

Cometto-Muñiz, J. E., & Cain, W. S. (1991). Nasal pungency, odor, and eye irritation thresholds for homologous acetates. *Pharmacology Biochemistry and Behavior, 39*(4), 983–989.

Dalton, P. (2003). Upper airway irritation, odor perception and health risk due to airborne chemicals. *Toxicology Letters, 140–141*, 239–248.

Domisseck, S., & Salesse, R. (2015). Le spectateur olfactif: La réception de la pièce Les Parfums de l'Âme par le public. In C. Jaquet (Ed.), *L'art olfactif contemporain* (pp. 287–298). Paris: Editions Classiques Garnier.

Gottfried, J. A. (Ed.). (2011). *Neurobiology of sensation and reward*. Boca Raton, FL: CRC Press/Taylor & Francis.

Jaquet, C. (2010). *Philosophie de l'odorat*. Paris: Presse Universitaires de France.

Mainland, J. D., Keller, A., Li, Y. R., Zhou, T., Trimmer, C, Snyder, L. L. . . . Matsunami, H. (2014). The missense of smell: Functional variability in the human odorant receptor repertoire. *Nature Neuroscience, 17*(1), 114–120.

Malnic, B., Hirono, J., Sato, T., & Buck, L. B. (1999). Combinatorial receptor codes for odors. *Cell, 96*(5), 713–723.

Mitcheson, K. (2013). *Nietzsche, truth and transformation*. New York: Palgrave Macmillan.

Morita, K. (1992). *The book of incense: Enjoying the traditional art of Japanese scents*. Tokyo: Kodansha International.

Nietzsche, F. (1999). *Thus spake Zarathustra* (T. Common, Trans.). New York: Dover Thrift Editions.

Paquet, D. (2004). *La dimension olfactive dans le theâtre contemporain*. Paris: L'Harmattan.

Saint Augustine. (2008). *Confessions* (H. Chadwick, Trans.). Oxford: Oxford World's Classics.

Shepherd, G. M., Chen, W. R., & Greer, C. A. (2004). Olfactory bulb. In G. M. Shepherd (Ed.), *The synaptic organization of the brain* (5th ed., pp. 165–216). New York: Oxford University Press.

Shikibu, M. (2010). *The tale of Genji* (A. Waley, Trans.). Tokyo: Tuttle.

Smeets, M. A. M., & Dalton, P. H. (2005). Evaluating the human response to chemicals: Odor, irritation and non-sensory factors. *Environmental Toxicology and Pharmacology, 19*(3), 581–588.

Su, C.-Y., Menuz, K., & Carlson, J. R. (2009). Olfactory perception: Receptors, cells, and circuits. *Cell, 139*(1), 45–59.

PART VII

Smell Capture, Distillation and Diffusion

INTRODUCTION

Making Scents

Alexandra Horowitz

I write this on the cusp of spring, and the air around me smells vernal. Spring reawakens odors, releasing from hibernation those smells trapped beneath the frosts of winter. Our noses, too, revive. Each day, a caught smell ignites a memory – specters, the haze of a forgotten scene, a punch to the viscera. This sensory experience, shared by everyone with a nose, is also deeply personal and private: the recollection it triggers is yours alone. Smell's public face is more blunt: the binary *good smells* and *bad smells* that we seek and apply on our bodies, on the one hand, and turn up our noses at, on the other.

Research indicates that humans may have the noses to discriminate a trillion smells (Bushdid, Magnasco, Vosshall & Keller, 2014). The downright irrelevance of this number in our lived experience aside, what this extravagance serves to remind us is how *little* most people typically smell. What have you smelled today? Alternatively, how many smells can you name? Chances are, each number is tiny. Most people are not experiencing large quantities of smell in daily life.

The number of discriminable smells itself is almost certainly less important than the question of whether one actually bothers to sniff. It is not, after all, the number of colors the eye can take in that figures most prominently in your experience of seeing, but whether one has one's eyes open and actually *looks*. As the number of colors that our eyes can detect does not suffice to describe the experience of seeing the face of your own child, the number of smells one can detect is neither explanation nor substitute for the feeling of déjà vu that a wafting odor can provoke.

The 'sniff' is the beginning of the odorant molecule's journey from outside the body to inside the body; it needs to be sufficiently vigorous to send the molecule up the length of the nose to the olfactory epithelium. At this postage-stamp-sized tissue (at a point behind the bridge of the nose, midway between the eyes), some 6 million olfactory receptor cells lie, awaiting the molecule's arrival. Should the odorant bind to a cell and cause it to fire, the signal travels to the olfactory lobe – the quickest sensory route into the innards of the brain – and then to various other parts of the brain responsible for learning, memory, emotion and motor response.

This process of olfaction is an 'active process,' say scientists who have mapped the route of a molecule from the outside world into the inner workings of the nose (Mainland & Sobel, 2006). While smells – especially ones labeled 'disagreeable' – may seem to happen *to* us, we actually must do a fair bit to *make* those smells happen. We must sniff, and send the molecule racing to the brain. We must attend to the smell, and be alert to the associations it triggers in our minds. We must revisit the smell, and keep it in our memories.

And, we could add, we might enable smells – some novel, some familiar – to appear on demand. In this part, the authors – artists, designers, academics – describe the means by which they have manipulated and created odors, as well as the conceptual projects that implicate smell. They treat smell as a medium in which to play; as a substance to be engaged with. In these essays, each considers the technologies that allow them to so engage.

Artist Olivia Alice Clemence (Chapter 24), inspired to capture the scent of a beloved grandmother who had passed, creates the means to do so. She describes her design of a steam distilling kit that enabled her to extract the scent-of-person from the person's clothes – in this case, a jumper. There is something apt about capturing, in an invisible and intangible form, the scent of a person no longer visible nor tangibly present.

For Clemence, the end defined the means – but she has also extended the means to enable a kind of 'recreation' of odors of the past: a Victorian pub and a junk shop. Her process is not only extendable to the far reaches of the reader's imagination; her distillery can, as she describes, be made with 'ordinary kitchen equipment.' In one's kitchen are the tools to smell the past.

In Chapter 25, Brian Goeltzenleuchter describes three of his own exhibitions which have used scent, and the 'scent emission devices' that served to deliver the relevant odors. The technology, in Goeltzenleuchter's hands, is in part defined by, and also serves to define, the experience that the exhibition attendee or passer-by has. A project using atomizers enabled – and was enabled by – the explicit intent that patrons interact with each other. In another project, microencapsulation of odor, painted into the wall, allowed for – and allowed – a more subtle and diffuse experience. Scent suspended in a gel polymer released odor in time, and was experienced differently, one suspects, at different times. The kinds of scent widely available, and the various means of producing scent, add up to myriad possibilities for the artist or designer.

Debra Riley Parr focuses in Chapter 26 on three methods of creating scent, considered historically and thematically. She considers how current and past scientists and artists 'capture,' 'encode' or 'disperse' scent. From headspace technology to recent developments, like the oPhone, the means of capture are, in Parr's reckoning, sensory analogs to the gramophone or camera. From another angle, the body is encoded – through its scent and through its scenting. And, like Goeltzenleuchter, she examines the varying successes of dispersion techniques and performances.

However much smell, as a sense and as a sensation, is maligned and overlooked, these contributions, like the previous chapters in this volume, go a fair way to overturning that state of affairs.

References

Bushdid, C., Magnasco, M. O., Vosshall, L. B., & Keller, A. (2014). Humans can discriminate more than 1 trillion olfactory stimuli. *Science*, *343*(6177), 1370–1372.

Mainland, J., & Sobel, N. (2006). The sniff is part of the olfactory percept. *Chemical Senses*, *31*(2), 181–196.

24

A DIY GUIDE FOR EXTRACTING SCENT IN YOUR KITCHEN

Olivia Alice Clemence

Introduction

Imagine an ordinary day, and try to recall all the different smells you encounter: invigorating shower gel, freshly washed clothes, earthy fresh coffee, fragrant fresh fruit, sweet honey, a spritz of perfume . . . and your day has hardly begun. Just imagine a day without smell; without your nose inhaling whiffs of odor, perfume, aromas and even an unpleasant stench – your day would be fairly dull. The food you eat would be tasteless, and the same goes for drinks; it would be a very bland day.

We continuously smell, whether consciously or unconsciously, and our brain recognizes almost all of the individual components of each aroma: ". . . we are able to detect thousands of different compounds and despite the great variation in chemical structure and molecular weight of these volatile compounds, we can differentiate between them and identify their individual smells" (Bakker & Clarke, 2012, pp. 23–24).

We regularly smell many things that are simultaneous individual scents. Coffee, for example, has up to 500 volatile compounds. Trained perfume chemists can distinguish the various chemicals in a given fragrance, but this may be explained by their extensive knowledge of the mixture of odorants needed to produce different fragrances – nevertheless, a very impressive skill to obtain (Adelman, 1987).

As a species, humans are highly dependent on sight and hearing to gather information about our surroundings. Smell is often the unsung sense; it is ephemeral, will linger in the air waiting to be sniffed, and is often very unpredictable. A whiff is inhaled through the nose to the olfactory bulb, and along the olfactory nerve directly to the brain, where the path divides into two. One path passes into the olfactory cortex, located at the front of the brain where identification and differentiation between odors occurs; the other passes into the limbic system, which is in the center of the brain (Pybus & Sell, 1999). All of this happens in seconds, and our brain almost instantly recognizes the hundreds of compounds and tells us what we are sniffing. With just one inhalation of a whiff, you can be instantly transported back to a place or situation, and this is due to its close connection to our limbic system (where all of our emotions and memories are stored). Smells are also essential signals in social bonding, and this is evident in babies. They can recognize the odor of their mothers soon after birth. Adults can identify their children or spouses by scent. In an experiment, women and men were able to distinguish t-shirts worn by their marriage partner – among dozens of others – by smell alone (Classen, Howes, & Synnott, 1994).

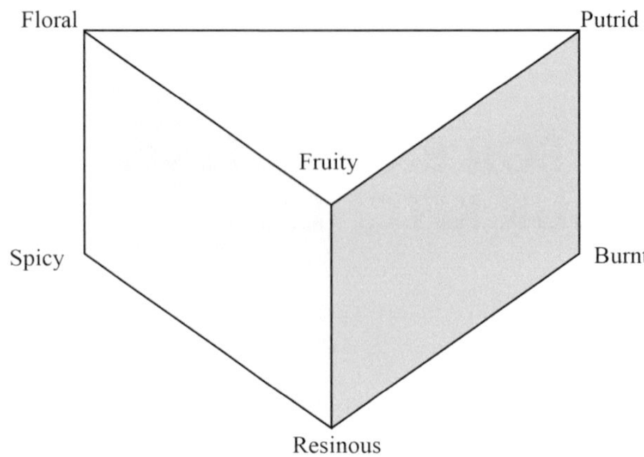

FIGURE 24.1 Henning's smell prism.

After Henning (1916).

Smell is a highly elusive phenomenon. There is no alphabet for smell; 'It smells like . . .' is how we would often begin to describe a smell using a range of metaphors. And to make it even harder for our nose receptors, smells are forever changing as one comes across new things not smelt before. In 1916, philosopher Hans Henning introduced the 'smell prism' (see Figure 24.1), in which six odor primaries were distinguished: fragrant (or floral), putrid, fruity, spicy, burnt and resinous. Though not bad descriptors, these, like all, have a subjective quality to them (Turin, 2006). Perfumes can also have many tones: for example, a rose top note, an earthy midtone and a burnt leather undertone. One could indeed argue that no smell is composed of a single molecule. Therefore, Henning's smell prism at least acts as a starting point for creating an olfactory alphabet or a range of smell descriptors.

Eating is an essential part of our lives; we eat around three meals a day, and can distinguish between a whole array of different foodstuffs. However, we only taste bitter, sweet, salt, sour and umami. The rest of what we taste is to down to smell. Smelling food is not always about the experience of eating, it is often a way of determining if the food has degraded. Without this instinct, we would not survive very long. Deprived of smell, eating food would be insipid and the act of eating would become rather laborious.

Try this experiment: using a pack of fruit sweets, close your eyes and hold your nose sealed tight. Reach for a sweet, and begin to eat. Can you guess the flavor? You should find this near impossible to guess until you release your nose! A wave of smell will come rushing through your nostrils, reaching your limbic system, where your brain will recognize the flavor.

Grandma's Jumper

I studied Design at Goldsmiths, University of London. During my final year, I began to explore the link between scent and memories. The project was named *Grandma's Jumper* (see Figure 24.2). It explored the life (and death) of a garment that is left behind when someone passes away. When keeping an item of clothing that belonged to a deceased loved one, it acts as an instant reminder; it can evoke memories and a sense of nostalgia. Some people may feel an obligation to keep an item simply because of its associations, even though it may be disliked aesthetically and never worn again.

This project focused on extracting the essential elements that are embedded within the garment – such as scent, color and DNA – so that the original form becomes obsolete. By deconstructing

FIGURE 24.2 The *Grandma's Jumper* project.

Image: author's own.

the clothing and extracting its component parts, we might be able to keep the associated memories without having to retain the physical object. The by-products of this process are a colored liquid, a steam-distilled scent and DNA-rich ashes. These substances exist as a new set of *memento mori* that can be distributed amongst relatives or utilized in new ways.

Distilling

> Distilling is beautiful. First of all, because it is a slow, philosophic, and silent occupation, which keeps you busy but gives you time to think of other things, somewhat like riding a bike. Then, because it involves a metamorphosis from liquid to vapor (invisible), and from this once again to liquid; but in this double journey, up and down, purity is attained, an ambiguous and fascinating condition, which starts with chemistry and goes very far.
>
> *Levi, 1984, pp. 57–58*

In order to extract the scent from the garment, I began to look into steam distilling. Distilling is not a new process; it has been used for many years in the fragrance industry to extract high-quality essential oils. It is slow, labor-intensive and requires vast quantities of raw ingredients to distil just a few milliliters of oil. To distil an essential oil, the raw ingredient needs to contain some oil. But what if the object is a jumper that has been worn?

During the distilling process, a large amount of by-product is produced, which is scented water. For example, in order to distil rose oil, the petals should be handpicked at sunrise, and to ensure a high-quality scent, distilled that same day. When distilling rose petals, you should not steam-distil, as this process often loses some of the more delicate tones. The best method is to allow the petals to float freely in the water, similar in a way to boiling vegetables. The still, which is usually made of copper, holds approximately 1500 L of water and 500 kg of rose petals. Steam is then introduced into the base of the still, causing the water to boil, and the process begins. The distillate (rose water) is captured in a beaker to the side, whilst the rose oil (also known as attar or otto) floats to the top. This occurs because oil is lighter than water. During this process, high-quality rose oil is captured, and the scented water by-product is also produced and can be used in cocktail making, for example. This steam-distilling process has been used for many years in the perfume industry. My work looked at whether it could be slightly adapted to distil the scent from a jumper. If so, it would suggest a process that could extract the scent from virtually anything.

The Steam-distilling Kit

I designed my steam-distilling kit to be made from glass so I could see what was happening (see Figure 24.3). However, I have since found that insulating the glass improves the yield. At the base is a 5 L borosilicate glass beaker, approximately 20 cm in diameter, which has a flat, ground rim. (Having a ground rim enables a better seal to form when the funnel is placed on top.) The beaker also has three indents in the side, 20 cm from the base, to enable the metal gauze to sit on it; this creates a shelf inside the beaker (the shelf is where you should place the object you want to distil). The glass funnel, which is hand-blown to fit onto the top of the beaker, should be approximately 30 cm in height with a fitting for a condenser at the top. The condenser needs to sit alongside this

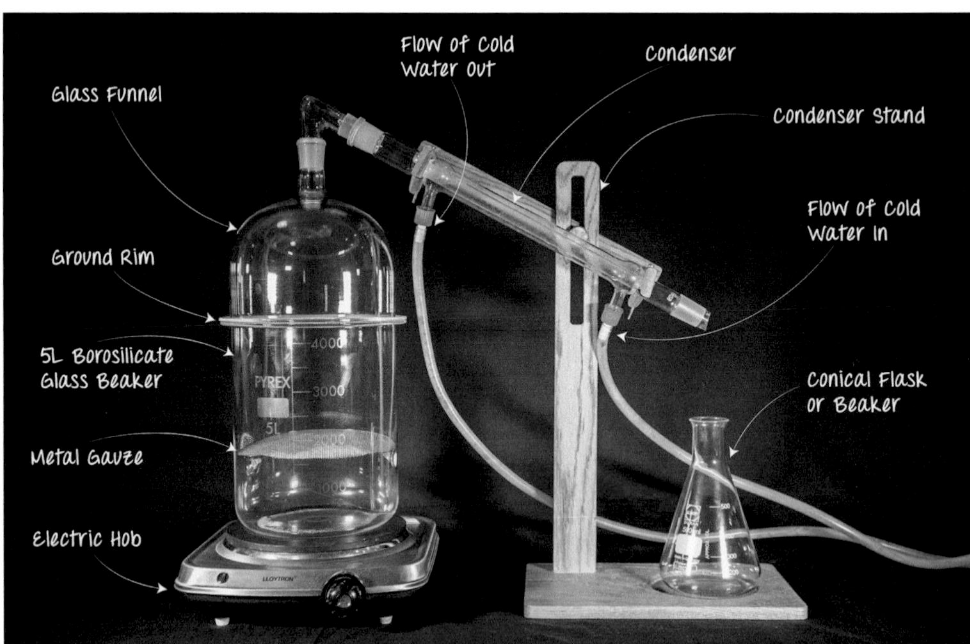

FIGURE 24.3 The steam-distilling kit.

Image: author's own.

stack at a 45° angle. A condenser has a jacket around a pipe, and the jacket fills with cold water, entering at the base and exiting at the top, which cools and condenses the steam. If you are using a tap for the flow, the water temperature will need to be under 20 °C, and the flow rate will need to be 2 L per minute. If the water is colder (5–10 °C), the flow rate changes to 1 L per minute. Alternatively, you can use a water pump in a bucket (I use a fish tank pump) to circulate the water. A bucket with water and ice cubes also works very well too, and you would require approximately two to three bags of ice per 100–200 ml of scent.

Finally, at the base of the condenser you need a beaker in which to capture the scent. You will also need a heat source to heat the water in the beaker. A gas hob will heat the glass in one area very well, but from experience, this can cause the glass to crack. I would advise you use an electric hob or a laboratory hotplate.

This still is designed to be open to the atmosphere, meaning that steam pressure cannot build up in the system, making it a much safer process to undertake at home. It will also ensure that the distillation process takes place at exactly 100 °C, and no higher. When using the still to distil the scent from a jumper, this is what happens: As water boils in the base of the glass beaker, it begins to create steam. The vaporized water rises through the wire gauze and engulfs the jumper. As the steam rises through the jumper, the scent molecules become encapsulated within the vaporized water. The steam now carrying the scent molecules continues to rise up the equipment, through the remaining garment, through the funnel and into the condenser. The condenser, which has a flow of cold water around the outside, condenses the steam back to water. The now scented liquid drops out into the beaker below. This process takes approximately one hour to yield 100 ml of scented liquid.

To improve on the design and to gather a better yield, here are a few adjustments you could make:

1. Use two borosilicate boiling glasses, one to sit on the hob for the water to boil inside. The other boiling glass that has the object you would like to distil within it fits on top of the other (see Figure 24.4). This will ensure that the object under distillation will never burn or overheat, and should therefore increase the quality of the yield.
2. A simple adjustment would be to get a different condenser: one that has more plates, or one that has a coil inside. This increases the surface area of the cold-water flow and cools the steam quicker and more efficiently.
3. Using a partial vacuum (e.g. an aspirator) to reduce the pressure, hence decreasing the boiling temperature of the water, speeds up the process but does not necessarily increase the yield.
4. Adding ethylene glycol (antifreeze) or a glycerol/water mixture helps reduce the vapor pressure of the water, and this enables more delicate scents to be distilled.
5. You can also use salts, such as sodium chloride, to increase the boiling point of water, but be careful to use ones that do not participate in the reaction, as using the wrong salt could ruin the scent.
6. Pumping superheated steam directly into the object you are distilling enables a better yield of scented distillate, but requires far more technical equipment that cannot be found in the average kitchen.
7. Another adjustment would be to use an ice bath surrounding the distillate-collecting beaker. This will ensure that the scent is fully locked into the liquid before it warms back to room temperature.

There are many different adjustments you could make, and I have just listed a few here.

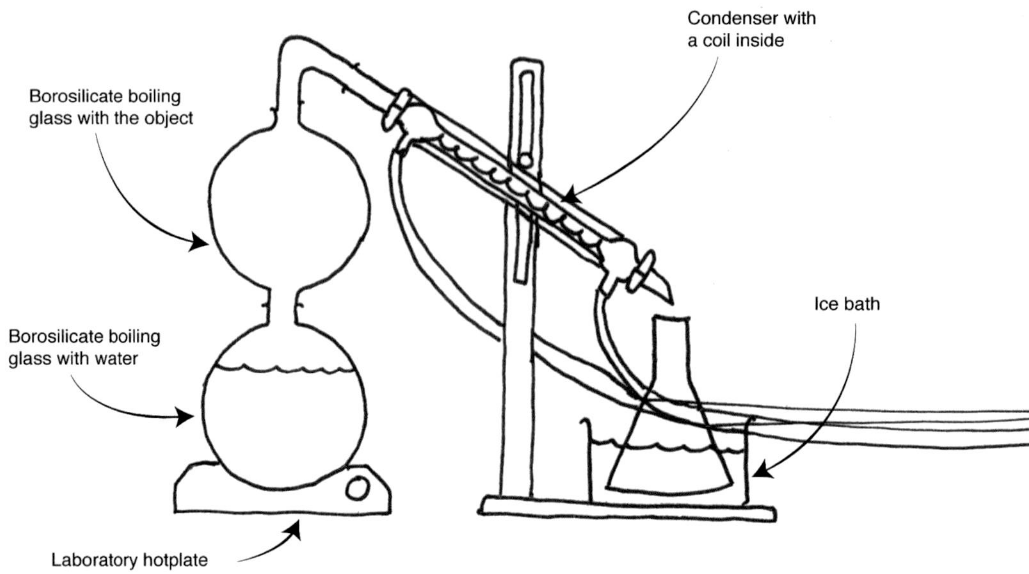

FIGURE 24.4 Alternative design for steam-distilling using two boiling glasses.

Image: author's own.

Distilling in Your Kitchen

The steam-distilling kit described above is mostly specially blown or altered by glassblowers, which of course makes it costly. However, you can distil at home using ordinary kitchen equipment. The results will not be as high-quality as the built glass kit, but it is a great starting point. You will need the following equipment:

- electric hob;
- saucepan steamer set (two tiers is sufficient);
- metal funnel large enough to fit onto your saucepan;
- silicone baking tray;
- cable ties;
- clear PVC tubing (two lengths that will reach from the hob to the tap and 1 m extra);
- condenser and a rubber bung to fit;
- glass beaker (a drinking glass will suffice);
- large syringe;
- chemistry stand and clamps (or an alternative method of holding the condenser at an angle of 45°).

Method of assembly (see Figure 24.5 as a guide):

1. Place the saucepans on the electric hob. The saucepan on the base will hold the water; the saucepan above should sit on top of this saucepan and have holes in the base, allowing the steam to rise through.
2. Cut a length off the silicone tray: approximately 4 cm in width and long enough to fit around the edge of the saucepan. Place the silicone around the join of the saucepan to make a seal. Secure this into position using the cable ties, ensuring that the cable tie sits on the silicone,

and not the metal saucepan. (The silicone will insulate the cable ties, therefore they should not melt.)

3. Pour approximately 300–400 ml of water into the base saucepan. Place the object you want to distil into the top saucepan.
4. The metal funnel should sit above the top saucepan. Attach this using the cable ties if you can.
5. Attach a section of the PVC tubing to the top of the funnel, with enough length to reach your condenser. Seal using the silicone and cable ties.
6. Using the chemistry stand and clamps, angle the condenser at 45°. Drill a hole into the rubber bung the size of the PVC tube. Thread the tube through the hole, and place the bung into the top of the condenser.
7. The condenser will usually come with a method of attaching a tube to the ends. Using the lengths of PVC tubing to reach the sink, attach these to the two nozzles at the side of the condenser.
8. The pipe that is attached to the bottom of the condenser should be connected to the tap; the syringe can help with this part, as there are so many different tap shapes and sizes – it should create a universal adaptor. I found that taping the syringe to the tap and attaching the pipe to the end of the syringe worked best. I put the tap on a low trickle; however, pressure can change, so when distilling, keep an eye on the syringe levels.

   Ensure the other tube coming from the top of the condenser is sitting in the sink ready for the flow of water to drain away.
9. Place your collecting glass or beaker at the end of the condenser.
10. Turn the hob onto a medium heat, and sit back and wait.

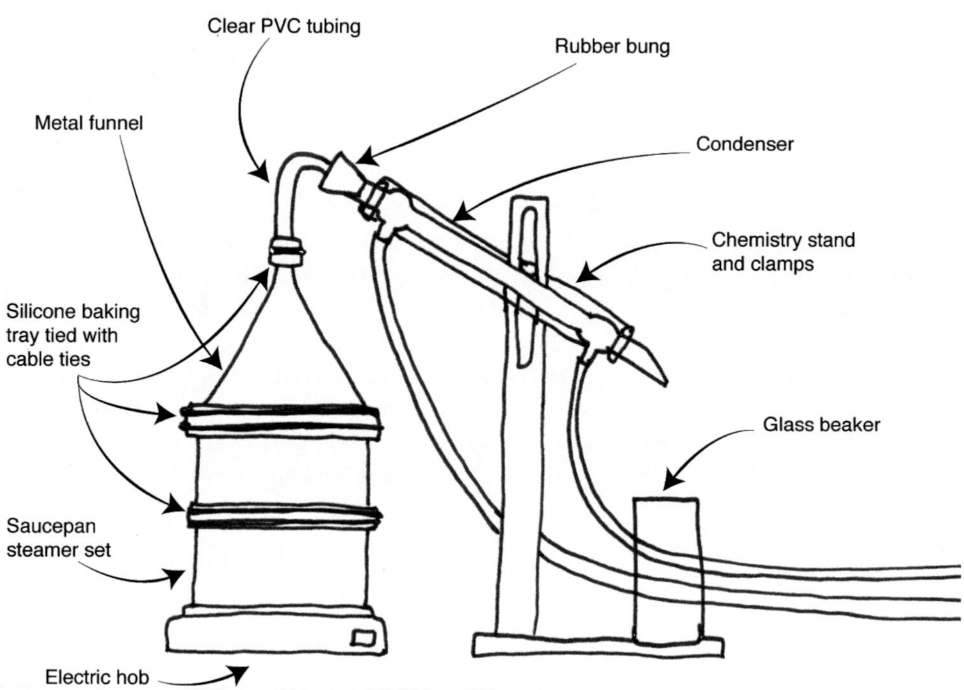

FIGURE 24.5 A home distilling setup.

Image: author's own.

This process is ideal for a beginner, and will not cost a fortune. There are many websites where you can purchase the scientific parts, and the rest you should find at your local homeware store. The technique is best suited to distilling natural matter; however, I have tried many items, including garments, shoes, gravel, napkins and natural materials such as basil, thyme, tomatoes and lavender. A word of warning: do not attempt to extract the scent of plastic, as when plastic melts, it can release toxic fumes. The same goes for household chemicals: please do not distil these items, as you might be at risk of damaging your health.

Conclusion

I designed and developed the distilling kits detailed above over three years ago, and have used them in many different ways. I began with extracting the scent from deceased loved ones' garments, which is where all of this began. I have since used the process to help recreate scents from the past. One project involved capturing the scent of a Victorian pub. On entering the pub, you smell a gin-infused bar – citrusy and heavily alcoholic. Flowing up the stairs was the scent of soapy laundry. At the top of the stairs were three rooms: one damp, mothball-ridden storage room; a musky, smoke-filled office where the master once ran the business; and the bedroom once occupied by prostitutes, infused with citrus, orange and lemon aromas. (Prostitutes during Victorian times attempted to hide the scent of sex by heavily dousing themselves in citrus-based perfumes.)

Another project captured the odor of a junk shop in Portobello Market. The resultant scent of a musty, damp space full of mothballs and vintage objects on top of the aroma of printed paper made the junk shop come alive through smell. For *Milan Design Week 2013*, I collaborated with scent mapper Kate McLean to distil Milan. The aromas, stenches and smells in and around the city were distilled live and placed back into a model of the city, creating juxtaposition between reality and memories. During the show, the audience were invited to participate in the scented model city by adding their own thoughts and comments about the smells encountered in Milan, in the form of graffiti written directly onto our model buildings.

Most recently, I have been exploring how food and scent are linked. When cooking food, the more delicate tones of scents are often lost. I have been exploring how you could incorporate this scent back into the foods using a distilled scent to enhance the dish. For example, what if the dish was bland, but the aromas in the surrounding space conjured up the taste? By using scent diffusers, could you scent a restaurant to smell of each dish throughout the evening – would this add to the sensory experience of each dish? The aim would be to work alongside a chef to incorporate extra scents both directly into the dishes and also into the atmosphere. For example, Bompas & Parr, who are renowned for their creative culinary skills and pop-up experiences,[1] often use a scent cloud, involving a combination of liquid nitrogen, boiling water and a splash of scented oil. Similarly, one of the most famous multisensory chefs, Heston Blumenthal, regularly experiments with scent, flavor, sound and setting. He has served dishes that include scent on the menu of his Michelin-starred restaurant the Fat Duck in Bray, UK. These have included a nitro-poached green tea and lime mousse served with spritz of lime grove scent from an atomizer, and a jelly of quail dish which comes on a bed of oakmoss, accompanied by a specially created scent of oakmoss that is dispersed by dry ice. The next phase of this project is to research and undertake several experiments to see if and how smell, texture, taste and the atmosphere in the room can affect how we eat.

Note

1 http://bompasandparr.com/ Retrieved March 1, 2017.

References

Adelman, G. (1987). *Sensory systems II: Senses other than vision*. Boston, MA: Birkhäuser.
Bakker, J., & Clarke, R. J. (2012). *Wine: Flavour chemistry*. Chichester: Blackwell.
Classen, C., Howes, D., & Synnott, A. (1994). *Aroma: The cultural history of smell*. London: Routledge.
Henning, H. (1916). *Der Geruch*. Leipzig: J. A. Barth.
Levi, P. (1984). *The periodic table* (R. Rosenthal, Trans.). New York: Schocken.
Pybus, D. H., & Sell, C. S. (1999). *The chemistry of fragrances*. Cambridge: Royal Society of Chemistry.
Turin, L. (2006). *The secret of scent*. London: Faber and Faber.

25

SCENTING THE ANTISEPTIC INSTITUTION

Brian Goeltzenleuchter

Introduction

The turn towards the olfactory in my art practice was in part fueled by a particular antagonism I was feeling towards the visual. It was not so much a desire to negate the visual, but rather to call into question the cultural complacency with which art unselfconsciously caters to a certain class of visual objects. I concede that this type of *complacent visuality* might serve as an organizational lube for disseminating art cultures, both contemporary and historical. No doubt it also allows historians and curators to catalog, archive and display art in a quasi-scientific way. It provides the default approach to the study of art history via books and journals that feature images and text. Likely, it even functions as a default template around which museum architecture is imagined and built. But it also presupposes a set of expectations and behaviors that structure how we experience and understand art itself.

For me, olfactory art throws a wrench into this system. The nature of smell tends to carry with it a certain immateriality. When we breathe, we take in volatile molecules through our nostrils. In some cases, smell is preceded by a complementary sense, such as seeing the bouquet of roses before registering its smell. But more often, smell asserts itself without warning, such as walking into a clean kitchen and smelling fried fish. Modern distillation processes and techniques have amplified this immateriality by removing smells from their traditional source objects. Thus, through its volatility and immateriality, olfactory art inherently challenges commodification, collection and archiving.

It is this invisible, involuntary nature of smell that prompted me to use it in my art. Whereas with other senses we make a conscious decision to look, listen, taste or touch, with smell it is often just an inherent aspect of existence – *if you're breathing, you're smelling*. As an artist who indulges in the legacies of conceptual art, I find great pleasure in how this adds context to Duchamp's historical claim to have shifted occupations from artist to 'respirateur' (Tomkins, 1996, p. 13).

While designing evocative smells might appear to be the olfactory artist's primary creative challenge, my chapter highlights another challenge which is often overlooked for its technical and creative potential: namely, the hardware infrastructure that is used to transmit the scent. Far from being an afterthought, the choice of a specific method for scent emission is a key element in a dispositif, which includes the institution, its architecture and contextual narratives, all of which shape how we understand scent in olfactory art. Using three recent art projects as case studies,

I will survey a variety of scent emission devices, and present technical, poetic and conceptual criteria that I used to determine the best device to use in a specific context.

Project: *c (pronounced /k/) Boutique*

Site: Museum of Contemporary Art San Diego (MCASD)
Context: Exhibition of region's top emerging artists
Scent delivery device: Fine mist sprayers on fragrance blotters; cold air diffusion units
Scent: Four different fragrance profiles

The Experience

A viewer wanders through a museum of contemporary art, experiencing a selection of work by emerging artists. One of the last galleries she approaches has been walled off, made to look like a storefront with massive windows and a retail door, which hermetically seals off the environment from the rest of the museum. Opening the door, she breaks the threshold and takes in a decisively different sensory environment. Techno music reverberates through the floor. Pink neon light casts a cool glow on the walls. A spicy, sweet aroma fills the air, which somehow feels cooler than the rest of the museum. A hip-looking store clerk greets her, "Welcome to c (pronounced /k/), the boutique wing of Contraposto Home Décor." She nods and begins to explore the inventory: a garden sculpture made of birdseed cast in the form of Venus de Milo; Rodin's Kiss reimagined as a candle with a 'patina' of gold leaf; and a display of bottles. Perfume? The clerk continues, "You're looking at The Genre Series of Home Fragrances. These are the olfactory equivalent of the great periods of Western painting – Cubism, Expressionism, Impressionism and Fauvism." She picks up a bottle marked 'TESTER,' and the clerk quips, "Best not to spray them all at once, lest the store begins to smell like a museum."

The Context

The Museum of Contemporary Art San Diego is a West Coast contemporary art museum located literally on the edge of the Pacific Ocean. I was invited to participate in an exhibition featuring work of regional artists who occasionally receive token support from the institution. The exhibition curator, Lucia Sanroman, and I reviewed several possible contributions to the exhibition. We settled on the *c (pronounced /k/) Boutique* in part because it spoke to my skepticism toward contemporary art museums and the roles they play in incubating artists and disseminating new art. My boutique was positioned in the heart of the museum, so as not to be confused with its gift shop. It was sealed off from museum patrons by a conventional storefront; two massive windows allowed patrons to look inside, while a commercial door remained locked unless I or one of my employees were there to run the boutique, which had sporadic hours. The boutique's logo – a 'c' inside a circle, which looked curiously like a copyright logo that had succumbed to gravity's force (see Figure 25.1) – branded many of the objects, including the perfume blotters that patrons used to sample *The Genre Series of Home Fragrances*. This logo was juxtaposed effectively against MCASD's logo, a bright pink 'x' which brands everything from the museum's welcome mats to the admission stickers patrons must wear. When patrons entered the boutique through the tight seal of the commercial door, they were confronted with a considerably different sensual experience than they experienced in the rest of the museum. A sonic mash-up of techno, world music, ambient sounds and storytelling provided the soundscape; the museum's conventional track lighting system was augmented by the pink glow of a neon 'c' sign positioned at the center of the boutique; and the smell of *The Genre Series* provided an unmistakable contrast to the antiseptic non-smell of the rest of the museum.

FIGURE 25.1 *c (pronounced /k/) Boutique*, views from environment as installed at the Museum of Contemporary Art San Diego, 2010.

Image: author's own.

The Device

The Genre Series of Home Fragrances was bottled and sold as a set of four 6 oz (178 ml) fine mist sprayers. The series adopted a straightforward branding scheme that playfully reduced the dynamic visuality of four iconic modern art 'isms' – Cubism, Expressionism, Impressionism and Fauvism, respectively – into a single graphic representation of each, which then became each product's label (see Figure 25.2). While this approach was engaging from the perspective of visual marketing, it proved challenging for use of testers of the fragrance in a museum environment. From an early stage, it was evident that the sprayers emitted mist particles that were too large to vaporize in the air. The result was an inconsistent olfactory environment, with an accumulation of residue amassing on the museum's terrazzo floor directly below the display shelf. To counter this, I made three adjustments to the installation. At the end of every business day, I mopped the floor to prevent the fragrance residue from building up and oxidizing. Second, I introduced oversized fragrance blotters to be used with the testers. Instead of spraying the fragrance into the air, patrons were invited to douse the blotter with an aroma, which greatly reduced the amount of overspray. Embossed with a 'c' logo, these blotters doubled as a free souvenir. Finally, I discreetly placed a scent diffuser in each corner of the boutique. Each diffuser had a chamber that stored a cartridge filled with aroma beads. The beads were tiny scent-free crystals that absorbed about one-third of their weight in liquid fragrance. The beads are typically used in potpourri applications. I placed them in cartridges that had a threaded pipe fitting on one end and a mesh screen on the other. Pressurized room-temperature air blew a steady stream of fragrance through the cartridges and into the space. The result was dry, micron-sized particles that lightly and consistently fragranced the space. Because these scent diffusers were hidden, and given that the fragrance was ultimately

FIGURE 25.2 *The Genre Series of Home Fragrances*, essential oil blends bottled in 6 oz mist sprayers.

Image: author's own.

the same as the one in the tester bottles on the display counter, customers assumed that the scent was coming from these.

The Scent

The commercial website for *The Genre Series* promised the 'olfactory equivalent of the great periods of Western painting' – a tall order, no doubt. In all honesty, these blends were relatively standard formulas that I borrowed from aromatherapy applications. However, I chose and adapted them using what I call 'synesthetic' or 'cross-modal empathy.' I am not a true synesthete, but I do find it useful to describe smell in nonverbal ways, specifically by using color and form as olfactory descriptors. With *The Genre Series*, I wanted to empathize with the figurative colors and forms of various iconic movements in painting via the use of natural olfactory essences or simulacra, however perverse that might sound. So, for instance, Impressionism subtly shifts between sinuous botanical and spice notes, while Cubism contrasts earth-toned, angular spices against dark, heavy florals. Though both Expressionism and Fauvism share a colorful aroma palette, the former resolves into a rough-edged citrus-spice accord once the sharp petitgrain top note evaporates, while the latter uses a shockingly persistent eucalyptus to penetrate a corrugated anisic-wood accord. What I learned was that the juxtaposition of the designed scents against art historical, visual and textual narratives created an imaginary link that either amused or troubled patrons.

Project: *Institutional Wellbeing: An Olfactory Plan for the Oceanside Museum of Art*

Site: Oceanside Museum of Art (OMA)
Context: Museum successfully completes capital campaign to quadruple its size, yet leaves staffing essentially unchanged; morale lags; exhibition functions as critique of the museum's management and board of directors for being thoughtless and provincial
Scent delivery device: Scent-infused gel with fan-assisted diffusion; microencapsulation
Scent: Ozonic: fresh, marine accord with salty, aldehydic top notes

The Experience

An entire wing of a museum is devoted to what appears to be documents from a corporate retreat. A diligent patron walks past the signage reading, "Institutional Wellbeing: An Olfactory Plan for the Oceanside Museum of Art." She is confronted by a pair of memos addressed to the museum's staff and board of trustees. One memo requests their participation in private wellness tests; the other requires their use of an environmental wellness fragrance made for the workplace. Both represent an attempted coordinated response to a lag in morale. Having dutifully read the memos, the patron then sees a suite of photographs illustrating a wellness test and a glossy advertisement for the wellness fragrance. The patron rounds a corner and is confronted by a floor-to-ceiling organization chart that visualizes the wellness profile of each affiliate of the museum. Rounding another corner, the patron walks into a darkened room with a video projected onto the wall. There appears a severe upshot of a man in a suit pounding his fist against a podium while almost shouting the following:

> You cannot underestimate the abrupt and predominantly negative shift in personal and collective equilibrium that occurs when the dynamics of one's personal and social environment change. Here we have a desirous institution, the Oceanside Museum of Art, who wants more for itself; who comes to terms with its ambition and successfully completes a capital campaign to quadruple the size of its facilities. And what does it get for its efforts? A board of trustees that micromanages its museum director? A museum director who micromanages his staff? A staff that cannot make a decision without committee consensus? In short, it gets a collective spirit of poverty and desperation.
>
> *Goeltzenleuchter, 2009*

Unnerved and slightly embarrassed, the patron exits the darkened room and is led into a small cube flooded with blue light – a meditation room, she is told (see Figure 25.3). She sits on a lone bench, breathes deeply, and drinks in the blue light. The air smells like a salty ocean breeze. She thinks she hears the wind blowing, until she notices a small fan-operated unit blowing the synthetic fragrance from the corner of the ceiling. She exits the meditation room, and suddenly the whole exhibition looks different. It is cast in a sickly red hue. While contemplating the scent in the blue meditation room, she forgot that her eyes had succumbed to the phenomenon of afterimage.

FIGURE 25.3 *Institutional Wellbeing: An Olfactory Plan for Oceanside Museum of Art*, exhibition view of the Meditation Room, 2009.

Image: author's own.

The Context

Institutional Wellbeing is an olfactory art installation that was exhibited over an eight-year period in seven museums and art galleries in four countries. While the conceptual premise was the same, each installation posed a unique challenge for designing scent experiences. Unless otherwise noted, my remarks herein will reference the 2009 incarnation of the project, as installed at the OMA and curated by Teri Sowell.

Institutional Wellbeing is designed to use smell as a way to reveal the structure, politics and social dynamics of cultural institutions. *Contraposto Living*, my company-as-conceptual artwork, identifies cultural institutions undergoing substantial changes in structure and/or identity and writes a letter to the institution acknowledging those challenges, and offers to create an 'environmental wellness fragrance' to encourage a productive, harmonious workplace. The outcome of the project is a multisensory exhibition that problematizes how fragrance can function as both olfactory brand and gentle critique. It should be said at the outset that this project requires the curator to perform the comedic role of the 'straight man,' which allows me to behave eccentrically without causing alarm among the staff and board, who tend to follow the curator's lead.

The process of creating the fragrance begins up to six months in advance of the exhibition's opening. I arrange to individually meet with all staff and board members of the institution. These meetings are known as 'wellness tests.' A major feature of these tests involves using natural essences and synthetic odor molecules in conjunction with techniques of applied kinesiology – a pseudoscience – to determine which individual fragrance notes make personnel strong, and conversely, which notes make personnel weak. According to the tenets of applied kinesiology, as outlined in Hawkins (2002), each test involves the subjects holding an essence, contained in a glass bottle, near the solar plexus while I test an indicator muscle – usually the shoulder. The muscle tests give me a baseline understanding of which substances make the most people strong and which make the most people weak. Each test can be visualized in a 'wellness profile' graph. This is a major simplification, but at the conclusion of the wellness tests, I superimpose all the graphs to determine if there are essences that make the majority in the institution strong. I basically isolate those essences – as a painter might lay out a set of colors on a palette – and then attempt to resolve them in the creation of an evocative fragrance blend. The resulting blend becomes that institution's environmental wellness fragrance. The fragrance is designed to be used by all those affiliated with the institution to cleanse the ambiance, stimulate consistent degrees of positive energy, and encourage high levels of productivity and enthusiasm. The process climaxes with a product launch unveiling the new fragrance.

The physical exhibition exists to substantiate the creation of the environmental wellness fragrance, which is bottled and sold out of the museum's gift shop. As opposed to the hyperbole implicit in most contemporary advertising, the exhibition design of *Institutional Wellbeing* exemplifies themes of transparency. For instance, I create a wall-sized organizational chart portraying the hierarchy of the institution (board and staff) using the wellness profile graphs for each cell in the chart. I exhibit all internal correspondence so that patrons can get a feel for how I discuss these issues with the institution's board and staff members. There is also a reading room that contains a small library of texts related to inter-disciplinary scholarship on olfaction. A series of wall-mounted, button-activated 'scent stations' encourage direct interaction with scent (see Figure 25.4). All of this is to substantiate the central narrative that X institution is unhealthy for Y reasons and therefore Z fragrance has been created (and is both bottled for sale and released environmentally in a meditation room) to mitigate the situation.

FIGURE 25.4 *Institutional Wellbeing: An Olfactory Plan for Oceanside Museum of Art*, exhibition view of scent stations, 2009.

Image: author's own.

The Device

The scent station device used in the OMA exhibition has since been used in many other projects. A simple box is flush-mounted to the wall. Inside the box, the desired scent is suspended in a gel polymer. As the gel evaporates, scent molecules are released into the box, which has a fan system directing the scented air out through slots on either the top or bottom. The box comes in two variations. One is a cordless, battery-powered unit that allows users to release the fragrance at the push of a button. The other is an AC-powered unit designed for a steady stream of air. The former is used for single-person scent experiences, while the latter is used for environmental applications. The gel units were designed to be low-maintenance so that museum preparators could maintain a consistent scent experience over the duration of a museum exhibition lasting several months. Working with museum staff, I have been able to allay fears of over-scenting the museum by testing several different concentrations of scent-infused gel. The only recurring challenge involved translating the fragrance, as I smelled it in my lab on perfume blotters, to an equivalent smell in the infused gel. Usually, some reformulation was necessary to compensate for the loss of the smaller scent molecules known as 'top notes.'

To complement the theme of transparency, and to further the pedagogical intentions of the exhibition design, I displayed a suite of four scent stations in close proximity to the meditation room. I deconstructed the OMA fragrance into its three primary accords across three of the scent stations – an accord being when two or more fragrance notes blend together to take on a new olfactory identity. The fourth station housed the complete composition. A simple wall text outlined the deconstruction process, and docents were trained to offer an augmented explanation. The arrangement and design of the scent stations maintained the look and feel of the exhibition's narrative, while they added a crucial introduction to how the fragrance was constructed. Many patrons told museum personnel that this was the first time they had considered, let alone smelt, a fragrance deconstructed.

The meditation room was the literal and figurative center of the exhibition. A 16-foot (1.8 m) unadorned cube, the room was punctured by a single entryway that allowed a monochromatic blue light and an ozonic aroma to spill out into the museum. The light came from medical-grade bilirubin phototherapy bulbs. These bulbs are known to many, if not by name, then by sight, as the light used to treat jaundice in babies born prematurely. In this context, they functioned as a gentle reminder of the critical essence of the *Institutional Wellbeing* project; so often I was working with museums which made decisions prematurely and were left struggling to grow into those decisions.

In other incarnations of this project, when such a full-scale meditation room could not be built, I designed a long, narrow meditation corridor through which a similar experience could be achieved. However, the limited cubic space coupled with dual entryways meant that scent could not be contained. The result entailed using microencapsulation to surround tiny droplets of fragrances in a non-permeable membrane that only released the fragrance when physically rubbed or scratched. The microencapsulated scent was mixed into the museum's wall paint and simply painted onto the interior walls of the corridor. When patrons walked through the narrow corridor, they inevitably rubbed up against the wall, releasing the fragrance. Simultaneously, the intense blue light still caused the optical phenomenon of afterimage, which, upon exiting the corridor, temporarily distorted the viewer's color perception.

The Scent

Every incarnation of *Institutional Wellbeing* affords the creation of a new scent. The scent I designed for OMA was an aldehydic ozone smell. Two of the three accords that comprised the scent could be classified as ozone or marine in nature. One, with salty minerality, paired well with the other, a sweet marine ozone with a clean dry down. Together they evoked a summery day at the beach. The third accord was an aldehydic top note which added a glass and steel angularity to the blend, which was designed to evoke the modernist architecture of the museum, itself situated just blocks away from the beach. The values exemplified in the fragrance were intended to be OMA's olfactory brand. Indeed, the olfactory branding was so convincing that of the seven incarnations of the *Institutional Wellbeing* series, this was the only fragrance that sold out before the exhibition closed.

Project: *Sillage*

Site: Santa Monica Museum of Art (SMMoA)
Context: Olfactory portrait of the museum's demographic
Scent delivery device: Bottled synthetic atomized directly on patron's wrist
Scent: Eleven different scentscapes designed to represent Los Angeles neighborhoods

The Experience

"Hi. I participated in a survey about the smells of various Los Angeles neighborhoods."
"Right this way. What part of town are you from?"
"Beverly Hills."
"I need to spray this on your wrist. Here you go."
"Whoa! It's so fresh. Like air and grass."
"When that wears off it's supposed to smell like the sweat of a trophy wife."
"Oh . . . so can I smell the other neighborhoods?"
"No. Well, yes. But you have to walk around the museum and ask the other patrons if you can smell them."
"Oh . . . OK. Thanks."

The Context

Sillage began as a collection of 11 fragrances that represented the major regions of Los Angeles. The scents were based on data collected from a survey asking Los Angeles residents to describe the smells they associate with various neighborhoods. Each fragrance can be thought of as a *scentscape*: a fragrance in which a distinctive background smell describing the sky and ground creates a foundation, on top of which episodic smells reveal themselves over time. At the SMMoA, patrons approached the *Sillage* booth and identified the region in which they lived to the museum's staff, who logged the data and sprayed a corresponding scent on the patron's wrist (see Figure 25.5).

Curator Asuka Hisa commissioned this project to coincide with an exhibition of large-scale, polychrome, non-objective paintings. Over the course of the one-day event, patrons wandered around the exhibition, talking to and smelling each other. The paintings served as a photogenic backdrop as the collective scent and enthusiastic chatter of the museum's demographic filled the space. The event effectively performed a figure–ground inversion on the conventional museum experience, wherein the patron loses him/herself in the quiet contemplation of the artwork. Upon closing, I used data from the event to formulate a bespoke perfume – the demographic smell of the museum's patron base – which was donated to the museum.

The Device

With *Sillage*, scent dissemination was comparatively easy. Whereas in previous projects I was interested in creating a multisensory environment, *Sillage* was, on one hand, heavily slanted toward the olfactory. I needed the scent to be atomized in micron-sized particles which, like perfume,

FIGURE 25.5 *Sillage*, olfactory artwork launched at the Santa Monica Museum of Art, 2014.
Image: author's own.

dried quickly on the skin. I also wanted to downplay any visual branding of the project in an effort to avoid associations with perfume culture and advertising. (There would be no perfume sold in the gift shop.) So I chose a nondescript perfume bottle with an atomizing spray pump that suited my needs. Simple vinyl lettering labeled the various neighborhoods represented in the project. The bottles were lined up in a row on an otherwise bare table set up near the front of the museum. Other than spraying the appropriate scent on a patron, the museum personnel only needed to log the user data so that a 'patron profile' could be recreated at a later date.

The Scents

I designed the scents at the concentration of 'extrait de parfum,' which contains the highest level of aromatic compounds. I wanted a classical dry down that revealed top, middle and base notes. Using the information I collected from the survey afforded me plenty of opportunity to layer each scent with evocative impressions of each neighborhood. The following are olfactory descriptions of each fragrance:

Downtown Los Angeles: hot asphalt penetrates traffic exhaust
East/Northeast Los Angeles: Mexican food mixes with orange blossom-infused air
Harbor Area: a breeze of salty ocean air mixed with heavy machinery
Hollywood: an old lady wearing cheap perfume, a kid sticky with cotton candy, and a hipster redolent of sweet tobacco
Los Feliz & Silverlake: dry air infused with night-blooming jasmine drifts across dry concrete
Northwest of Downtown: musty pond water fades into warm air
San Fernando Valley: rosemary gives way to bubblegum melting on the sidewalk
South Los Angeles: metallic heat and hydraulic fluid pulse over old asphalt
West Los Angeles: wet lawn gives way to dry air, which gives way to the clean sweat of a trophy wife
Wilshire Area: traffic exhaust and tar hover above hot concrete
Other (for visitors who lived outside of Los Angeles): suntan lotion on a sweaty tourist

I tried to make each scent equally diffusive. I had no idea how comfortable patrons would be with requesting to smell one another, so I added to each blend aromachemicals that augmented diffusivity (see Figure 25.6). In addition to stimulating conversation among the patrons about the construction of Los Angeles – its myths and realities – I wanted to generate an expansive environmental scent in the museum; what I referred to as the 'SMMoA patron demographic profile.' Given the massive cubic volume of the museum, the most effective way to do this was to use patrons as dynamic, kinetic scent diffusers who, by walking around, left fragrance trails in their wake.

Conclusion

Designing olfactory artworks for museums is a highly collaborative affair. Patrons register smells in very subjective ways, which vary with age, culture, health, subjective experience and genetics. In-house legal counsel and museum registrars must sign off on issues related to health, safety and conservation. Preparators must maintain the exhibit. Docents need to be instructed on how to talk about it. And all of that is predicated on a curator even taking an interest in it. Added to this is the fact that smell represents in a very different manner than do images and sound (Batty, 2010), so any artist who takes him/herself seriously should feel daunted to work in this medium. And yet the history of art points to numerous artists who, over the last century, have made scent-based

FIGURE 25.6 *Sillage*, olfactory artwork launched at the Santa Monica Museum of Art, 2014.
Image: author's own.

works (Osman, 2013). Though many of these artists are well known, their engagement with the sense of smell is not (Drobnick, 2012). Those experiments in olfaction existed outside the formal conventions of art display, historiography and conservation. They are a mystery yet to be solved, a territory yet to be explored. This alone should be reason enough for artists to inquire about this liminal space, olfactory art.

References

Batty, C. (2010). A representational account of olfactory experience. *Canadian Journal of Philosophy, 40*(4), 511–538.

Drobnick, J. (2012). Towards an olfactory art history: The mingled, fatal, and rejuvenating perfumes of Paul Gauguin. *The Senses and Society*, 7(2), 196–208.

Goeltzenleuchter, B. (2009). *Institutional Wellbeing: Product launch* [Art video]. Retrieved March 1, 2017 from www.bgprojects.com/iwoma

Hawkins, D. (2002). *Power vs. force: The hidden determinants of human behavior* (rev. ed.). Carlsbad, CA: Hay House.

Osman, A. (2013). *Historical overview of olfactory art in the 20th Century*. Retrieved March 1, 2017 from https://www.academia.edu/4608919/Historical_Overview_of_Olfactory_Art_in_the_20th_Century

Tomkins, C. (1996). *Duchamp: A biography*. New York: Henry Holt.

26

INDETERMINATE ECOLOGIES OF SCENT

Debra Riley Parr

Introduction

Almost every article on olfaction outlines as a starting point the denigration of the sense of smell over the past several centuries. In his article "Toposmia," Jim Drobnick (2002) begins with the observation that "the stigmatization of smell over the past two centuries and the resultant odorphobia still persists, as olfaction is often considered to be the most animalistic of the senses" (p. 31). Laura Marks (2008) sets out to correct "the neglect of touch, smell, and taste (and to some extent, hearing) in visual culture," which she argues arises from "the tendency to dismiss the proximal senses as inferior" (p. 123). It is a standard trope of much writing on olfaction to recount its history in terms of a lack of importance, a lack of standing in philosophical thought as a reliable source of knowledge, and an alignment with the animalic and lower registers of understanding. Gale Largey and Rod Watson (1972) point out that while humans produce and sense smells, sociologists historically have "ignored odors or regarded them as an insignificant dimension of human interaction" (p. 1021). Victoria Henshaw sums up many other accounts referencing the historical depreciation of smell, including the research of Constance Classen and Mirko Zardini (Henshaw, 2014, pp. 9–10). That all being recognized, lamented and surely mentioned in other chapters in this book, perhaps this is the space to think more positively about the future possibilities of designing with smell, and of moving the critical discourse to an enthusiastic state of anticipation of wondrous experiences that await.

It is also a trope of much writing on olfactory experiences that smells are processed directly by the limbic part of the brain, which governs emotions and memories. In his article "The uneducated nose," Alastair Gee (2015) distinguishes the different way the brain parses visual and scent information, noting the limbic system's "crude and unpolished" processing of smell. The description of the 2015 Museum Tinguely exhibition *Belle Haleine: The Scent of Art*, claims that "smell is apperceptual," can be "directly experienced" via the limbic system, and is closely associated with recollection (Museum Tinguely, 2015). Scent can move us easily; transport us to a moment in time, to a different space in a way that feels more immediate, more visceral, than other sensory perceptions precisely because it *is* more immediate and more visceral. The Japanese artist Mariko Mori, for instance, accelerated a moment of meditative contemplation in her exhibition entitled *Nirvana* (1997) by lightly spritzing an aerosol of jasmine into a gallery where a projected three-dimensional image of herself as beautiful bodhisattva, a being in search of enlightenment, beckons her viewers to join her. This slight scent transformed the experience of museum space into something different – into a spiritual space, or at least for a moment introduced a feeling of unexpected material bliss generated by the smell of jasmine, which, with its indolic quality, recalls all sorts of bodily smells.

Now that the eradication of noxious smells has often been accomplished in urban environments, and regimes of hygiene are well enough observed that we are protected from all sorts of smells generated by human and animal bodies, the designed re-introduction of smell into environments like that of the museum and other public spaces is at hand. But is it all roses and jasmine? Walking by the chemical cloud of scent wafting out of an Abercrombie & Fitch boutique is enough, perhaps, to give pause about the imagined pleasure of unexpected encounters with someone else's definition of a lovely smell. This aggressive use of scent as a marketing strategy raises ethical questions that stretch to the far more sinister military industrial deployment of smell, as in Israel's use of a foul odor called *Skunk* as a form of crowd control (B'Tselem, 2013). When we think of designing with smell, what is it that we are thinking of precisely, and what is the motivation to design with smell? What can be accomplished in demarcating a space – a human body, a museum gallery or a retail boutique – with an aroma? Can this marking be done in a discreet way that is not harmful to people and the environment, that recognizes and respects borders? Or is the way scent works, both in the atmosphere and in the body, antithetical to the way design works? Smells in the air and the olfactory system's mode of operation defy the ordering of a strict ecology – that is, the demarcating of a space with an impermeable border.

The scent opera *Green Aria*, performed at the Guggenheim Museum in 2009, demonstrated that aesthetic ecologies can be staged, as did the New York Museum of Art and Design's exhibition *The Art of Scent 1889–2012* (November 20, 2012–February 24, 2013), with its complex installation apparatus designed by architects Diller Scofidio + Renfro. But more ambitious, exciting, and perhaps troubling, are the technologies being used and developed by Sissel Tolaas, Roman Kaiser of Givaudan and David Edwards of Le Laboratoire.[1] Tolaas uses headspace technology to capture and store an archive of over 7,000 smells; Kaiser is using it to capture odors of flora on the endangered species list. Edwards has produced the oPhone, a device that can deliver scent through digital means. These endeavors, with all of their preservationist and archivist impulses, call for a re-assessment of scent as a category of experience, knowing and cultural discipline.

The 1970s cult film *Harold and Maude* features a smell generator, which functions as a signifier of Maude's *joie de vivre*, and of her unwillingness to follow all kinds of cultural logic and received notions of what is important and acceptable. This apparatus – fantastical and poetic – introduces the young Harold to Maude's 'odorifics,' as she calls them – smell experiences both exotic and mundane, which are all designed to expand his knowledge of worlds he has yet to encounter or consider, such as snowfall on 42nd Street or the New York Subway. Turning on the machine is a way of turning on Harold – exciting his desire for adventure and for libidinally driven sensations, feelings tamped down and policed by his puritanical upbringing. Perhaps this chapter could work along those lines: if not exactly a smell generator, then a modest apparatus that might prompt some errant sniffing, some questioning of how scent's indeterminate qualities can or should be harnessed.

The following three sections, "Capture," "Encode" and "Disperse," resituate earlier technical modes of creating and controlling ecologies of scent – the alembic, the atomizer, perfume, tinctures and herbal distillation – by looking at contemporary methods of capture and control, such as headspace technology and the oPhone. For centuries, scent has been captured, recorded and dispersed, but a major design problem remains in figuring out how scent can become data that can be digitally manipulated.

Capture

While there is a long (if relatively sidelined) history of smell, its ecologies and the seemingly innocuous technologies designed to control and extend the olfactory experience are not generally well known or appreciated. Friedrich Kittler's *Gramophone, Film, Typewriter* (1999) offers a method and model for thinking about media in our experience of sound, image and language. However, scent is still waiting for its phonograph, camera or typewriter (Kittler, 1999).[2] The technologies of scent capture lag a century behind those that record images, sounds and words, largely because it still remains unclear what communicative value odors have or what can be done with smell, beyond getting rid of it. Once difficult smells have been eradicated, and then replaced with the lovely and usually welcome smells of flowers, perfumes or delicate cuisine, what then?

Perfume itself may be thought of as a way of designing with smell: an olfactory capture technology or recording device. The earliest perfumes, dating back 4,000 years, were composed of vegetal aromatics – herbs and flowers – typically crushed in oil as a method of harnessing their smell for later use (Stamelman, 2006). The alembic, first used by alchemists in ancient Persia, is an apparatus that refines and distils vegetal and floral substances through a process of heating, vaporizing and condensing. This technical transformation and destruction of raw materials produces substances that are intense and, often, not immediately recognizable as referents to their sources. And yet the metamorphosis (via extraction or distillation) from flower to essence constitutes a preservation of the past life of plants and can be used to conjure up the memory of times past – of summer in the dead of winter. The production of perfume continues as a highly profitable practice within the fragrance industry and in DIY circles, where designing one's own perfume sits alongside home distilling (see Chapter 24), foraging and other artisanal activities. In the case of mass perfume production, and more scientific apparatuses like the still on display at the Fragonard Museum in Paris, the technologies that make such production possible remain esoteric and beyond the technical capacity of most people.

Headspace technology, crucial to capturing aroma, is rarely seen outside the chemist's laboratory, although it can be used in the field to record the scent of flowers and other aromatic materials. Developed during the 1980s, this equipment consists of various glass cones, tubes and a motorized mechanism for pumping inert gas into a central chamber fitted with a resin trap that will absorb molecules of scent found in the air around the object. The apparatus can fit inside a carrying case, but is still cumbersome and expensive. The term 'headspace' originally refers to the practice of brewing beer and sniffing the space above the head, or foam, to determine the level of fermentation (Newman, 1998, p. 90). In the study and analysis of smell, the area of scent around an object is also referred to as 'headspace,' and headspace technology works quite simply to absorb odors that can then be analyzed through gas chromatography. The captured data can be represented visually as a formula, or a 'recording' of the smell, which can then be replicated synthetically. This technology is more sensitive, and easier and quicker, than extraction and distillation. It also has the advantage of not destroying the object under analysis. The point is that smell, with instruments like headspace technology, is already being captured and coded – but there is a push to develop more portable technology to make it readily available and shareable in a way that we are accustomed to with images, sound and writing. Questions arise about why this is happening now.

The idea of scent as a stimulator of memory provides one answer. Most famously, if somewhat erroneously, this triggering capacity is often traced back to Marcel Proust's account in his novel *Remembrance of Things Past* (1982) of the madeleine – a delicate, crumbly pastry. Dipped in a cup of tea, the madeleine blooms: Proust muses on the startling effect of the taste, briefly mentioning the smell as part of the experience. He writes: "taste and smell alone, more fragile but more enduring," serve to activate from even a "tiny and almost impalpable drop of their essence, the

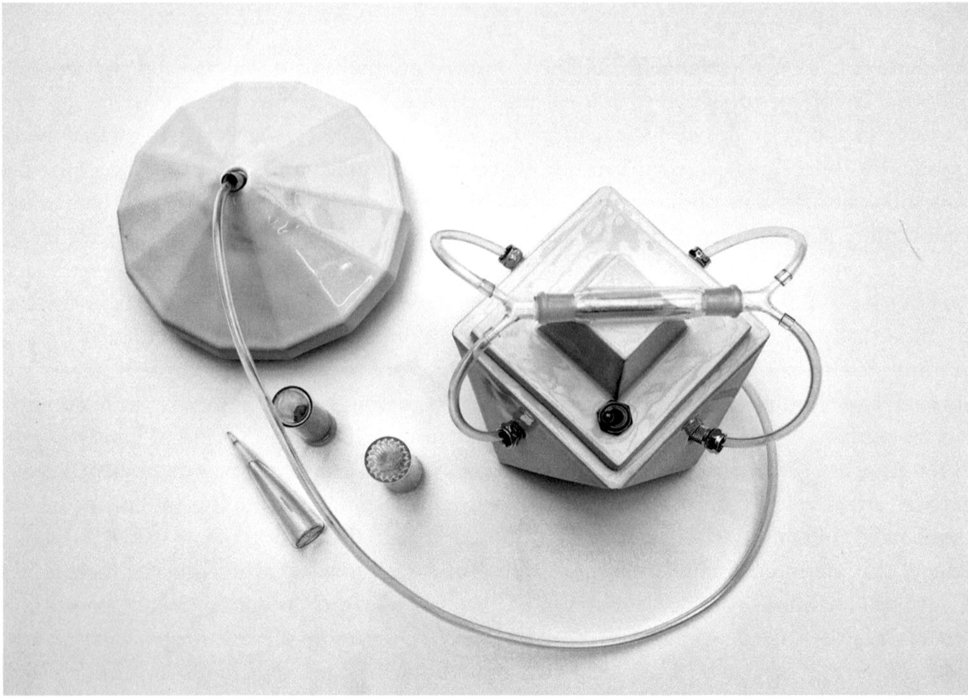

FIGURE 26.1 Madeleine, Amy Radcliffe, 2013.

Image courtesy of Amy Radcliffe.

vast structure of recollection" (pp. 50–51). Scent's power to prompt recollection of things forgotten and lost holds a great deal of potential in a number of arenas. The capacity scent has to transport us in time and place – and its ability to render one almost powerless to resist it, as seen in both Proust's *Remembrance of Things Past* and the film *Harold and Maude* – may be one of the motivations for capturing it, and for making it less random and subject to chance. London artist Amy Radcliffe references Proust in her conceptual prototype of a 'scent camera,' in an elegant little device she calls Madeleine[3] (see Figures 26.1 and 26.2).

Radcliffe's machine is actually an example of headspace technology. In her appropriation of this device, she imagines a machine that can produce smell samples of sentimental objects in small ampoules that are as shareable, at least theoretically, as contemporary digital photographs. Radcliffe writes:

> If an analogue, amateur-friendly system of odour capture and synthesis could be developed, we could see a profound change in the way we regard the use and effect of smells in our daily lives. From manipulating our emotional wellbeing through prescribed nostalgia, to the functional use of conditioned scent memory, our olfactory sense could take on a much more conscious role in the way we consume and record the world.[4]

Radcliffe's speculative olfactory sampling of objects redolent of memories underscores the fact that headspace technology is not widely used beyond its technical applications. It is used by governmental bodies in the United States, such as the Environmental Protection Agency, to determine the vinyl chloride content of wastewater, and by the US Food and Drug Administration

FIGURE 26.2 Madeleine in operation, Amy Radcliffe, 2013.

Image courtesy of Amy Radcliffe.

and other institutions and corporations to analyze the scents of various locations and environments, from coffee shops to battlefields.[5] Its most familiar form may be the Breathalyzers used by police forces to determine blood alcohol levels registered in the exhalations of people suspected of being drunk. Many corporations, mostly in the highly profitable industry of fragrances and flavors, have patented headspace technology: for example, Firmenich – NaturePrint; Givaudan – ScentTrek; Takasago – Aromascope. As yet, it is a highly specialized technology, and obviously not something you can carry around in your pocket like an iPhone.

Headspace technology, however, has been used to create enormous scent archives: 7,000 samples collected by Norwegian artist/scientist Sissel Tolaas, and over 300 gathered by scent chemist Roman Kaiser documenting flora on the endangered species list. The work of Tolaas and Kaiser is supported by two of the largest corporations in the scent industry: International Flavors & Fragrances and Givaudan. The research their archives may make possible is only beginning to be realized. Tolaas investigates a wide array of odors, from the smellscapes of cities to the yeasty elements of the human body. Her work does not shy away from the difficult aspect of smell; she insists that there is no good or bad smell, and is comfortable working in the area delimited by Kant (2006) in his theory of smell being linked to the experience of disgust. Tolaas pushes the boundaries of what smells are acceptable: intimate and difficult physical smell portraits of men, cheesy manifestations produced by the harvesting of human body odors, and the capture and reproduction of local urban smells of city neighborhoods (Poon, 2015). These last projects move her work into larger ecological concerns – the smell of shared public spaces that are defined by economics, ethnic communities and political divisions (Tolaas, 2011).

Kaiser, too, is engaged at a macro-level in his preservation of essences gathered from threatened plants. This work is an outgrowth of his role in analyzing smell at Givaudan. As early as the 1970s, Kaiser (2011) notes he began investigating "attractive scents that are not available as commercial essential oils or related products" (p. 7), but he cites larger ecological interests as motivations for this investigation, not least a conference held in St. Louis, Missouri in 1999, where he first learned of the global threat climate change and environmental degradation pose to plants. At that conference, Peter Raven outlined the projected loss of 100,000 of the known 300,000 plant species by 2050. Kaiser also sees his project as aligned with the call a decade later in 2010, by German Chancellor Angela Merkel, for immediate action to protect biodiversity *now*, rather than delaying until conditions seem even more dire (Kaiser, 2011).

So to summarize, headspace technology, while probably the best method of capture, remains out of reach for most people. It is still not really the typewriter, phonograph or camera scent needs in order to be more like language, sound and imagery, and to be captured and shared more widely.

Encode

Here, I turn to a private ecology of the body and how that is mediated by an encoding of scent. However narrowly enacted by advertisers, the perfume industry and olfactory artists, encoding requires specialized technical training – again, like headspace technology, outside of the purview of most people. The chemical analysis of smell – and even the accurate description of smell – remains a hidden, specialized activity for the most part: a symptom, perhaps, of the lack of general knowledge about, and the long-standing denigration of, the sense of smell. Victoria Henshaw (2014) outlines part of the problem, writing that "most modern European languages have an insufficiently developed vocabulary for smells" (p. 16). While we all have memories tied to scent, it is beyond most of us to recreate a remembered smell, whether in a chemical formulation or in language. Perfumers rely on distinct chemical terms in their work, but most of us use imprecise analogies to suggest what something smells like. A perfume called *RODIN bis*,[6] designed by independent perfumers D. S. & Durga for Linda Rodin, a high-end beauty specialist, demonstrates this dilemma.[7] The perfume is described in the press as based on her childhood memories of wrapping up in her mother's clothing, and redolent of "powder puff, a bit of toasted sugar with a touch of black pepper, lipstick (which used to have this intoxicating smell in the 1950s), whiffs of tobacco and Juicy Fruit gum" (Rodin, 2014). This private ecology of the maternal body is highly specific – a kind of homage to a bodily experience now past, irretrievable, and yet communicated in a way that is not specific to Rodin's mother's body at all. The scent generated is mediated, reliant on the perfumer's talent of matching chemical and natural materials to language, and to narratives of recollection.

Artist Martynka Wawrzyniak, likewise, collaborated with perfumers in her exhibition entitled *Smell Me*, which presented the aromas of her body as a kind of fragrant self-portrait. Givaudan perfumer Yann Vasnier and the Scent Director of 12.29, Dawn Goldworm, analyzed the smells gathered from Wawrzyniak's tears and sweat (collected by students at Hunter College over the course of a year) and translated them into an array of synthetic fragrances.[8] During the exhibition, a scent chamber permeated the gallery with the smells of the artist's armpit, her sweat, her tears and her hair. Goldworm is reported as saying of the project: "Her nightshirt smell is deep and warm like a small, delicate animal in hibernation . . . . Her hair smells like a memory from my childhood – civet, costus and musk with cumin, pepper, coconut water and spicy green galbanum" (Zalopany, 2012). Much of this description of the artist's scent draws its terms from the specialized vocabulary of the perfumer and the language of composed smells.

Scent critic Barbara Hermann notes that these synthetics distinctly relate to the smell of the human body, and as she explains, in Wawrzyniak's project, "the body comes back to perfume *as* its own perfume" (Hermann, 2013, p. 246). The power of the exhibition lies in the way the scent portrait denies the viewer the pleasure of the image of the artist, offering instead an experience of *becoming* the object of Wawrzyniak's odors, of being "passively penetrated – with all of the sexual and political connotations that verb carries" (Hermann, 2013, p. 243). For Hermann, the olfactory pleasure of the exhibition holds larger implications about the history of perfume, especially in the turn away from animalic scents after the 1980s, and the current rise in popularity of 'difficult scents' that challenge the nose with their odd synthetics or brute bodily references. Her reading of the whole project sees it as underscoring the history of perfume in its dialectical negotiations with the body as mask or enhancer.

Also relying on the encoding expertise of perfumers, the scent opera *Green Aria*, performed at the Guggenheim Museum in 2009, demonstrates that aesthetic ecologies of scent can be staged using aromas that are entirely chemical formulations. The opera, part of the Guggenheim's ambitious program *Works & Process*, played to sold-out audiences in New York, and then in Bilbao. The *New York Times* describes *Green Aria* as a "beguiling 30-minute work" resulting from a collaboration between the writer and director Stewart Matthew, composers Nico Muhly and Valgeir Sigurdsson, and the French perfumer Christophe Laudamiel (Tommasini, 2009). The opera, in a highly abstracted way, narrates the human struggle with nature – in the dark and without images – using a sequence of smells in conjunction with pre-recorded music, but no lyrics or human voices. The *Green Aria* team orchestrated sound and more than 30 odors to be experienced over the course of the performance. Some of the smells were strange, or hard to identify, while others conjured more familiar scents of fresh cut grass, industrial chemicals, charred wood and damp soil. The perfumer on the team, Christophe Laudamiel, conceived of the wide array of smells as characters in the opera, giving them names like *Chaos* (a fruity smell) and *Absolute Zero* (a cold fragrance).

Green Aria's theme, and indeed its very performance, is resonant with the anxiety of designing a world in which smell – one of the last wild senses – is tamed and turned into something industrially produced and orchestrated. Of course, the world of synthetic smells created in *Green Aria* is the world we already live in. However, partly because of the secrecy with which the corporate entities mixing up olfactory formulas operate,[9] one is not necessarily conscious of inhaling manufactured smells on a daily basis. Also, this lack of awareness stems from the fact that smell is perceived to be an integral part of things, while the reality is that most of the smells one experiences do not genuinely belong to the things they seem to belong to, even though they are precisely formulated to be mistaken as such. As Classen, Howes and Synnott (1994) indicate, "artificial odours are a sign without a referent," presenting us with "olfactory simulacra" (p. 205). Some of the scents in *Green Aria*, for example, may reference the natural world, in the form of 'green' smells (think plants) and those of water, but all of them are produced chemically in a laboratory setting far away from any body of water or field of grass. Laudamiel explains the complexity of formulating the smell of water by aligning it with the painting technique of making white seem whiter by adding a bit of black. The design of a watery smell entails amplifying the freshness with a woody note, and as Laudamiel explains, "[s]omething smells always of several things" (Lubow, 2009).

Disperse

The dispersal of scents also relies on complicated technologies, and the implications of dispersal – beyond the personal use of atomizers to transform liquid perfume to a more vaporous cloud – range from the collective experiences of crowd control to the aesthetic enjoyments of scented performances like *Green Aria*, and the recent innovations around digital scent delivery systems.

Green Aria's success depended on specific technologies of dispersal, as did the microencapsulation[10] component of artist Maria Loboda's outdoor installation for the 2013 *New York Frieze Art Fair*, and the New York Museum of Arts and Design's exhibition of perfume, with its hidden installation apparatus designed by architects Elizabeth Diller and Ricardo Scofidio. These are complex, industrial systems requiring technical and expert knowledge that is usually lacking amongst the general public, and therefore remaining out of reach of amateurs.

For the technical production of *Green Aria*, Matthew and Laudamiel worked with a ventilation company, Fläkt Woods, to design and manufacture a 'scent organ' that would play the 'scent notes' of the opera. Ideally, it would seem that a scent opera should feel unmediated, so that the audience can immerse themselves in the experience without being distracted by the technology making it possible. But in fact the apparatus, from the centralized 'scent organ' to the flexible 'scent microphones' attached to each seat in the theater, created an unavoidably awkward system by which audience members breathed in the scents being dispersed during the performance. The synchronizing of the smell and sound experience presented technical problems, particularly in creating the desired effect of each smell lasting for only a few seconds. It required the release of each aroma precisely on cue, according to exact calculations of the diffusion rates of the different chemical compositions in play.

Another problem in the dispersal of scents is related to the fact that smells resonate in the nose at different temporal rates, and also have a different relationship with the body compared to sound. So the desire to have the *Green Aria* audience experience an extended sequencing of smells, and yet have each of those smells distinct from each other and perfectly aligned with the unfolding of an aural composition, posed a significant challenge. Fläkt Woods solved this problem by designing a box of pressurized steel canisters containing an array of scented crystals that produce smells which evaporate more quickly than scented liquids or oils. These crystals allowed for more control of the temporal and spatial range of the scents as they were delivered through the 'microphones' to each person in the audience. It is telling that this term used to describe the tubes delivering the smells (via strategically timed puffs of unscented air) is derived from sound technology, even though they did not function like a microphone.

In her review of *Green Aria*, Alexandra Alter comments: "[m]ore is at stake than novelty and artistic success" (Alter, 2009). She goes on to note: "Messrs. Matthew and Laudamiel are marketing their new scent dispersal technology to hotels, movie theaters, videogame makers and other entertainment companies that want to pack a bigger sensory punch" (Alter, 2009). Arthur Lubow (2009), in his review, explains that *Green Aria* emerged from a highly technical research and development scenario. Specifically, in 2004, Matthew (who had long worked in the finance industry) founded SenseLab, a company interested in the potential intersections of art practices and science. With Laudamiel, Matthew collaborated four years later in the formation of a scent media company, Aeosphere, which focused explicitly on the research and development of fragrance as a form of communication.[11]

The *Green Aria* opera, and more precisely the technology that made the design of this opera possible, follows numerous attempts to add scent experiences to films throughout the 20th Century. Not all elements of the project were entirely new in terms of the olfactory concept. For example, it is suggested that in 1906, the audience of a movie theater in Forest City, Pennsylvania had the scent of rose essence blown over them using an electric fan whilst watching silent newsreel footage of flower-trimmed floats at Pasadena's New Year's Day Rose Parade (Gilbert, 2015). In the 1960s, the film *Scent of Mystery* relied on Smell-O-Vision, with a system of odors cued to be released at certain moments aligned with onscreen actions. Indeed, prior to working on *Green Aria* with Matthew, Laudamiel created several fragrances – edited by the French fashion designer Thierry Mugler – to accompany the film *Perfume*, based on Patrick Süskind's 1985 novel.[12] According to Lubow's (2009) review of *Green Aria*, Laudamiel dispersed the scents at a screening of the film in Paris, but because of the inadequacies of the technology, the experience was somewhat muddled. Yet despite the technical issues of delivery, when the virgin appeared on screen, her scent drifted through the cinema, and as Lubow (2009) notes, Laudamiel's assessment of this moment was that "the feeling was incredible."

Conclusion

The instrumentalizing of scent as a way of creating shared, breathable ecologies, and as a form of intimate or mass communication, still remains quirky and uneven, partly due to the lingering, long-held ideas about smell delivering a lower order of information than sight, as noted by many scholars working in the field (e.g. Classen et al., 1994; Henshaw, 2014). Recent developments in digital devices capable of delivering scent, however, reveal a steady interest, which will be highly important in the way olfaction might accompany, deepen and extend textual, visual and auditory digital experiences in everyday life. Of course, there is the dumb – if luxury – version of this in Chanel's iPhone case, which quite simply mimics the shape of the iconic *Chanel No. 5* bottle. But it is David Edward's team at Le Laboratoire whose prototype oPhone and oNote application steal this stage (see Figure 26.3), even if the Japanese company Scentee's simpler device is already available as an attachment to one's smartphone.[13]

The oPhone is a digital device that allows the user to compose over 300 different aromas into an olfactive communication (Stinson, 2014). The oNote is a message – text, image or sound – that is accompanied by an aroma. Edwards sent the first oNote to perfumer Christophe Laudamiel in 2014, an email message with a photograph and scent representing the city of New York.[14] While still in a developmental stage, there are plans to extend the number of aromas that the oPhone can deliver by pushing air over the surface of small chips in the device. On the market as of 2016, 'Cyrano' is the latest iteration, and with the tagline "Design your air. Design your mood," promises a personal scent system that will boost energy levels and promote relaxation.[15] Still, these technical advances, while promising, remain limited in scope, since there is still a need to network oPhone technology into existing communication platforms.

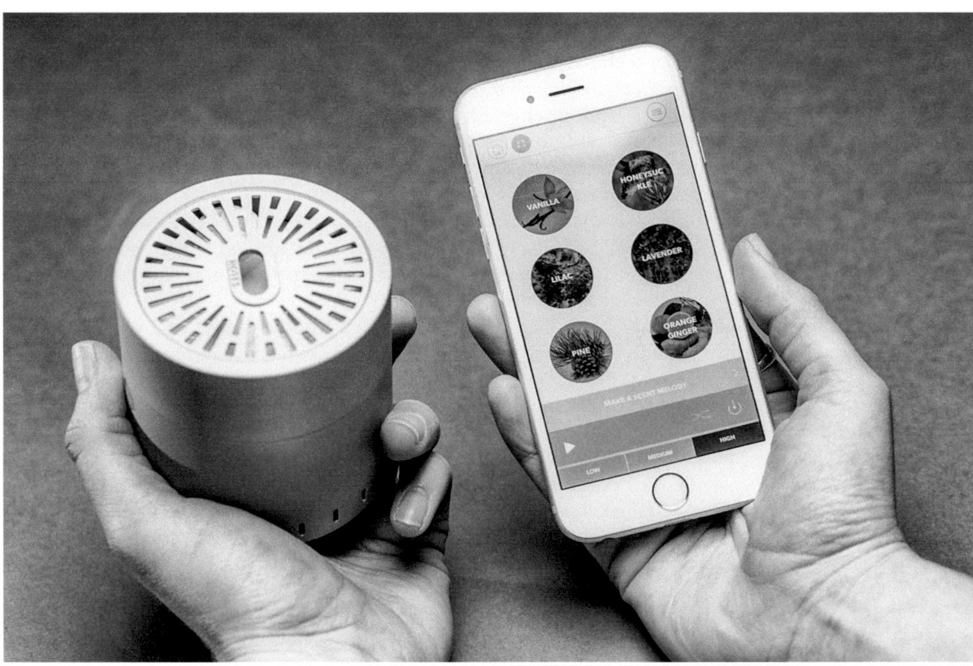

FIGURE 26.3 The oNotes application allows you to select scent tracks and play mood medleys in your personal space.

Image courtesy of Wayne E. Chinnock.

All of this is to suggest that ecologies of scent – and their technologies – remain indeterminate. Scent is emerging from, but still largely operating in, something of an olfactory unconscious, which Laura Marks attributes to the historical denigration and repression of the sense of smell. Despite recent forays into designing with smell, olfactory experience, she posits, is relegated to a private realm, and surrounded by taboos that make speaking about smell difficult and even an asocial act (Marks, 2008). As a field of critical inquiry and practical design, then, scent continues to unfold. There is an argument still to be made that scent is critical to understanding both intimate and shared ecologies of the body and of the air we breathe. Literature concerned with olfaction, as far back as the Ancient Greeks, alerts us to contemporary problems in establishing frameworks to help assess and amplify the value of scent in everyday life. It emphasizes the stakes involved in challenging the visual regime of the last century, and in the development of technologies and knowledge that would make scent, like images and sound, more democratically valued, reproducible and shareable.

Notes

1 Sissel Tolaas is a contemporary Norwegian artist who works with odors and has compiled a personal archive of smells. She has researched smell for companies such as Adidas and Sony, and has exhibited work at the San Francisco Museum of Modern Art and the National Art Museum in Beijing. Roman Kaiser is a scent chemist at Givaudan, the largest manufacturer of fragrances and flavors. David Edwards, a Professor of Biomedical Engineering at Harvard University, launched Le Laboratoire in 2007, an incubator for new initiatives in scent technologies and other projects. Most notable is his experimentation with the oPhone, a digital device that affords scent communication.

2 The reference here is to Friedrich Kittler's *Gramophone, Film, Typewriter*, which is described by the author as a collection of stories about "early and seemingly harmless machines capable of storing and therefore separating sounds, sights, and writing" (Kittler, 1999, p. xl).

3 See Amy Radcliffe's *How to Succeed with Your Madeleine* (http://vimeo.com/68778690 Retrieved March 1, 2017).

4 https://vimeo.com/68778690 Retrieved March 1, 2017.

5 For example, Entech Instruments currently offers several versions of headspace technology, which they describe as very sensitive (www.entechinst.com Retrieved March 1, 2017).

6 A clever name, since 'bis' is a prefix signifying two or twice, thus referring to the doubling of something – the second iteration; and it is also used in chemistry to indicate two identical but separated complex chemical groups appearing in one molecule. In music, a field to which perfumery is indebted for much of its lexicon, 'bis' signals 'again' as a direction in a musical score indicating that a passage is to be repeated.

7 D. S. & Durga is an independent perfumery in Brooklyn, New York.

8 Dawn and Samantha Goldworm own the New York olfactive branding firm 12.29, which has created scents for Rodarte and Jason Wu fashion runway shows, as well as for Design Miami and Lady Gaga's perfume launch at the Guggenheim Museum (https://www.wmagazine.com/story/dawn-samantha-goldworm-designer-scents-and-candles Retrieved March 1, 2017).

9 The secrecy of the perfume industry has a long history that can be noted in the title of early volumes of scent recipes and formulas such as *The Secrets of the Reverend Maister Alexis of Piemont*, published in the 16th Century. More recently, Cathy Newman (1998) notes the "deadbolt secret world of the fragrance industry" (p. 10) and the fact that formulas (often many pages long) are jealously guarded, partly because scents cannot be patented.

10 Microencapsulation is a "process by which individual particles or droplets of solid or liquid material (the core) are surrounded or coated with a continuous film of polymeric material (the shell) to produce capsules in the micrometer to millimeter range, known as microcapsules" (https://uqu.edu.sa/files2/tiny_mce/plugins/filemanager/files/4290121/MICROENCAPSULATION.pdf Retrieved March 1, 2017). Most popularly known in its scratch-and-sniff form, microencapsulated scent can be incorporated into the printing process.

11 The technological developments that made *Green Aria* possible have been built on by Christophe Laudamiel's startup, DreamAir, which emerged after internal disagreements brought about the end of Aeosphere. DreamAir has developed numerous projects, including *Air Sculpture* and *The Plume Scent Solution*, which materialize scent in designed spaces using "complex air-curtain systems and adjustable intensity levels" (http://dreamair.mobi Retrieved March 1, 2017).

12 It is no coincidence, then, that Thierry Mugler, the fashion, fragrance and design house, sponsored *Green Aria*.

13 https://scentee.com/ Retrieved March 1, 2017.
14 http://techtheday.com/the-worlds-first-scent-message-scent-sent-as-onotes-decoded-by-ophones/ Retrieved March 1, 2017.
15 www.onotes.com Retrieved March 1, 2017.

References

Alter, A. (2009, May 20). This opera hasn't even opened, but we know it smells. *Wall Street Journal*. Retrieved March 1, 2017 from http://online.wsj.com/news/articles/SB124277733367437141

B'Tselem. (2013, January). *Crowd control: Israel's use of crowd control weapons in the West Bank* [Press release]. Retrieved March 1, 2017 from www.btselem.org/publications/summaries/201212_crowd_control

Classen, C., Howes, D., & Synnott, A. (1994). *Aroma: The cultural history of smell*. London: Routledge.

Drobnick, J. (2002). Toposmia: Art, scent and interrogations of spatiality. *Angelaki: Journal of Theoretical Humanities*, 7(1), 31–47.

Gee, A. (2015, February 26). The uneducated nose. *The New Yorker*. Retrieved March 1, 2017 from www.newyorker.com/tech/elements/naming-scents-uneducated-nose

Gilbert, A. (2015). *What the nose knows: The science of scent in everyday life*. Seattle, WA: CreateSpace Independent Publishing Platform.

Henshaw, V. (2014). *Urban smellscapes: Understanding and designing city smell environments*. New York: Routledge.

Hermann, B. (2013). *Scent and subversion: Decoding a century of provocative perfume*. Guilford, CT: Lyons Press.

Kaiser, R. (2011). *Scent of the vanishing flora*. New York: Wiley-VCH.

Kant, I. (2006). *Anthropology from a pragmatic point of view* (R. B. Louden, Ed. & Trans.). Cambridge: Cambridge University Press.

Kittler, F. A. (1999). *Gramophone, film, typewriter*. Stanford, CA: Stanford University Press.

Largey, G. P., & Watson, D. R. (1972). The sociology of odors. *American Journal of Sociology*, 77(6), 1021–1034.

Lubow, A. (2009, May 24). The smell of music. *New York Art*. Retrieved March 1, 2017 from: http://nymag.com/arts/art/features/56912/

Marks, L. (2008). Thinking multisensory culture. *Paragraph*, *31*(2), 123–137.

Museum Tinguely. (2015). *Belle Haleine: The Scent of Art*, February 11–May 17. Retrieved March 1, 2017 from www.tinguely.ch/PDFGeneratorServlet?pageuuid=3d1c9673-d239-4e59-b091-54401184a188

Newman, C. (1998). *Perfume: The art and science of scent*. Washington, DC: National Geographic Society.

Poon, L. (2015, November 4). Wake up and smell your city. *The Atlantic City Lab*. Retrieved March 1, 2017 from www.citylab.com/tech/2015/11/wake-up-and-smell-your-city/413908/

Proust, M. (1982). *Remembrance of things past. Volume 1: Swann's Way* (C. K. Scott Moncrieff & T. Kilmartin, Trans.). New York: Vintage Books.

Rodin, L. (2014, April 1). Linda Rodin's memory of her mother, in a bottle. *New York Times Style Magazine*. Retrieved March 1, 2017 from https://www.nytimes.com/2014/04/01/t-magazine/on-beauty-linda-rodins-memory-of-her-mother-in-a-bottle.html?_r=0

Stamelman, R. (2006). *Perfume: Joy, obsession, scandal, sin – a cultural history of fragrance from 1750 to the present*. New York: Rizzoli.

Stinson, L. (2014, June 16). This magical scent-sending phone is now available to order. *Wired*. Retrieved March 1, 2017 from www.wired.com/2014/06/you-can-now-buy-this-scent-sending-phone/

Tolaas, S. (2011). An alphabet for the nose. *Journal for Artistic Research*. Retrieved March 1, 2017 from www.researchcatalogue.net/view/?weave=1036&x=875&y=0

Tommasini, A. (2009, June 1). Opera to sniff at: A score offers uncommon scents. *The New York Times*. Retrieved March 1, 2017 from www.nytimes.com/2009/06/02/arts/music/02scen.html

Zalopany, C. (2012, October 18). Vain glorious: Get a whiff of this girl. *T The New York Times Style Magazine*. Retrieved March 1, 2017 from http://tmagazine.blogs.nytimes.com/2012/10/18/vain-glorious-get-a-whiff-of-this-girl/

AFTERWORD

Jim Drobnick

Anthologies provide key opportunities to assess the current state of a field or topic. In the case of *Designing with Smell: Practices, Techniques and Challenges*, the subject is relatively new and the anthology takes an adventurous step forward in specialization. Up until now, compilations regarding smell and design have done so through the general context of the senses in particular niches, such as the city (Zardini, 2006; Diaconu, Heuberger, Mateus-Berr & Marcel Vosicky, 2011) or the museum (Levent and Pascual-Leone, 2014). In a similar way, anthologies on smell have featured some chapters on design, but in the midst of art, literature and other cultural forms (Drobnick, 2006; Jaquet, 2015).[1] *Designing with Smell*, by contrast, not only concentrates on smell, it does so in a broad, inclusive way to cover a full range of design applications: events, installations, exhibitions, buildings, landscapes, schools, workspaces, museums and public space, among other areas. With this pioneering anthology, one can take in a breath and sniff the prevailing winds, so to speak, of the diverse examples of olfactory research currently present in design practice.

A striking feature foregrounded in *Designing with Smell* regards the protean nature of olfaction. To take stock of the many practices utilizing smell in the previous pages, a partial list would include entertainment, art, social commentary, education, remediation, therapy, narrative, consumption and history. Clearly, the era has passed when smell was held to be inconsequential or frivolous; now scent's versatility and strengths bring forward possibilities that force a reconsideration of many accepted design practices. Since the introductions to the parts do such an incisive job of introducing their themes and chapters, my comments here will operate more laterally and pick up on some of the general threads that run throughout the anthology. In particular, there are three topics that deserve emphasizing: technology, literacy and ethics.

Technology is an essential consideration for those producing smell-based artworks, installations and events. Even though olfactory experiences can seem primal and immediate, they still need to be constructed for audiences, though that construction may be as subtle as a well-hidden nozzle. Such aromatic technologies are often new to museums, galleries and theaters, and so it falls upon the artists themselves to research, install and deploy diffusers, atomizers, fans and other apparatuses. Technology is also involved in the analyzing and synthesizing of scents, which occur far away and behind the scenes in chemical laboratories and perfume workshops. A strength of *Designing with Smell*, then, is the number of practitioners writing about their process with olfactory technology and installation methodologies. First-person accounts by those writing Chapters in Parts I, VI and VII (olfactory art, historic and theatrical smellscapes, and capture, distillation and diffusion) detail the many strategic choices made to procure and develop scents, to figure out the best means to convey them, and to most effectively engage their audiences. Even when visitors experience a smell seemingly arising out of nowhere, an ingenious use of technology renders this possible.

Many of the individuals featured in this volume exemplify the full scope of the term 'artist-researcher,' for their activities bring together and hybridize elements of engineers, fabricators, scientists and perfumers to that of being artists and designers.[2] Having access to adequate technology – both in the creation of scents and in their delivery – has been a major impediment to the flourishing of olfactory art and performance. Both in late 19th-Century theater and in mid-20th-Century cinema, the inability to release odors and clear them out in a timely and uniform way hindered the wider application of scent to plays and movies. Now that perfume companies and olfactory labs are providing access to their scent libraries, manufacturing capabilities and hi-tech diffusion technologies, independent artists and designers have gained an unprecedented level of agency and opportunity to utilize smell creatively. In the contemporary post-media age, when it is expected that aesthetic experiences incorporate multiple types of media and present multisensory effects, smell has achieved marked prominence, in large part because the technology now exists to facilitate olfactory experimentation and adventure.

With the proliferation and increasing sophistication of recent olfactory art, performances and interventions, the question arises how audiences are equipped to not only experience smells in new ways, but also to process and understand them. Concomitant with the *production* of fragrant illusions and provocations, then, is the need to enhance the *reception* of scent's nuances and complexities. This brings me to the second virtue of this anthology, the focus on what Thompson and Barratt call 'smell literacy,' which is accentuated in the chapters in Part V on learning, pedagogy and academia. Improving reception and literacy operates on a number of different levels. Besides increasing people's exposure to a broader range of specially highlighted olfactory events, there is the exercising of appreciation and the fostering of connoisseurship (McBride & Nolan, Chapter 19); the linking together of sensation and cognition, embodiment and epistemology in experiential education (Henshaw, Chapter 16); the development of scent-oriented programs in schools of art and design (Osman, Noppeney & Endrissat, Chapter 17; Norman and Pozzani, Chapter 18); historical re-enactment (Verbeek, Chapter 20); and the cultivation of self-teaching tools that encourage ongoing, independent inquiry (McBride & Nolan, Chapter 19). Too often, sensory training is deemed useful for children (think of Montessori schools), but not substantive enough to warrant serious attention for secondary students or adults. The apprenticeship model of perfumers and the training in a scientific field can be too insular or time-consuming for many who would otherwise be interested in smell. Given the lack of educational opportunities, olfactory artists in previous decades have resorted to relying on alternative methods, such as aromatherapy workshops, or low-cost do-it-yourself tactics, such as sniffing at department store perfume counters. Literacy also involves discourse, and in Chapter 17, Osman, Noppeney and Endrissat rightly point out how the domains of physiology and perfume have thus far overly controlled how smells are considered, spoken about and thus perceived. Breaking the determinism implicit in the former (whereby smells inculcate specific physiological effects) and the commercialism of the latter (whereby smells promote or become the object of consumerism) will raise new possibilities for olfactory endeavors.

To the need for expanding the opportunities for olfactory discourse and education, I would add critical thinking. In my teaching about the senses in art and design programs at university level, I have found that while practice is important, so too is framing the senses historically, culturally and theoretically. The burgeoning literature generated by the 'sensory turn' has contributed trenchant analyses about sensory hierarchies, ocularcentrism and how the senses bring forward complex (and conflicting) attitudes toward the body, personal identity and social affiliation. Students have been fascinated by the chance to focus on the senses in a sustained and in-depth manner.[3] Whatever the student's medium or specialization – for artists, performance, sculpture, painting, new media, printmaking and so on; for designers, graphic, industrial, environmental, advertising and so on; as well as future art critics, curators and art historians – considering the senses has opened up content and issues unapproachable by conventional means. To those from a somatophobic (and odorphobic) Western milieu, valuing the senses initiated new ideas about life and society; for those from diverse populations, giving the senses meaningfulness validated cultural practices that differed from the West's deodorized norm. Educational programs involving the senses in general, and smell in particular, then, can accomplish several outcomes: a fuller and more conscious experience of scents, a greater sense of agency to investigate and do things with scent, and a facility to develop new olfactory methodologies, concepts and theories.

Scents also exist in social contexts, which brings me to the third strength of *Designing with Smell* – its foregrounding of the ethics of smell. While some scholarship on perception reduces the senses to mere conduits that transmit the 'raw data' of the world to the brain for processing, such an abstracted view ignores the inherent embodiment of the senses and how life experiences and societies act upon those bodies. McBride and Nolan (Chapter 19), for instance, ask the incisive question 'Whose senses?' in their argument to consider the cultural significance of the situatedness of smell in particular subjectivities. Just as normative hierarchies can exclude or marginalize some of the senses, so too can society disenfranchise those who challenge conventional assumptions of neurotypicality or bodily ability.[4] For Warren and Riach (Chapter 15), issues of power infuse any practice of 'aroma management' – that is, the deliberate alteration of the air by organizations to influence the emotional states, behavior and sociality of people, be they workers, consumers, prisoners and so on. At stake is the exercise of biopolitics at the level of smell, whereby notions of dominance, exclusion and conformity can be instilled subliminally and without consent.

Conversely, smells contain the possibility to operate in empowering ways, such as for remediation, therapy or urban renewal. Given the essentialness of the life-sustaining medium of smell (air) and its global enveloping presence (the atmosphere), a holistic concern for human and planetary well-being arises in a number of the chapters in *Designing with Smell*. Smellwalks and smell maps sensitize people to the variations of scent within their environments and locate places to appreciate or improve (Henshaw, Chapter 16; McLean, Chapter 7). Malodors identify instances for recuperative efforts and provide testing grounds for new olfactory techniques (Kwon, Chapter 10). Fragrances also spawn opportunities for collaboration as designers attend to the needs of the sick, elderly, children and anyone suffering pollution's effects (Xiao, Tait & Kang, Chapter 8; Lygum, Chapter 9). Such restorative practices offer models of response to what is, for many, humankind's greatest crisis – climate change – and its root cause in the damage to the fragile composition of the atmosphere. Songster's *GhostFood* (Chapter 5) deserves special mention here for its use of scent in both an experiential *and* political manner: not only are the effects of climate change made immediately tangible, but the project constructs a polemic for both personal and societal action. New forms of olfactory technology are being used to remediate older forms of polluting technology and the deleterious results they have wrought. The challenge in moving forward with a convincing olfactory ethos, then, is to avoid considering smell just as a masking or coping strategy: rather, an olfactory-conscious sensibility could serve as the framework for rethinking the sustainable basis of all technology.

Given the potential magnitude of smell, should olfactory studies be considered a discipline? My own experience in researching smell gives some perspective of its growth. In 1996, when I first began investigating the cultural dimension of smell in earnest for my doctoral studies, there were only a few choice sources, such as Corbin (1986), Rindisbacher (1991) and Classen, Howes and Synnott (1994). To find anything useful, I had to scrounge for brief passages about smell in the general expanse of books and articles (e.g. Freud's notorious footnote on smell).[5] There was, of course, a solid literature on the neuroanatomy of smell and the science of olfaction, but the links to cultural practices were not clear. Ten years later, when I edited *The Smell Culture Reader* (Drobnick, 2006), the difference was notable. Not only was there a much greater number of texts distinctly about smell, the discussions were more theoretical, in-depth and sophisticated. At the time, I felt reasonably satisfied that the anthology was able to bring together a representative overview of the cultural research into smell through excerpts and citations (if a text could not be included, I intentionally sought to at least mention it in my introduction and prefaces). Overall, about 250 books and articles are referenced in *The Smell Culture Reader*, which gives an approximate baseline about the state of olfactory research at the time.

Now, in 2016, the literature on the social, cultural and aesthetic aspects of smell is vast by any measure. Purely in terms of the numbers, I have noted at least 100 substantial texts on scent just in the first half of this year. Whereas previously writers on smell would draw together insights from a number of different fields, it is now possible to solely draw upon the writings in one's specialty, whether it be literature, history, art, religion, linguistics, ethnography, geography, urban studies, education, disability studies and so on. If a decade ago one could entertain the idea that all of the diverse writing about the socio-cultural aspects of smell could be gathered and assimilated, today it is nearly impossible to keep up with the plethora of notable publications.

Does this flourishing research merit its own discipline, such as 'olfactory studies,' as Thwaites (Part V Introduction) and Osman, Noppeney and Endrissat (Chapter 17) suggest? Many new disciplines start off as an offshoot of a more established one, and the obvious trunk for olfactory studies would be sensory studies. Yet sensory studies itself is a still-emerging discipline, and probably not yet established enough in the academy to justify splitting itself into new fields, even with its successes in publications, grants and various forms of institutional support. Yet there are factors supporting olfactory studies as a discipline. For instance, Maria Manzon (2011, pp. 15–16) surveyed the literature on disciplinarity and outlined the basic epistemological and sociological criteria necessary to define a discipline (which I have abridged and modified):

- clear objects or materials of inquiry;
- identifiable attitudes, values or a point of view;
- interconnected truths or theories;
- a particular set of skills, procedures or methods;
- distinct concepts and analytical tools;
- pertinent applications and results;
- a discernible historical tradition with canonical references;
- recognition by academic institutions, professional organizations and other established disciplines.

Some of these aspects of a discipline are fulfilled in the current state of olfactory studies research: smells comprise the objects of study, smellwalks and smell maps are distinct methodologies, olfactory remediation is a pertinent application. For some of the other criteria, the actualization is an ongoing process. Still, the trend is obvious that through the accumulation of publications, olfactory studies is moving towards specialization as a discipline.[6] One might ask, then, what would be gained? Initially, status and credibility, certainly, and more institutional support for valuable work to be done, whose potential is only now being understood.

As much as anthologies like *Designing with Smell* indicate that olfactory studies is reaching a critical mass of research that might justify the designation of a discipline, I would argue for extending the moment when definitions are imprecise and subjects open to different angles of approach. Perhaps it is the amorphousness and unpredictability of smell that favors the disciplinary messiness and permeability that have become fruitful strategies for olfactory research. In this anthology alone, one can find artists conducting science, scientists making exhibitions, and new forms of hybridity and collaboration being generated to include a diversity of researchers. The in-depth specialization fostered by a discipline might impel a silo mentality that could jeopardize the cross-over inter-disciplinarity energizing so much of current researchers' analyses. At this stage in olfactory studies, concentrating on innovative research is more interesting than expending effort to construct and defend disciplinary boundaries.

With or without a dedicated discipline, *Designing with Smell* exemplifies the dynamism coursing through contemporary research into olfaction. What is striking to note from the range of chapters is the performativity and versatility of smell: there is nothing preventing smell from being put to almost any kind of use, be it experiential or political, instrumental or aesthetic, manipulative or liberating, affective or conceptual. With the advent of new technologies and creative ingenuity, the so-called intrinsic traits of smell can be modified, reimagined, transformed. Rather than simply serving as an extraneous element added to products, buildings or environments, smells can fundamentally alter conventional practices of architecture and urban space towards those that are sustainable, healthy, and treat the atmosphere in an ethical way. Even in its incipient stages, the field of olfactory studies has made an important contribution by shifting the discourse on smell. To paraphrase a valuable insight by Warren and Riach (Chapter 15), smells are not just physical phenomena acting upon passive bodies, they are things that people and societies purposefully engage with. By demonstrating the protean nature of smell, the texts in this anthology not only interrogate assumptions about olfaction, they reconceive and broaden its cultural utility and significance.

Notes

1 This is not to say that there have not also been a number of key solo- or dual-authored books discussing design and smell, such as Malnar and Vodvarka (2004) or Franck and Bianca Lepori (2007). My discussion here, though, regards anthologies.

2 See Drobnick (2015). I've also written about olfactory art and the artists Brian Goeltzenleuchter, Sissel Tolaas and Jenny Marketou, who are mentioned several times in this anthology (Drobnick, 1998, 2010, 2011; Drobnick & Fisher, 2008).

3 My course *The Senses in Artistic Practice* covers the traditional five senses, plus more hybrid phenomena such as altered states, intoxication, synesthesia, neurodiversity, paranormal experiences and sensory prostheses.

4 Even further, Korsmeyer (1999) and Smith (2006) have demonstrated how the supposed differences in the use of the senses provided evidence for intolerant individuals to justify their (and society's) misogyny and racism.

5 Also available at that time were a number of popular histories of perfume. While useful in a general sense, their emphasis on perfume limited the fuller articulation of smell in culture.

6 My comments about the emerging discipline of olfactory studies pertain to research in the humanities, rather than in the physical and biological sciences, which already maintain well-defined olfactory specializations.

References

Classen, C., Howes, D., & Synnott, A. (1994). *Aroma: The cultural history of smell*. London: Routledge.

Corbin, A. (1986). *The foul and the fragrant: Odor and the French social imagination*, Cambridge, MA: Harvard University Press.

Diaconu, M., Heuberger, E., Mateus-Berr, R., & Marcel Vosicky, L. (Eds.) (2011). *Senses and the city*. Berlin: Lit Verlag.

Drobnick, J. (1998). Reveries, assaults and evaporating presences: Olfactory dimensions in contemporary art. *Parachute*, *89*, 10–19.

Drobnick, J. (Ed.) (2006). *The smell culture reader*. Oxford: Berg.

Drobnick, J. (2010). *Air*chitecture: Guarded breaths and the [cough] art of ventilation. In P. di Bello & G. Koureas (Eds.), *Art history and the senses* (pp. 147–166). London: Ashgate.

Drobnick, J. (2011). The city, distilled. In M. Diaconu, E. Heuberger, R. Mateus-Berr, & L. Marcel Vosicky (Eds.), *Senses and the city* (pp. 257–275). Berlin: Lit Verlag.

Drobnick, J. (2015). Smell: The hybrid art. In C. Jaquet (Ed.), *L'art olfactif contemporain* (pp. 173–189). Paris: Université Sorbonne.

Drobnick, J., & Fisher, J. (2008). Odor limits. *The Senses & Society*, *3*(3), 349–358.

Franck, K. A., & Bianca Lepori, R. (2007). *Architecture from the inside out*. Chichester: Wiley.

Jaquet, C. (Ed.). (2015). *L'art olfactif contemporain*. Paris: Université Sorbonne.

Korsmeyer, C. (1999). *Making sense of taste: Food and philosophy*. Ithaca, NY: Cornell University Press.

Levent, N., & Pascual-Leone, A. (Eds.) (2014). *The multisensory museum: A cross-disciplinary perspective on multiple modalities of a museum experience*. Lanham, MD: Rowman and Littlefield.

Malnar, J. M., & Vodvarka, F. (2004). *Sensory design*. Minneapolis, MN: University of Minnesota Press.

Manzon, M. (2011). *Comparative education: The construction of a field*. London: Springer.

Rindisbacher, H. (1991). *The smell of books*. Ann Arbor, MI: University of Michigan Press.

Smith, M. M. (2006). *How race is made: Slavery, segregation and the senses*. Chapel Hill, NC: University of North Carolina Press.

Zardini, M. (Ed.) (2006). *Sense of the city: An alternative approach to urbanism*. Montreal: Canadian Centre for Architecture.

INDEX

Entries in **bold** denote tables; entries in *italics* denote figures.

For Product Safety Concerns and Information please contact our EU representative GPSR@taylorandfrancis.com Taylor & Francis Verlag GmbH, Kaufingerstraße 24, 80331 München, Germany

Batch number: 10777126

Printed by Printforce, the Netherlands